Springer Series on

Wave Phenomena 2

Springer Series on

Wave Phenomena

Editor: Leopold B. Felsen

Rayleigh-Wave Theory and Application

Proceedings of an International Symposium
Organised by The Rank Prize Funds
at The Royal Institution, London 15–17 July, 1985

Editors
E. A. Ash and E. G. S. Paige

With 239 Figures

Springer-Verlag
Berlin Heidelberg New York Tokyo

Professor Eric A. Ash, F.R.S.

Department of Electrical Engineering, University College London,
Torrington Place, London, WC1E 7JE, England

Professor Edward G.S. Paige, F.R.S.

Department of Engineering Science, University of Oxford,
Parks Road, Oxford OX1 3PJ, England

Series Editor:

Professor Leopold B. Felsen, Ph.D.

Polytechnic Institute of New York, 333 Jay Street, Brooklyn, NY 11201, USA

ISBN-13:978-3-642-82623-8 e-ISBN-13:978-3-642-82621-4
DOI: 10.1007/978-3-642-82621-4

2153/3130-543210

The Lord Rank, J.P., LL.D., 1888-1972

Lord Rank, who died in 1972, was an outstanding industrial personality in the United Kingdom. He consolidated his family's flour milling and bread business, Ranks Hovis McDougall, into one of the major British food manufacturers. He also built up The Rank Organisation, starting in films and now incorporating a wide range of activities, as well as having a close relationship with the Xerox Corporation of the United States through its shareholding in Rank Xerox.

Lord Rank was much concerned with the quality of life and provided for many charities during his lifetime. Just before his death, Lord Rank established The Rank Prize Funds to encourage scientific work in the fields of Nutrition and Opto-electronics.

Preface

There is no respect for mere age in Science or Technology. Yet the centenary of the discovery, by the third Lord Rayleigh, that elastic waves can be guided by a surface, is memorable for the contradictions which it encompasses: Rayleigh's assessment of his classic 1885 paper as a rather minor mathematical development with a potential value only in seismology on the one hand; on the other the rediscovery of the subject in a totally different field – that of electronic signal processing – which has led to its explosive growth over the last twenty years.

The new surface acoustic wave (SAW) devices have proved to be pervasive electronic components. Today almost all television receivers contain one or more SAW filters, sophisticated radars depend on SAW signal processing, and SAW devices have been incorporated in ultra-high performance opto-electronic systems.

With the objective of surveying these developments and attempting to predict those that can be foreseen, as well as marking the centenary of Lord Rayleigh's discovery, The Rank Prize Funds' Committee for Opto-Electronics organised a symposium in London, in July 1985. It was held in the Royal Institution, London, where Rayleigh himself once worked and often lectured. The present volume is the tangible outcome of the symposium, in which international experts surveyed their various fields, and presented the most exciting recent developments.

Following a brilliant historical vignette, the book covers the main topics of surface wave propagation and generation; SAW devices and systems incorporating them; the role of SAW in the new techniques of acoustic microscopy and in the technologically important field of non-destructive testing; the involvement of SAW in acousto-optical interaction. The book can be read as a snapshot in time of a rapidly developing subject. It provides a concise guide to the now voluminous literature. It could also serve well as a one-semester course on surface acoustic wave electronics in an M. Sc. course. The editors trust that it will provide a useful source for workers in the field as they extend their studies into the second Rayleigh Wave century.

London, July 1985

E.A. Ash
E.G.S. Paige

Preface

[illegible] of the discovery [illegible] Lord Rayleigh, [illegible] elastic waves [illegible] in a totally different [illegible] of electronic [illegible] has led to [illegible] devices.

[illegible] surface acoustic wave (SAW) devices [illegible] to be an [illegible] of electronic components. Today almost all television receivers contain one or more SAW filters; sophisticated [illegible] on SAW signal processing, and SAW devices have been incorporated in ultra-high performance microelectronic systems.

With the objective of surveying these developments and attempting to predict those that can be foreseen, as well as marking the centenary of Lord Rayleigh's discovery, The Rank Prize Funds' Committee for Opto-Electronics organised a symposium in London in July 1985. It was held in the Royal Institution, London, where Rayleigh himself once worked and often lectured. The present volume is the tangible outcome of the symposium, in which international experts reviewed their various fields and presented the most exciting recent developments.

Following a brilliant historical vignette, the book covers the main topics of surface wave propagation and generation, SAW devices and systems incorporating them, the role of SAWs in the new techniques of acoustic microscopy and in the technologically important field of non-destructive testing, the involvement of SAW in acousto-optical interactions. The book can be read as a snapshot in time of a rapidly developing subject. It provides a concise guide to the now voluminous literature. It could also serve well as a one-semester course on surface acoustic wave electronics in an M.Sc. course. The editors trust that it will provide a useful source for workers in the field as they extend their studies into the second Rayleigh Wave century.

London, July 1985 E.A. Ash
E.G.S. Paige

Contents

Part I

History

Some Sketches of Rayleigh

J.N. Howard

Applied Optics, 7 Norman Road, Newton Highlands, MA 02161, USA

Several short sketches are presented of Lord Rayleigh, to show his method of working and his interaction with his fellow scientists. The topics discussed are: his research on the blue of the sky (Rayleigh scattering); his rescue of Waterston from near-oblivion; his research on surface acoustic waves (Rayleigh waves); his collaboration with Agnes Pockels; and his research on blackbody radiation (the Rayleigh-Jeans Law).

1. Introduction

This paper was presented at the conference marking the centennial of the publication by Rayleigh in 1885 of his paper on surface acoustic waves. The oral version of the talk was rather informal and was intended simply as a historical introduction and a glimpse of Rayleigh as a scientist and as a person. For this written version of the talk we have instead substituted several short vignettes of Rayleigh that illustrate some of his characteristics. His son (the fourth baron) has written the most complete biography of Rayleigh, and there are detailed biographies by R.B. Lindsay in the Dictionary of Scientific Biography and in Lindsay's introduction to the Dover edition of the Theory of Sound. We have omitted here his most widely publicized discovery, the identification of argon in the atmosphere. Rayleigh was not flamboyant or aggressive; he was in general calm, patient and thorough in his analyses, and understanding and generous in his dealings with his fellow scientists.

2. Rayleigh and the Blue of the Sky

Soon after he left Trinity College, Cambridge University, in 1868, John William Strutt converted into a laboratory part of the former stables at Terling Place, the home of his father, the second Baron Rayleigh. He then pursued experimental studies on acoustics, electricity, photography and optics. One of his earliest interests was a study of human color vision, which he began in 1870, perhaps after hearing a public lecture on this topic at the Royal Institution, given in 1870 by James Clerk Maxwell. Strutt read several earlier papers by Maxwell on color and color vision, and constructed a device - called a color top - that Maxwell had used to make quantitative measurements specifying any color (including white) in terms of three primary colors: red, green, and violet. He wrote at the time that he had commenced this study more for his own satisfaction rather than the hope of adding anything new to a subject already so fully and ably treated by Maxwell. By mid July 1870 he had repeated most of Maxwell's published experiments, but one set of data did not agree with the others. As he wrote: "One set of observations made with the same care, and apparently as satisfactory as any of the others, puzzled me for some time on account of the great discrepancies with the others which it exhibited. I have no doubt that the cause lay in the different character of the light on the day in question (July 23), which came from the unusually blue sky which sometimes accompanies a high wind. On the other days the light came principally from clouds . . . I am not aware whether the difference of sky and cloud light has ever been made the subject of direct investigation . . ."

It did not take John Strutt long to devise an experiment to study the difference between the light from the zenith sky on a bright cloudless day and diffuse light from the sun. He arranged a simple spectroscope to look towards the sun with a piece of opal glass or thin white paper inserted in the beam to diffuse the light. He then inserted a mirror set at an angle of 45 degrees, so that instead of looking at sunlight the spectroscope saw the zenith sky. He then plotted the intensity of both of these spectra from the red to the blue ends (using Fraunhofer lines for wavelength calibration). Meantime he also pondered the theory of possible mechanisms of refraction of sunlight by the atmosphere. He reasoned that small particles in the atmosphere somehow interact with and scatter the sunlight incident on them. "The problem is to compare the intensities of the scattered and the incident rays, for clearly these will be proportional. Let i be the ratio of the two amplitudes. This is a function of the following quantities: V, the volume of the disturbing particle; r, the distance of the point under consideration from it; λ, the wave length of the light; c, the velocity of propagation of light; D and D' the original and altered densities. Of these quantities the first three are dimensions only of length; the fourth depends on length and time, and the last two introduce the consideration of mass. There are no other elements of the problem except mere numbers and angles, which do not depend on dimensions of length, time and mass. Since the ratio i, whose expression we seek, has no dimensions in mass, it follows at once that D and D' can occur only as a ratio D:D', which is a simple number and can therefore be omitted. It remains to find out how the ratio of amplitudes, i, varies with V, r, λ and c.

Now, of these quantities, c is the only one depending on time; and since i has no dimension in time, c cannot occur in its expression. We are then left with V, r and λ and from what we know of the dynamics of the question we can be sure that i varies directly with V and inversely with r. Since the ratio i is dimensionless, and the volume V has three dimensions in space, i must therefore be proportional to $\frac{V}{\lambda^2 r}$.

We thus have the important law: When light is scattered by particles which are very small compared with any of the wavelengths, the ratio of the amplitudes of the vibrations of the scattered and incident light varies inversely as the square of the wavelength, and the intensity of the scattered radiation therefore varies as the inverse fourth power."

John Strutt then proceeded to derive mathematically the complete expression for the scattering of light by small particles, after he had already shown by straightforward dimensional analysis that the intensity of the scattered radiation varies inversely as the fourth power of the wavelength. He then turned the same dimensional reasoning to two earlier hypotheses on why the sky is blue: an argument by Newton that sunlight is refracted by particles high in the atmosphere in the form of thin plates (ice crystals?) which bend the sunlight like little Newtonian prisms; and a much more complicated theory of water bubbles in the atmosphere postulated by Rudolf Clausius. Strutt was able to show that both of these theories predicted only an inverse square variation of scattering with wavelength, instead of the inverse fourth power that he had just derived. These results, published in 1871 as his paper "On the light from the sky, its polarization and colour," derive the expressions for what we now call Rayleigh scattering, and properly explain why the sky is blue and the sunset red.

3. Rayleigh and Waterston on Kinetic Theory of Gases

In 1845 a Scottish scientist, J.J. Waterston, sent to the Royal Society for publication a detailed manuscript on the kinetic theory of gases: "On the physics of media that are composed of free and perfectly elastic molecules in a state of motion." The manuscript received a mixed reception from the reviewers; one of them said it was nonsense, unfit for reading before the Society. The other was kinder, but not enthusiastic, and did not recommend publication. Instead he suggested that the manuscript be deposited in the Archives of the Royal Society. If the manuscript had been rejected the author could have submitted it elsewhere; instead it lay buried and unnoticed in the Archives. Waterston in one or two subsequent papers in other journals attempted to draw attention to his earlier paper, but the original paper remained absolutely unnoticed until Lord Rayleigh, in 1891, while writing a paper on the theory of gases, came across a reference to the Waterston paper. Since Rayleigh was at that time Secretary of the Royal Society, on his next visit to the Society he examined Waterston's paper in the archives. He saw at once that a great injustice had been done in not publishing the manuscript when it had been submitted in 1845. He said "The omission to publish it at the time was a misfortune, which probably retarded the development of the subject by ten or fifteen years." Waterston had anticipated some calculations of molecular velocity (published independently by Joule in 1848 and by Clausius in 1857) and had sketched much additional theory that was later developed and published by Maxwell in 1860.

Rayleigh then arranged to have the entire memoir by Waterston published in Phil. Trans. Roy. Soc. 183 1 (1892), with an introduction by himself pointing out the importance and priority of Waterston's paper. Rayleigh also made an effort to locate Waterston. He found from relatives that Waterston, after his manuscript had been ignored, had largely withdrawn from scientific matters and had taken a position in India as a naval instructor to the Indian Navy. He retired in 1856 and returned to Edinburgh. He was very fond of walking along the seawall at Leith near Edinburgh, and one day in 1884 he had simply vanished without a trace, presumably having slipped off the seawall into the ocean.

4. Rayleigh and Surface Acoustic Waves

The year 1985 marks the centennial of a particularly important paper by Rayleigh: his work on surface waves on elastic solids. In his years as a student in physics at Trinity College, Cambridge (1864-68) John Strutt and his fellow students were steeped in mathematical techniques to study the stresses and strains of bodies subjected to forces and the vibrations induced in elastic solids by deforming forces. The great British scientists of theoretical classical physics - Kelvin, Maxwell, Stokes, Rayleigh - when thinking about any practical problem in mechanics or dynamics, immediately set up a mathematical model, perhaps limited to one dimension or to small deformation for simplicity, and patiently and happily worked out the problem as a challenging mental puzzle. For example, Lord Kelvin would attend scientific meetings and listen while the next speaker was introduced, but soon he produced from his coat pocket a little green notebook, and he then worked busily on his own problem while half listening to the speaker. (Kelvin managed to publish over 600 papers in his lifetime by seldom taking time off.)

One of Rayleigh's earliest interests was acoustics, and in the time period 1872-1878 he wrote his treatise "The Theory of Sound," in which he worked out powerful mathematical techniques for studying acoustic waves. He also realized immediately that many of these mathematical formalisms derived for acoustics also had close counterparts in optical wave generation and propagation, and also in the propagation of disturbances in elastic solids. His colleague George Howard Darwin (who had been a fellow student at Trinity with Rayleigh, and who was also a son of Charles Darwin) wrote a major paper in 1882 on the deformations of an elastic solid as part of his research on the mathematical description of the tides of the earth. Horace Lamb, best known for his work in hydrodynamics, wrote a paper on the vibrations of an elastic sphere in 1882. Rayleigh had already published a dozen papers in hydrodynamics and several on elastic solids when in 1885 he decided to investigate the behavior of waves on the plane free surface of a homogeneous isotropic elastic solid. The mathematical analysis was similar to that for deep-water waves, which he had studied in 1876. The problem is complicated, but he introduced various simplifications and boundary conditions (the solid is incompressible; certain components of tangential stress must vanish at the free surface, etc.) and quickly determined that the wave component into the depth of the solid was rapidly damped, and a two-dimensional wave would then propagate along the surface with a velocity in the acoustic region. He further determined that these relations must still hold even for natural bodies with finite compressibility.

At the end of this short (7 page) mathematical paper (which was published in the Proceedings of the London Mathematical Society) Rayleigh observed: "It is not improbable that the surface waves here investigated play an important part in earthquakes, and in the collision of elastic solids. Diverging in two dimensions only, they must acquire at a great distance from the source a continually increasing preponderance."

Others, such as Horace Lamb, Sir James Jeans and more recently Sir Harold Jeffreys, later worked out the paths of propagation of the Rayleigh waves induced by earthquakes that propagate along the interface of elastic media in the earth's crust. By measuring the intensities and times of arrival of these Rayleigh waves with a seismograph it is possible to calculate the location where the earthquake occurred. The presence of these Rayleigh waves in earthquake shocks was verified almost immediately; with the sensitive apparatus used now Rayleigh waves have been detected after having made as many as six circuits around the earth. In 1911 A.E.H. Love greatly extended the theory of surface waves in elastic media.

In the fields of optics and electronics there is another important use of these surface acoustic waves: as delay lines. An optical signal is converted to an electronic signal and fed into a tiny wafer of quartz or silicon a few millimeters long. The electrical disturbance propagates along the solid as a surface acoustic wave, taking say milliseconds to traverse the wafer. To obtain a delay time of milliseconds with an optical signal propagated at the velocity of light would require optical paths hundreds of kilometers; but with Surface Acoustic Waves a pathlength of only a few millimeters is needed. Such SAW devices have important uses in many electronic and optical systems.

5. <u>On Sealing Wax and String</u>

Many of Rayleigh's contemporaries remarked on his remarkable ability to obtain useful and interesting experimental results with very simple, crude, homemade apparatus. J.J. Thomson once visited the laboratory at Terling —

where lenses and mirrors for optical studies were sometimes held in the desired position with a dab of sealing wax — and said "It looks like everything is done with sealing wax and string!" What mattered to Rayleigh was not the appearance of the apparatus, but rather its calibration, in which he took great pains. For example, in the late 1880's he turned his attention to surface physics: surface tension, viscosity and capillary phenomena. In 1889, at the hardware store in nearby Witham, he bought a sponge-bath tub, 33 inches in diameter, and filled it with water. He knew that tiny shreds of camphor actively bounce around on a clean surface of water (as a small amount of the camphor dissolves and changes the surface tension), but that the dancing stops if the water surface is even slightly contaminated by dust or oil. He reasoned that a drop of olive oil on the water surface would spread out until it covered the surface as a film, and he observed a series of smaller and smaller drops of oil on clean water surfaces until he found the size that would not quite cover the surface and camphor particles would again dance. He reasoned that at that point the oil film was only one molecular thickness on the surface, and he then calculated the molecular thickness of olive oil as 1.6 nanometers (which he called micro-millimeters), a length only 1/400 of the wavelength of yellow light. This was the first direct experimental measurement of the size of a molecule, and Rayleigh's result is remarkably close to later, more precise values obtained by I. Langmuir, N.K. Adam, and others, who refined the apparatus. Using that same tub of water Rayleigh also made some useful studies of the polarisation of light when reflected near the Brewster angle from liquid surfaces.

6. Rayleigh, Agnes Pockels, and Surface Tension

In 1890 Rayleigh published in Phil. Mag. (his favorite journal) several papers relating to liquid surfaces and surface tension. Even with simple apparatus he was able to draw many interesting conclusions about surface physics.

In mid-January 1891 Rayleigh received a long letter (in German) from Miss Agnes Pockels (1862-1935) in Braunschweig. Agnes Pockels had read Rayleigh's papers in her brother's journals and wanted to tell him of some experiments on surface forces that she had carried out in the kitchen sink. Her father and her brother were physicists, and she too was deeply interested in science, but at that time no mechanism for the formal university training of women existed in Germany, so her only education (after the Municipal High School for Girls) was self-training, with help from her brother Friedrich (who is now remembered for his studies of the Pockels effect, used in electro-optical modulators).

At the age of 18, in 1880, she had begun a series of researches on the effect of oil films on the surface tension of water. She had divided the water in her kitchen sink into two areas, and had devised a means of skimming the surface of contaminants and then measuring and comparing the surface forces. To measure these forces quantitatively, she had suspended a horizontal disk (later identified as a porcelain button) from one arm of a balance scale until the disk just touched the water surface, and then added weights to the other arm until the disk just separated from the water. (She had invented what we now call the surface film balance.) She had written to various German scientists about her work, but they showed no particular interest. Then her brother Friedrich saw Rayleigh's papers and advised her to write to him. Rayleigh had the letter translated and saw at once that she had made several fundamental advances to the understanding of the physics involved, even beyond his previous paper. He then published her paper in Nature in

1891, with an introduction by himself commenting on the importance of her work. This public recognition by one of the leading British scientists of the time brought instant recognition and fame to Agnes Pockels. Her later papers were treated by journals with deferential respect. Professor Voigt of the Physical Institute in Braunschweig offered her facilities in his institute; she attended meetings and visited Quincke, Riecke, Ostwald and Weber (all active in surface physics). She later was given an honorary doctorate by the University of Braunschweig (Brunswick) and was even elected to the Hannover Academy of Science. Her rising fame and recognition even helped her brother to obtain tenure in his previously precarious appointment at his university. (The American researcher Irving Langmuir adopted essentially her scheme for measuring surface forces and his subsequent efforts in surface physics won him the Nobel Prize in 1932.)

7. Rayleigh, Jeans and Blackbody Radiation

Towards the end of the Nineteenth Century the attention of physicists turned towards understanding the behavior of blackbody radiation. Thermodynamicists had determined that all bodies give off radiation according to their temperature (and emissivity) and if a body strongly absorbed incident radiation of a certain wavelength, then that same body also radiated strongly at that wavelength to its surroundings. If a body were perfectly black - that is, if it completely absorbed all radiation of all wavelengths incident on it - then that body would also emit radiation (according to its temperature) to its surroundings with an emissivity of one, and there was considerable interest in determining the precise wavelength distribution of blackbody emitted radiation as a function of temperature. Stefan and Boltzmann had derived an expression for the total radiation from a blackbody and Willi Wien had established "Wien's Law:" that the product of the absolute temperature and the wavelength of maximum energy density is a constant. He also had proposed — on rather unsatisfactory grounds — an expression that fitted experimental blackbody data on the short wavelength side of the maximum, but not at long wavelengths.

In 1900 Lord Rayleigh turned his attention to possible mechanisms for explaining blackbody radiation, from the point of view of classical physics. He considered a cavity surrounded by walls at a uniform temperature (a "Hohlraum") and then assumed that the "complete radiation" in the cavity would be a summation of all of the possible standing waves of light radiated from the walls. He thus obtained an expression (Rayleigh's "Law") that nicely represented the curve of blackbody radiation at long wavelengths. However, it "blew up" at short wavelengths (because there was a wavelength term in the denominator). This failure in the theory was called "ultraviolet catastrophe."

In mid-1900 Max Planck, a young physicist in Berlin, tried to devise an empirical formula that would reduce to Wien's expression at short wavelengths and to Rayleigh's expression at long wavelengths. Almost by serendipity he found such a formula, which he showed to Professor Rubens. Rubens tried the formula on some recent experimental blackbody data measured by Lummer and Pringsheim. He returned to Planck and said "This is more than empirical; it fits everything perfectly - it must be right." Planck then spent several weeks trying to derive this formula on a theoretical basis. He was finally able to do this by postulating that radiation is not emitted continuously, but rather in small steps of energy, which he called "quanta of action." He published his formula in 1901, thus beginning the age of quantum physics.

Rayleigh returned to this topic in 1905 in some letters to Nature. He noted Planck's theoretical formula, "which seems to agree very well with the experimental facts." He then made a comparison of the total energy per cubic centimeter of blackbody radiation from a gas under standard conditions, and found that using his formula ("Rayleigh's Law") he obtained eight times as much energy as that calculated from Planck's formula. Almost immediately Rayleigh received a letter from a very bright, 28-year-old scientist, James Jeans (1877-1946), who had recently graduated from Trinity College, Cambridge, and had become a lecturer in mathematics there. He pointed out to Lord Rayleigh that when Rayleigh counted all of the possible modes of standing waves in his isothermal cavity he had introduced a redundant factor 8 by counting both positive and negative values of x, y, and z. (In other words, to count all of the possible modes, one needs only to count those modes designated by the octant of a sphere containing positive x, y, and z). Rayleigh at once wrote to Nature "I hasten to admit the justice of this correction," and the relation Rayleigh had derived in 1900 became the Rayleigh-Jeans Law. The young James Jeans gained instant recognition by having his name thus linked with Lord Rayleigh, who was then President of the Royal Society and had just won the Nobel Prize in 1904, as the first Englishman to win a Nobel Prize in physics.

8. Rayleigh and Suggestions for Research

In his published papers Rayleigh often mentioned thoughts that occurred to him but that he did not pursue. He might remark as an aside that it would be interesting to investigate the effect of higher temperature, or pressure (or some similar parameter), on the phenomenon under consideration, but that he did not have the apparatus to pursue this at the moment. Remarks such as this proved to be gold mines for graduate students in physics and engineering. For years many graduate students' theses began with the sentence, "In 1887 (or some similar date from 1880 to 1915), Lord Rayleigh suggested that . . . " and then the student proceeded to elaborate on the point suggested by Rayleigh. As an example of this sort of prescience by Rayleigh, we can observe his monumental paper of 1895 on argon (Paper 214, for which Rayleigh received the Nobel Prize in physics in 1904). After describing in detail the inertness of this noble gas, argon, to aqua regia, caustic soda, sodium peroxide, and almost every conceivable reagent, Rayleigh remarked: "It will be interesting to see if fluorine also is without action, but for the present that experiment must be postponed, on account of difficulties of manipulation." For many years it was assumed that the noble gases could not be made to react, but in the middle 1960s a team of researchers at the Argonne National Laboratory found that fluorine reacted with xenon and argon to form hexafluorides (and at remarkably ordinary temperature and pressure). This astonishing result was almost anticipated in 1895 by Rayleigh.

9. A Brief Bibliography

Every ten years or so Rayleigh collected into a bound volume his Scientific Papers, which were published in six volumes by Cambridge University Press between 1899 and 1920. This set was re-published in 1964 as a three volume set by Dover, New York, with additional material by J.N. Howard.

Charles H. Giles has written in detail of Rayleigh's contributions to surface physics and his interaction with Agnes Pockels in Chemical Industry (London) 1971 p. 42; 1971 p. 324 and 1979 p. 469.

Martin J. Klein has related Rayleigh's influence on the original introduction of quantum theory in the Archives for History of the Exact Sciences 1 459 (1962) and the Natural Philosopher 1 83 (1963).

The book by his son, Robert John Strutt: Life of John William Strutt, 3rd Baron Rayleigh (London, Edward Arnold and Co. 1924) is the most detailed biography of Rayleigh, particularly with respect to his personality and family life. This volume has been reprinted in 1968 by the University of Wisconsin Press in an expanded version, with annotations by the fourth baron and foreword by J.N. Howard.

Part II

Propagation

Rayleigh Wave Propagation

B.A. Auld

Edward L. Ginzton Laboratory, Stanford University,
Stanford, CA 94305, USA

(I) The Early Period

It is now a century since LORD RAYLEIGH published his theoretical proof of the existence of elastic waves propagating along the plane surface of an elastic solid [1]. In reading this paper one is struck by the fact that, except for minor changes in notation, the formulation established by Rayleigh is still followed today in the theory of elastic guided wave propagation. To illustrate this point vividly, the following summary of Rayleigh's derivation paraphrases his approach with only a few notational substitutions: $\boldsymbol{u}_\ell$ and $\boldsymbol{u}_s$ for the dilatational (longitudinal) wave and distortional (shear) wave displacements; ω for the frequency;

$$\omega/V_\ell, \alpha_\ell, \beta$$

$$(\omega/V_s, \alpha_s, \beta)$$

for the longitudinal (shear) wavenumber magnitudes and components, perpendicular and parallel to the surface.

Rayleigh began with the equations for elastic displacements in an isotropic solid described by its Lamé constants. That is,

$$\rho \frac{d^2 u_x}{dt^2} = (\lambda + \mu)\frac{d\theta}{dx} + \mu \nabla^2 u_x, \text{ etc.} \tag{1}$$

where

$$\theta = \text{div } \boldsymbol{u} \tag{2}$$

is the dilatation. Assumption of the time dependence $\exp i(\omega t)$ and some algebraic manipulation converted (1) to

$$\{\nabla^2 + (\omega/V_\ell)^2\}\theta = 0. \tag{3}$$

With $z = 0$ taken as the surface of the solid (Fig. 1), θ was assumed to be of the form

$$\theta \sim \exp(-\alpha_\ell z + i\beta x) \tag{4}$$

where

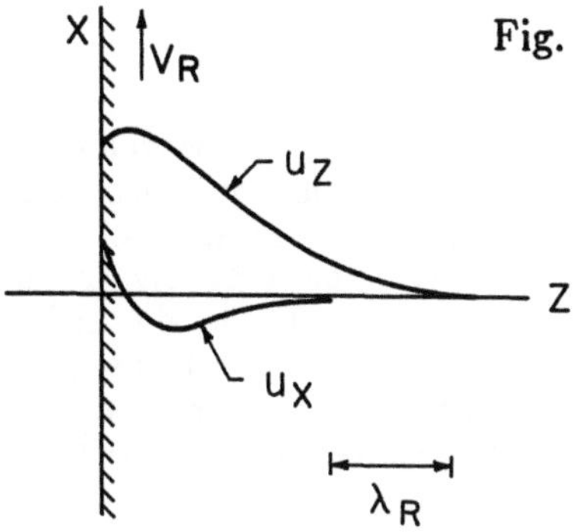

Fig. 1

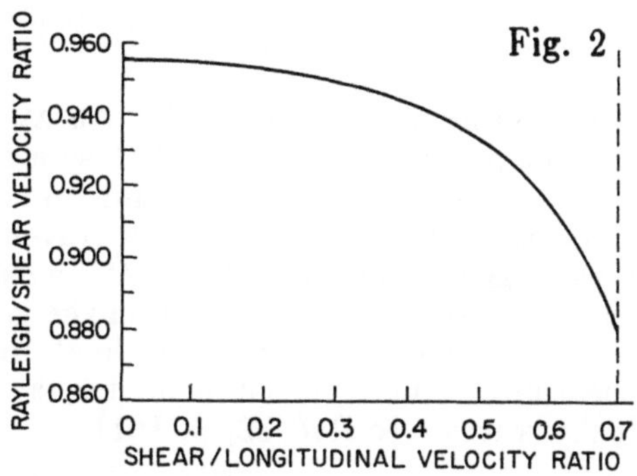

Fig. 2

Fig. 1. Displacement components of a progressive wave in two-dimensions propagated along the plane surface of an elastic solid

Fig. 2. Rayleigh wave velocity as a function of the ratio of shear and longitudinal velocities

$$\alpha_\ell^2 = \beta^2 - (\omega/V_\ell)^2 \tag{5}$$

from (3). The corresponding displacement u_ℓ was obtained using a method given by Lamb in a paper on the Vibrations of an Elastic Sphere, *Math Soc. Proc.*, May 1882. This gave

$$(u_\ell)_x \sim -i\beta\,; \quad (u_\ell)_z \sim -\alpha_\ell. \tag{6}$$

For $\theta = 0$, the corresponding displacement u_s was shown to satisfy

$$\{\nabla^2 + (\omega/V_s)^2\}u_s = 0 \tag{7}$$

and, assuming a wave function

$$u_s \sim \exp(-\alpha_s z + i\beta x) \tag{8}$$

the relation

$$\alpha_s^2 = \beta^2 - (\omega/V_s)^2 \tag{9}$$

was deduced. From the condition $\theta = 0$, the displacements for this solution were found to be

$$(u_s)_x \sim \alpha_s\,; \quad (u_s)_z \sim i\beta. \tag{10}$$

Finally, the stresses associated with (6) and (10) were calculated and the two solutions superposed to give zero tangential stress and normal traction at $z = 0$. This gave the equation

$$\{2\beta^2 - (\omega/V_s)^2\}^2 = 4\alpha_\ell \alpha_s \beta^2 \tag{11}$$

determining the period of the vibration as a function of the wavelength and the properties of the solid. Rayleigh evaluated the velocity of propagation for an incompressible solid and one with a Poisson ratio σ of 1/4, showing that the surface wave travels more slowly than a purely transverse elastic plane wave (Fig. 2). In calculating the velocity, he discarded the complex solutions to (11) on the ground that, for these solutions, part of the displacement grows without limit for increasing penetration into the solid. This type of behavior is now known to be characteristic of leaky waves (see Ref. 32, p. 161).

For a progressive wave the real part was taken of the combined (complex exponential) wave functions ((6) and (10)), for $\sigma = 0$ (incompressible solid) and $\sigma = 1/4$, or $V_\ell/V_s = \sqrt{2}$ and $\sqrt{3}$. In the second case

$$\left.\begin{aligned} u_x &\sim (\exp -\alpha_\ell z - 0.5773 \exp -\alpha_s z)\sin(\omega t + \beta x) \\ u_z &\sim (0.8475 \exp -\alpha_\ell z - 1.4679 \exp -\alpha_s z)\cos(\omega t + \beta x) \end{aligned}\right\} \tag{12}$$

showing that the motion takes place in elliptical orbits and that the displacement parallel to the surface goes through zero at a finite depth. At $z = 0$ the $\sigma = 0, 1/4$ solutions give two points on the general curve of Fig. 3, where

$$\left.\begin{aligned} u_x(z=0) &\sim F_x \\ u_z(z=0) &\sim F_z. \end{aligned}\right\} \tag{13}$$

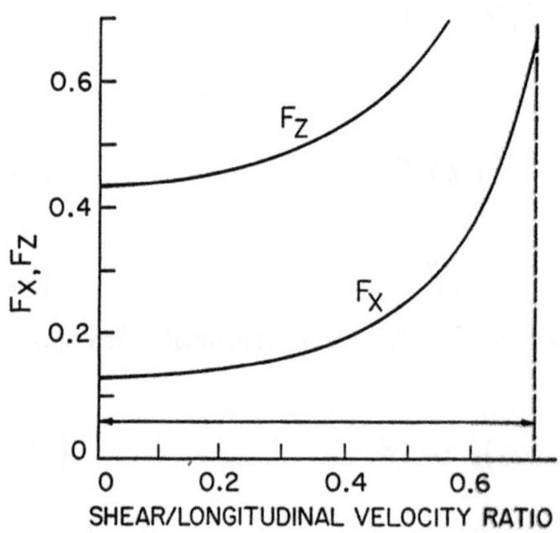

Fig. 3. Perpendicular (u_z) and parallel (u_x) displacement components as a function of the ratio of shear and longitudinal velocities

Rayleigh concluded his paper with the following paragraph:

> "It is proper to remark that the vibrations here considered are covered by the general theory of spherical vibrations given by Lamb * in the paper referred to. But it would probably be as difficult, if not more difficult, to deduce the conclusions of the present paper from the analytical expressions of the general theory, as to obtain them independently. It is not improbable that the surface waves here investigated play an important part in earthquakes, and in the collision of elastic solids. Diverging in two dimensions only, they must acquire at a great distance from the source a continually increasing preponderance."

thereby setting the course of elastic surface wave research for decades to come.

In a footnote to his surface wave paper [1], Rayleigh pointed out that his analysis might also be used in deducing the complete solution applicable to a plate, bounded by parallel free surfaces, but that he had not obtained any results which seemed worthy of quotation. His analysis of the plate problem was not published until four years later [2]. In that paper he noted that the results, for thin plates, had already been long-known [3-5]; but that the method proposed might not be without interest because of difficulties in rigorously treating thin plates, and also because the new solution was not limited to thin plates. He also commented that his 1885 paper was a particular case of the plate wave analysis.

Rayleigh approached the plate wave problem using the same mathematical formalism as that of his surface wave paper. The plate was taken to have surfaces at $z = \pm z_1$ and the waves were considered to propagate along x. He resolved the vibrations into a first kind (with displacement u_y only) and a second kind (with displacements u_x, u_z); in each category, he separated the vibrations into classes having odd and even symmetrics with respect to the middle surface of the plate. In doing this, he discovered the vibrations later known as SH plate waves and Lamb plate waves. For SH waves, he gave the frequency equations

$$\left(\frac{\omega}{V_s}\right)^2 = \beta^2 + \left(\frac{q\pi}{z_1}\right)^2 \quad \text{even} \tag{14a}$$

$$\left(\frac{\omega}{V_s}\right)^2 = \beta^2 + \left(\frac{(q+1/2)\pi}{z_1}\right)^2 \quad \text{odd} \tag{14b}$$

and, for Lamb waves, the frequency equations

$$4\beta^2 \alpha_\ell \alpha_s \tanh \alpha_\ell z_1 = \left\{\left(\frac{\omega}{V_s}\right)^2 - 2\beta^2\right\}^2 \tanh \alpha_s z_1 \quad \text{even} \tag{15a}$$

*See citation after (5) above.

$$4\beta^2 \alpha_\ell \alpha_s \coth \alpha_\ell z_1 = \left\{\left(\frac{\omega}{V_s}\right)^2 - 2\beta^2\right\}^2 \coth \alpha_s z_1 \quad \text{odd} \qquad (15b).$$

Rayleigh noted that (15a) and (15b) describe vibrations in which the middle plane undergoes, respectively, extension without bending and bending without extension. Equations (l5a) and (15b) are now known as the Rayleigh-Lamb frequency equations, because of a contemporary paper by LAMB on the flexure of an elastic plate [6]. It is not clear when this terminology was first adopted; the designation of elastic surface waves as Rayleigh waves was certainly already established in 1925 (see Ref. 13). In any case, the linkage of Rayleigh and Lamb waves through (15a) and (15b) is not of purely historical interest; it is of direct physical significance in modern applications of these waves to nondestructive evaluation and to signal processing devices.

No account of the origins of Rayleigh waves would be complete without reference to earlier work by the great nineteenth-century elasticians. Research by CAUCHY and POISSON on thin elastic plates has already been mentioned [3-5]. In connection with modern applications of surface and plate waves, it is interesting to note that in 1829 Cauchy was already considering anisotropic plates. POISSON [5] was the first to show that elastic vibrations, in general, consist of both dilatational and distortional waves. Subsequently, Lamé (1852) expressed the general solution in terms of scalar and vector potentials. Lamb's investigation of elastic vibrations in a sphere, as noted above, was referenced in Rayleigh's 1885 paper. There also existed a considerable work on elastic waves in isotropic cylinders, such as the paper by L. Pochhammer in 1876. This paper, and a later one by CHREE [7], laid the foundations for later investigations of Rayleigh waves on curved surfaces.

(II) Geophysics

It is clear from publications appearing early in the twentieth century that Rayleigh's paper on elastic surface waves immediately stimulated an intense interest in the theory of earthquakes. A paper by Bromwich in 1898 dealt with the vibrations of an elastic globe and, in particular, with waves on a layered half-space. His calculation was for layers small compared to the wavelength. LOVE [8] extended this work to layers of arbitrary thickness, in seeking to explain the observed properties of earthquake waves. At this time the first long-period seismographs, which measured horizontal motion only, showed that earthquake waves contained an intense horizontal component of motion. Love, during his investigation of the layered half space [8], discovered a horizontally polarized surface wave, later to bear his name. His studies were also directed toward the effect of thick layers on the propagation of Rayleigh waves. In this work he discovered that some solutions for the thick layer depended on the properties of *both* media. This led to the later clarification by STONELEY [9] that these are surface waves guided by the interface. Layer surface waves were subsequently studied in great detail by SEZAWA [10,11] and others.

During the same period, intensive studies were made of the transient excitation of Rayleigh and other waves by surface and buried sources. The seminal paper was by LAMB [12], who confirmed Rayleigh's supposition that the surface wave should be dominant at large distances. Other studies of what came to be known as Lamb's problem followed, beginning with NAKANO [13] and continuing through the 1930's. A key advance in this area was the publication in 1939 of a Laplace transform method by CAGNIARD [14] for solving transient problems of the elastic half space. This method is exhaustively reviewed by MIKLOWITZ [15].

In view of the wide-ranging earthquake research stimulated by the 1885 paper on elastic surface waves, it is difficult to imagine that Rayleigh's study did not also influence the early investigations of electromagnetic surface waves excited on the earth by radio antennas. Workers in this field (Coller, Uller, Zenneck, Sommerfeld, etc.) pursued their research at the same time (1900–1910) as Lamb and Love.

(III) Ultrasonic Resonators and Delay Lines

During the 1930's the use of pizeoelectricity to excite ultrasonic waves in solids led to a revival of interest in the theory of elastic vibrations in finite bodies. Prior to this time the center of activity was the half-space problem, although LAMB in 1917 [16] first analyzed the lowest even and odd symmetry modes of the free plate and showed that they become Rayleigh surface waves in the short wavelength limit. This renewed attention given to finite elastic bodies finally culminated, after about twenty years of research, in a complete understanding of the spectrum of the Rayleigh-Lamb frequency equations (15a,b).

The motivation for initiating ultrasonic wave propagation and vibration studies came from the application of ultrasonic techniques to rods and slabs, and the use of these structures as transmission media [17,18]. During and after World War II, ultrasonic propagation in solids began to be applied in delay line devices for pulse storage purposes in radars, computers, and pulse decoders [19]. In addition, ultrasonic vibrations in thin piezoelectic plates were finding applications as high Q electromechanical resonators for the frequency control of radio transmitters.

Both the resonator and delay line applications of ultrasonics were linked to Rayleigh's pioneering work on surface and plate waves. This connection is obvious for delay line devices, but less so for resonators. The key relationship was first explored in detail by EKSTEIN [20], referencing earlier work by Sykes and Mason. In a thin crystal plate used as an electromechanical resonator the desired resonance is a *thickness mode*, consisting of a standing elastic plane wave between the free faces of the plate. Such a mode is an exact solution for an infinite plate; but the free edges of a finite plate perturb the thickness solution and couple it to standing plate waves in the *plane* of the plate. These *lateral modes*, so named by Ekstein, introduce spurious responses in the frequency spectrum of the resonator and must be avoided by proper design. Ekstein's paper, for monoclinic crystal plates, was the first attempt at a fundamental solution to this important problem, but, owing to its complexity, it was not

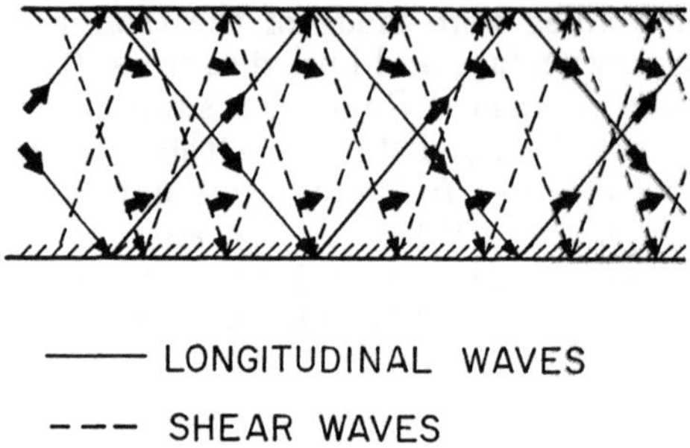

Fig. 4

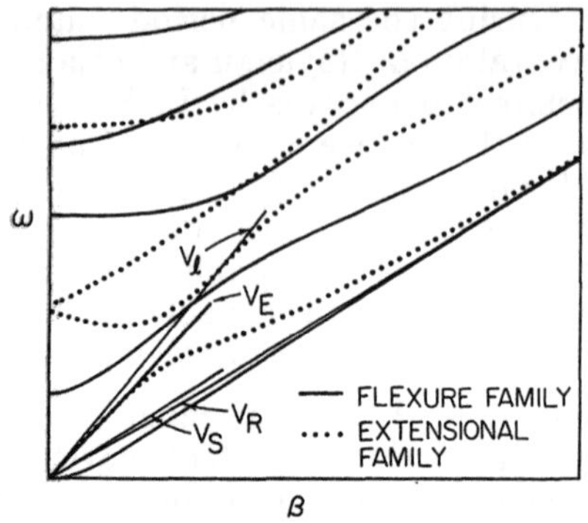

Fig. 5

Fig. 4. Lamb waves (Rayleigh plate vibrations of the second kind, 1889) constructed by superposing longitudinal and shear partial waves

Fig. 5. Typical dispersion curves (ω versus β) for Lamb waves in an isotropic plate

pursued further. Lateral resonances of thin plates were finally analyzed by MINDLIN [21], using the expansions in a series of powers of the thickness coordinate first developed by CAUCHY [3,4] and POISSON [5].

Despite this renewed interest in plate waves—for both resonator and delay line applications—parts of the spectrum defined by the Rayleigh-Lamb frequency equations remained unexplored, even after publication of HOLDEN's paper in 1951 [22]. Nevertheless, this paper introduced a fundamental concept—simplified bounding curves for the plate wave dispersion curves—that was later elaborated and extended to allow construction of the full Rayleigh-Lamb dispersion curves (ω versus β) without requiring detailed numerical computations. Mindlin based an analogous system of bounds on the concept of longitudinal and vertically-polarized shear waves reflecting back and forth between the plate faces (Fig. 4). This concept, explored in great detail by TOLSTOY and USDIN [23], is a more physical presentation of the longitudinal-shear decomposition of plate waves proposed by RAYLEIGH [2]. Mindlin's original contribution was to introduce special boundary conditions at the plate surfaces to decouple the longitudinal and shear partial waves. Dispersion curves for these decoupled waves were then shown to provide physically meaningful bounds for the exact Rayleigh-Lamb dispersion curves. Using this method, ONOE [24] and Mindlin showed that the exact dispersion curves for the Lamb modes of an isotropic plate are of the form shown in Fig. 5. The two lower branches reduce at the upper right to Rayleigh waves, as previously proved by LAMB [16]. Imaginary branches of the dispersion curves were later computed by LYON [25] and complex branches by SHERWOOD [26] (Fig. 6). This completed, some seventy years after the first publication of the Rayleigh-Lamb equations, a physical interpretation of their spectra. Dispersion spectra for elastic waves on rods were also constructed at this same time, with results later reviewed by THURSTON [27]. An interesting aside, relating to the finite plate resonator problem, is the interpreta-

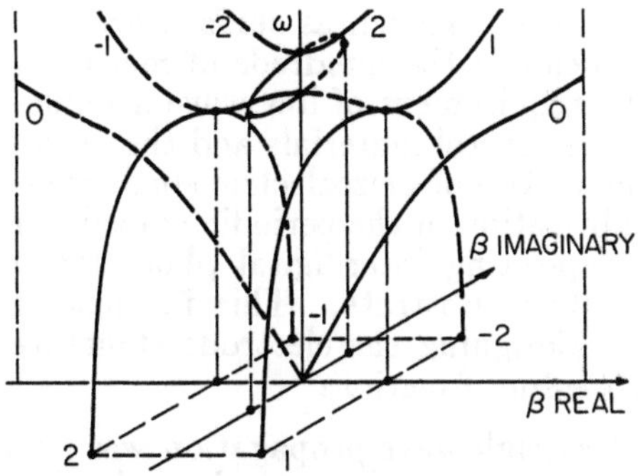

Fig. 6. Complex branches of the Lamb wave dispersion curves

tion of localized resonances at a finite plate edge as standing Rayleigh waves across the edges [28].

(IV) Rayleigh Wave Nondestructive Evaluation and Signal Processing

The development of ultrasonic technology soon attracted the attention of the material inspection community [29,30]. Rayleigh waves were particularly attractive because of their sensitivity to surface imperfections; but Lamb waves were also quickly identified as ideal for inspecting sheet materials. One of the most common excitation methods, the wedge transducer (Fig. 7a) was described in Refs. 29, 30; and the first version of the periodic transducer (Fig. 7b) was proposed by Sokolinskii in 1958. A detailed investigation of transduction techniques was published in 1961 by VIKTOROV, who later wrote the first book on ultrasonic Rayleigh and Lamb waves [31]. This book describes not only the applications in material inspection but also points out the advantages of Rayleigh and Lamb wave delay lines in signal processing, including the concept of the dispersive line. Details of nondestructive evaluation and signal processing applications are given in other papers at this symposium and will not be considered here. Signal processing applications, however, led to the introduction of new physical concepts and an extended understanding of fundamentals, particularly with respect to surface waves on anisotropic and piezoelectric materials.

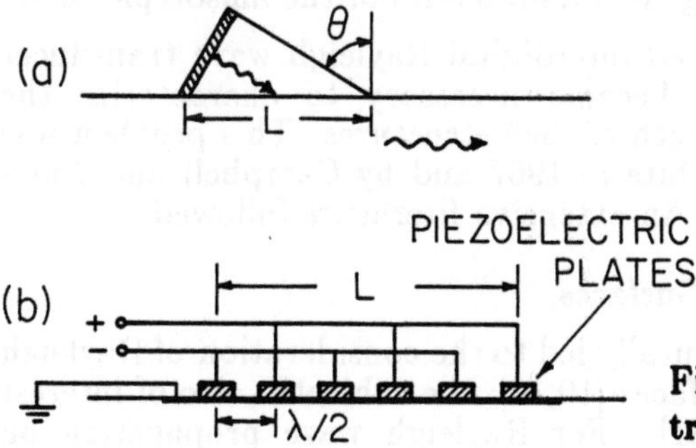

Fig. 7. (a) Wedge and (b) Periodic transducers for Rayleigh waves

In Rayleigh wave signal processing devices—known as surface acoustic wave (SAW) devices—operating frequencies in the hundreds of megaherz were targeted from the beginning. The requirement of low wave attenuation led automatically to the use of single crystal materials and the necessity of addressing the anisotropy problem. Use of piezoelectric single crystals also made it possible to simplify fabrication of the periodic transducer structure (Fig. 7b)—i.e., by directly depositing interdigital photoetched electrode structures onto the piezoelectric substrate. This fabrication technique was later widely exploited in designing the electrode structure to provide a wide variety of signal processing functions.

The effect of material anisotropy on Rayleigh wave propagation was first approached, in the late 1950's, as an exercise in theoretical mechanics. With increasing availability of high-speed computers, this pioneering work was extended numerically after 1960 by many researchers. There was an initial period of controversy over the existence of Rayleigh wave solutions for all orientations of a crystal substrate. This question was finally resolved by Toning and Ingebritsen in 1967 and by Lim and Farnell in 1968, showing that forbidden directions of propagation do not exist on crystal substrates. The difficulty arose from the fact that, on crystal substrates, the surface waves (now called generalized Rayleigh waves) have displacements that vary with depth as decaying sinusoids, rather than as simple exponentials ((4) and (8)). For certain crystal directions the generalized Rayleigh wave degenerates into a generalized surface skimming shear wave. At this same point a true surface wave, with a significantly different propagation velocity, appears. At directions in the neighborhood of this isolated point, the second solution still exists; but it radiates (or leaks) energy into the substrate, except at the isolated point where the generalized Rayleigh wave becomes a skimming wave. (Solutions to (11) with leaky wave characteristics were noted by Rayleigh and rejected.) These so-called psuedosurface waves were first observed experimentally by Engan *et al.* in 1967 and by Rollin *et al.* in 1968. Further details concerning the properties of generalized Rayleigh waves are available in Ref. 32. Reference 33 reviews the subject of elastic wave propagation on anisotropic layered half spaces. Although SAW devices are based on Rayleigh wave propagation, they are fabricated on plates of *finite* thickness. Spurious effects arising from acoustic waves multiply reflected from the bottom and the top of the plate have been accurately modeled by WAGERS [34], using the Lamb modes of the anisotrpic plate.

With the advent of the deposited interdigital Rayleigh wave transducer for piezoelectric substrates, it became necessary to characterize the electromechanical coupling strength of such structures. This problem was first addressed by Tseng and White in 1967 and by Campbell and Jones in 1968 (see Ref. 32, page 111). An extensive literature followed.

(V) Rayleigh Waves on Curved Surfaces

The early geophysicists were naturally led to the consideration of Rayleigh wave propagation on curved surfaces [10,11]; the subject is also of interest in nondestructive evaluation [31]. For Rayleigh wave propagation on the circumference of a circular cylindrical surface, one must distinguish between convex surfaces, where the Rayleigh wave is on the same side as the radius of curvature, and concave surfaces, where the Rayleigh wave is on the opposite side. In the second case the Rayleigh wave has a velocity greater than the shear velocity of the solid medium and therefore radiates energy into the solid. It is therefore a leaky wave. (See Ref. 31, pages 29–46.) Another type of Rayleigh wave on a curved surface appears in the mode spectrum of a cylindrical rod [35]. In the short wavelength limit the lowest order longitudinal mode becomes an axially-propagating Rayleigh wave on the cylindrical surface. Similarly the lowest order mode in each azimuthally-different family of flexural modes becomes a spiralling Rayleigh wave in the short wavelength limit. Propagating on surfaces of variable curvature is discussed by RULF [36], including the effect of energy shedding when the radius drops below a certain critical value. Another phenomenon of interest is the propagation of additional modes confined to the surface of a curved solid-vacuum interface [37]. These waves have been called "whispering gallery" waves after RAYLEIGH's first investigation of this problem in acoustics [38]. Figure 8 illustrates the horizontal-shear version of whispering gallery modes, using the ray representation commonly followed in acoustics. In (a) the ray representation is shown for an SH mode on a bent plate. Part (b) shows the typical ray representation for a whispering gallery mode of pure shear type. For the longitudinal-shear type of mode, which also occurs, there are two types of rays [37].

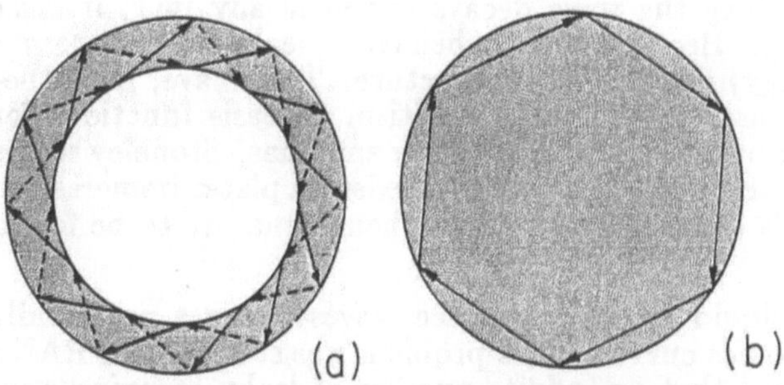

Fig. 8. (a) Ray representation of an SH mode in a bent plate; (b) Pure shear whispering gallery modes on the surface of a cylindrical rod

(VI) Solid-Liquid Interfaces

In 1924 solid-solid interface surface waves were first identified by STONELEY [9]. For a *plane* solid-liquid interface it was shown by Gogoladze in 1948 and Brekhovskikh in 1957 that there exist two waves traveling along the interface. One, called the Stoneley wave, decays exponentially away from the interface in *both* media and has most of its energy in the liquid. It is a true surface wave. The other has most of its energy in the solid and radiates energy into the liquid. It is a leaky

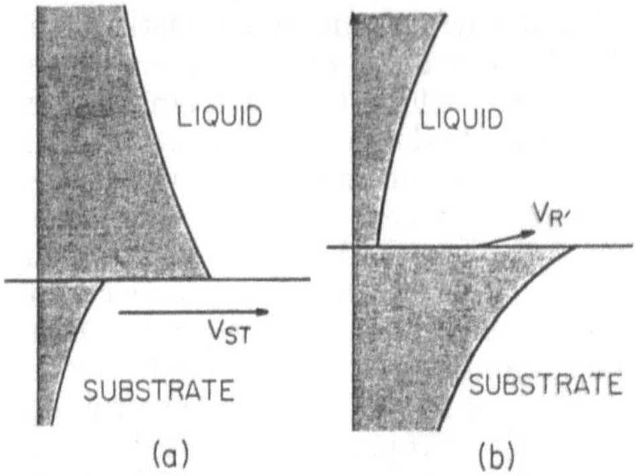

Fig. 9. Waves at a plane solid-liquid interface. (a) Stoneley wave; (b) Leaky Rayleigh wave. (after Ref. 40)

Rayleigh wave and therefore propagates along the interface with attenuation. Figure 9 illustrates schematically the wave functions for these two solutions. Part (b) shows that the field strength of the leaky Rayleigh wave grows with increasing penetration into the liquid, precisely the type of behavior noted by Rayleigh in connection with the complex roots of (11). The reason is simple. As shown by the arrow marked V'_R, the energy radiates obliquely into the liquid, so that the radiation field at larger and larger values of y originates from points on the surface at more and more negative values of z. Since the wave decays exponentially in z, it must grow exponentially in y. Because of this behavior, leaky waves are not proper (i.e., finite energy) modes of the structure. They have, nevertheless, been found to be useful in electromagnetism as basis functions for analyzing the radiation patterns of surface wave antennas. Stoneley waves and leaky Lamb waves are, similarly, found to exist on plates immersed in a liquid. Further details concerning all these phenomena are to be found in Refs. 31 and 39.

The nature of solid-liquid interface surface waves changes profoundly when the interface becomes curved. This problem was studied by FRANZ in 1954 [40], who showed that an infinite number of leaky *creeping wave* solutions now exist. These correspond to the "trapped" Stoneley surface wave for the plane solid-liquid interface (Fig. 9(a)). In addition to the creeping (or Franz) waves, there also exist leaky Rayleigh waves on curved solid-liquid interfaces, corresponding to Fig. 9(b). All of these waves are now of practical importance in underwater sound and nondestructive evaluation [40,41,42].

(VII) Period Composites (Elastic Superlattices)

The design and fabrication of artificial materials for special purposes is a rapidly growing technology with applications in many areas—for example: mechanical structures, piezoelectric transducers, and electronic devices. Such materials frequently take the form of composites, in which the nature and proportion of the component phases are chosen to maximize a figure of merit for some particular application. Composite materials are increasingly being fabricated with strictly periodic structures, notably in the case of PZT-epoxy piezoelectric transducer materials [43]. Currently, these are constructed on a *macroscopic* scale; but the perfection of molecular beam epitaxy techniques for fabricating microscopically-layered materials now makes it possible to create new materials with tailored *microscopic* properties. In elasticity theory these materials may all be described as elastic superlattices.

The availability and growing use of such periodic composite materials for sonics and ultrasonics is now driving new basic studies of Rayleigh and other types of elastic waves on superlattices. At wavelength much larger than the structural period, an elastic superlattice behaves as an artificial anisotropic material, with controllable elastic parameters. It is, however, also possible, by suitably choosing the period, to locate the Brillouin zone boundaries at arbitrary frequencies in the ultrasonic range. This feature substantially modifies the wave solutions discussed in the preceding sections and introduces some completely new phenomena.

In LOVE's analysis of waves guided by a layer on a half space [8], he showed that a necessary condition for horizontal shear (or Love) wave "trapping" was that the shear velocity of the layer be slower than that of the half space. If the half space is a one-dimensional periodic composite with layers parallel to the free surface, horizontal shear surface waves exist over a restricted frequency range, regardless of the velocity ratio between the surface layer and its immediate neighbor [44,45]. The effect is based on the existence of stopbands for horizontal shear waves propagating obliquely in the layered composite half space.

A Floquet-type analysis applicable to elastic wave propagation in general three-dimensional periodic composites has also been formulated [46,47], and detailed solutions have been developed for two-dimensional periodic arrays of isotropic rods in an isotropic matrix. This is a fundamental analysis of the equations of elasticity, with material parameters that vary periodically in space. Because the continuity conditions at boundaries between the component phases are automatically taken into account in this approach, it is easily applicable to the most general unit cell structures. The analysis is performed by expanding the elastic wave function in a series of space harmonics, following the method first presented by Brillouin for solving the scalar wave equation with spatially-varying coefficients. In the elastic case, each space harmonic contains the three allowed elastic wave polarizations, which must be separated by an orthogonality property. The result is shown to be the same as in Brillouin's analysis, an infinite set of coupled linear equations for the space harmonic amplitudes, which can then be solved numerically to any desired accuracy or can be solved approximately by the coupled-mode approximation, using only two space harmonics. The coupled-mode approach is appropriate for calculating the stopband (or forbidden zone) behavior of the superlattice, a major goal of the references cited. (A second analysis of elastic wave propagation in an infinite periodic composite has just recently appeared [48]. This is a hamiltonian formulation and was applied only in calculating effective elastic moduli of the composite in the long wavelength limit.)

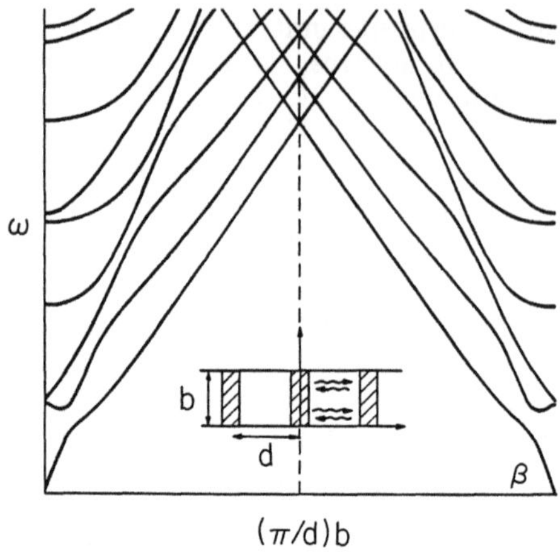

Fig. 10. Brillouin diagram for even-symmetry Rayleigh-Lamb wave propagation on an isotropic composite plate.

The Floquet method was first implemented using plane wave space harmonics. This correctly predicted the experimentally observed elastic wave function distributions at the Brillouin zone boundaries [46,49,50]. However, it did not take into account free boundary conditions at the surfaces of the composite plates used in the experiments. This was done in another analysis, where the space harmonics were expanded in terms of the Lamb wave modes having a plate density, and a compliance equal to the spatial averages of these quantities for the composite material [51]. These composite piezoelectric plates are of growing importance in ultrasonic transducer technology [52,53,54].

Figure 10 shows the 0 and -1 space harmonic curves on the Brillouin diagram obtained using Lamb wave basis functions. When space harmonic coupling is introduced by periodicity of the material, stopbands appear at all crossing points of the curves in the figure. It can be seen from this that the stopband spectrum of the plate is much more complicated than that of an infinite elastic composite, where the space harmonic curves on the Brillouin diagram are pairs of straight lines, one for longitudinal and one for shear. Another feature of Fig. 10 also deserves attention. It was first noted by LAMB [16] that the lowest branches of the Lamb wave curves become Rayleigh waves in the short wavelength limit. The appearance of a stopband gap at the crossover point for the lowest Lamb wave branch in Fig. 10, suggests the existence of evanescent Rayleigh surface waves in the stopband gap. Such evanescent surface waves are known to exist on the periodic surface gratings used in SAW resonators. In this case, which is currently being studied by a variant of the Floquet approach described above, the evanescent surface waves lie below the longitudinal and shear "sound lines" of the Brillouin diagram. Consequently, none of the space harmonics are radiating, and the solution is a true surface wave. Stated differently, the bulk waves excited by scattering at each tooth of the grating are phased so as to exactly cancel in the radiation field. On this basis, the same phenomenon might be expected in Fig. 10, but this hypothesis remains to be tested by further analysis and experiment.

References

1. Lord Rayleigh: "On Waves Propagating Along the Plane Surface of an Elastic Solid," Proc. Lond. Math. Soc. **17**, 4 (1885).
2. Lord Rayleigh: "On the Free Vibrations of an Infinite Plate of Homogeneous Isotropic Elastic Matter," Proc. Lond. Math. Soc. **20**, 225 (1889).
3. A. L. Cauchy: "Sur l'Equilibre et le Mouvement d'une Plaque Solide," Exercises de Mathématique **3**, 328 (1828).
4. A. L. Cauchy: "Sur l'Equilibre et le Mouvement d'une Plaque Elastique dont l'Elasticité n'est pas la Même dans tous les Sens," Exercises de Mathématique **4**, 1 (1829).
5. S. P. Poisson: "Mémoire sur l'Equilibre et le Mouvement des Corps Elastiques," Mém. Acad. Sci. Ser. 2 **8**, 357 (1829).
6. H. Lamb: "On the Flexure of an Elastic Plate (Appendix)," Proc. Lond. Math. Soc. **21**, 85 (1889).
7. C. Chree: "The Equations of an Isotropic Solid in Polar and Cylindrical Coordinates, Their Solutions and Applications," Trans. Cambridge Phil. Soc. **14**, 250 (1889).
8. A. E. H. Love: *Some Problems in Geodynamics*, Cambridge University Press, London, 1911, 1926.
9. R. Stoneley: "Elastic Waves at the Surface of Separation of Two Solids," Roy. Soc. Proc. London, Series **A 106**, 416 (1924).
10. K. Sezawa: "On the Propagation of Rayleigh Waves on Plane and Spherical Surfaces," Bull. Earthquake Res. Inst. (Tokyo) **2**, 21 (1927).
11. K. Sezawa: "Dispersion of Elastic Waves Propagated on the Surface of Stratified Bodies and on Curved Surfaces," Bull. Earthquake Res. Inst. (Tokyo) **3**, 1 (1927).
12. H. Lamb: "On the Propagation of Tremors over the Surface of an Elastic Solid," Phi. Trans. Roy. Soc. (London) **A 203**, 1 (1904).
13. H. Nakano: "On Rayleigh Waves," Japan. J. Astron. Geophys. **2**, 233 (1925).
14. L. Cagniard: "Réflexion et Réfraction des Ondes Séismiques Progressives," (Thèse), Gauthier-Villars et Cie, Paris, 1939.
15. J. Miklowitz: *The Theory of Elastic Waves and Waveguides* (North-Holland, Amsterdam 1978).
16. H. Lamb: "On Waves in an Elastic Plate," Proc. Roy. Soc. A **93**, 114 (1917).

17. D. Bancroft: "The Velocity of Longitudinal Waves in Cylindrical Bars," Phys. Rev. **59**, 588 (1941).
18. G. E. Hudson: "Dispersion of Elastic Waves in Solid Circular Cylinders," Phys. Rev. **63**, 46 (1943).
19. D. L. Arenburg: "Ultrasonic Solid Delay lines," J. Acous. Soc. Am. **20**, 1 (1948).
20. H. Ekstein: "High Frequency Vibrations of Thin Crystal Plates," Physical Review **68**, 11 (1945).
21. R. D. Mindlin: "Thickness-shear and Flexural Vibrations of Crysal Plates," J. Appl. Phys. **22**, 316 (1951).
22. A. N. Holden: "Longitudinal Modes of Elastic Waves in Isotropic Cylinders and Plates," BSTJ **30**, 956 (1951).
23. I. Tolstoy and E. Usdin: "Dispersive Properties of Stratified Elastic and Liquid Media: A Ray Theory," Geophysics **18**, 844 (1953).
24. M. Onoe: "A Study of the Branches of the Velocity-Dispersion Equations of Elastic Plates and Rods," Report of the Joint Committee on Ultrasonics—Institute of Electrical Communications Engineers and the Acoustical Society of Japan, 1955.
25. R. H. Lyon: "Reponse of an Elastic Plate to Localized Driving Forces," J. Acous. Soc. Am. **27**, 259 (1955).
26. J. W. C. Sherwood: "Propagation in an Infinite Elastic Plate," J. Acous. Soc. Am. **30**, 979 (1958).
27. R. N. Thurston: "Elastic Waves in Rods and Clad Rods," J. Acous. Soc. Am. **64**, 1 (1968).
28. E. A. G. Shaw: "On the Resonant Vibrations of Thick Barium Titanate Disks," J. Acous. Soc. Am. **28**, 38 (1956).
29. E. G. Cook and H. E. Valdenburg: "Surface Waves at Ultrasonic Frequencies," Am. Soc. Testing Mater. Bull. **198**, 81 (1954).
30. C. Minton: "Inspection of Metals with Ultrasonic Surface Waves," Nondestructive Testing **12**, 13 (1954).
31. I. A. Viktorov: *Rayleigh and Lamb Waves*, (Plenum, New York 1967).
32. G. W. Farnell: "Properties of Elastic Surface Waves," in *Physical Acoustics*, ed. by W. P. Mason and R. N. Thurston, Vol. VI (Academic Press, New York 1970, Ch. 3.)
33. G. W. Farnell and E. L. Adler: "Elastic Wave Propagation in Thin Layers," in *Physical Acoustics*, ed. by W. P. Mason and R. N. Thurston, Vol. IX (Academic Press, New York 1972, Ch. 2.)
34. R. S. Wagers: "Plate Modes and Surface Acoustic Wave Devices," in *Physical Acoustics*, ed. by W. P. Mason and R. N. Thornton, Vol. XIII (Academic Press, New York 1977, Ch. 3).

35. T. R. Meeker and A. H. Meitzler: "Guided Wave Propagation in Elongated Cylinders and Plates," in *Physical Acoustics,* ed. by W. P. Mason, Vol. IA (Academic Press, New York 1964, Ch. 2).

36. B. Rulf: "Rayleigh Waves on Curved Surfaces," J. Acous. Soc. Am. **45**, 493 (1968).

37. L. M. Brekhovskikh: "Surface Waves Confined to the Curvature of the Boundary in Solids," Soviet Physics–Acoustics **13**, 462 (1968).

38. Lord Rayleigh: " Whispering Galleries," Section 287 *The Theory of Sound 2* (MacMillan, London 1896, 2nd Ed.), p. 126.

39. H. Überall: "Surface Waves in Acoustics," in *Physical Acoustics,* ed. by W. P. Mason and R. N. Thurston, Vol. X (Academic Press, New York 1970, Ch. 1).

40. W. Franz: "Über die Greenschen Fonctionen des Zylinders und der Kugel," Z. Naturforsch. **9A**, 705 (1954).

41. R. E. Bunney, R. R. Goodman, and S. W. Marshall: "Rayleigh and Lamb Waves on Cylinders," J. Acous. Soc. Am. **46**, 1223 (1969).

42. H. Überall, L. R. Dragonette, and L. Flax: "Relation Between Creeping Waves and Normal Modes of Vibration of a Curved Body," J. Acous. Soc. Am. **61**, 711 (1977).

43. T. R. Gururaja, W. A. Schulze, T. R. Shrout, A. Safari, L. Webster, and L. E. Cross: "High Frequency Applications of PZT/Polymer Composite Materials," Ferroelectrics, **39**, 1245 (1981).

44. B. A. Auld, G. S. Beaupre, and G. Hermann: "Horizontal Shear Surface Waves on a Laminated Composite," in *Modern Problems in Elastic Wave Propagation,* ed. by J. Miklowitz and J. D. Achenbach (Wiley, New York 1978), p. 459.

45. G. Hermann, G. S. Beaupre, and B. A. Auld: "Applicability of Floquet-type Solutions to Bounded Layered Composites," in *Mechanics Today,* ed. by S. Nemat-Nasser (Pergamon Press, Oxford and London 1980), p. 83.

46. B. A. Auld, H. A. Kunkel, Y. A. Shui, and Y. Wang: "Dynamic Behavior of Periodic Piezoelectric Composites," in 1983 IEEE Ultrasonic Symposium Proceedings, ed. by B. R. McAvoy (IEEE, Piscataway, N.J. 1983), p. 554.

47. B. A. Auld, Y. A. Shui, and Y. Wang: "Elastic Wave Propagation in Three-dimensional Periodic Composite Materials," J. de Physique, Colloque C5, supplément au n°4 **45**, 159 (1984).

48. R. Tao and P. Sheng: "First Principle Approach to the Calculation of Elastic Moduli for Arbitrary Periodic Composites," J. Acous. Soc. Am. **77**, 1651 (1985).

49. T. R. Gururaja, W. A. Schulze, L. E. Cross, B. A. Auld, Y. A. Shui, and Y. Wang: "Resonant Modes in Piezoelectric PZT-polymer Composites with Two-dimensional Periodicity," Ferroelectrics **54**, 183 (1984).

50. T. R. Gururaja, W. A. Schulze, L. E. Cross, R E. Newnham, B. A. Auld, and Y. Wang: "Resonant Modes in Piezoelectric PZT Rod-polymer Composite Materials," in 1984 Ultrasonic Symposium Proceedings, ed. by B. R. McAvoy (IEEE, Piscataway, N.J. 1984), p. 523.

51. B. A. Auld and Y. Wang: "Acoustic Wave Vibrations in Periodic Composite Plates," in 1984 Ultrasonic Symposium Proceedings, ed. by B. R. McAvoy (IEEE, Piscataway, N.J. 1984), p. 528.

52. T. R. Gururaja, W. A. Schulze, L. E. Cross, R E. Newnham: "Ultrasonic Properties of PZT Rod-polymer Composites," in 1984 Ultrasonic Symposium Proceedings, ed. by B. R. McAvoy (IEEE, Piscataway, N.J. 1984), p. 533.

53. W. A. Smith, A. A. Shaulov, and B. M. Singer: "Properties of Composite Piezoelectric Materials for Ultrasonic Transducers," in 1984 Ultrasonic Symposium Proceedings, ed. by B. R. McAvoy (IEEE, Piscataway, N.J. 1984), p. 539.

54. A. A. Shaulov, W. A. Smith, and B. M. Singer: "Performance of Ultrasonic Transducers made from Composite Piezoelectric Materials," in 1984 Ultrasonic Symposium Proceedings, ed. by B. R. McAvoy (IEEE, Piscataway, N.J. 1984), p. 545.

The Relationship Between Surface Wave Displacement and Anisotropy on Selected Crystal Structures

E. Dieulesaint and D. Royer

Laboratoire d'Acoustoélectricité de l'Université
P. et M. Curie E.S.P.C.I., 10, rue Vauquelin,
F-75231 Paris Cedex 05, France

I - INTRODUCTION

The influence of anisotropy on conditions of propagation of elastic waves in solids is known. As regards the bulk waves, three waves (one quasi-longitudinal, two quasi-transversal) can propagate in an anisotropic solid, whereas only two propagate in an isotropic solid. Moreover in an anisotropic solid, the phase velocity and the Poynting vector of each wave change with the direction of propagation. As regards the Rayleigh surface wave, the possible differences, besides the variation of the velocity and of the Poynting vector with the propagation direction in a given plane, are *i)* a mechanical displacement with three components instead of two for an isotropic solid, *ii)* an oscillation of each component with depth and along a distance from the surface greater than 2λ, *iii)* the existence of pseudo surface modes.

The object of this paper is to examine the two last features. It is shown that the existence of a pseudo surface wave can be predicted from the velocity variation of the bulk waves which is related to the value of the anisotropy factor and from the concept of wave coupling. For simplicity, this comment is applied to the case of a crystal having a cubic symmetry. However, it is valid for crystals belonging to other symmetry classes. The effect of anisotropy on the decrease versus depth of the mechanical displacement is shown analytically when the Rayleigh wave displacement has only two components.

Before developing these points, let us mention that analyses on surface acoustic wave propagation in anisotropic solids were carried out before the advent of computers, by STONELEY [1, 2], SYNGE [3], DERESIEWICZ and MINDLIN [4], GAZIS [5]. Later on, many scientists in particular CAMPBELL and JONES [6], FARNELL [7], have applied numerical techniques to this problem.

2 - ANISOTROPY AND PSEUDO-SURFACE WAVE IN A CUBIC SYMMETRY CRYSTAL

As in a semi-infinite medium the absence of matter above the surface decreases the stiffness of the solid, the phase velocity of Rayleigh waves propagating in an isotropic solid is slightly less (about 10 % less) than the velocity of the shear wave.

In a crystal of cubic symmetry i.e. described by three constants c_{11}, c_{12} and $c_{66} \neq (c_{11} - c_{12})/2$, a vertically polarized transverse wave (TV) and a horizontally polarized quasitransverse wave (TH) can propagate in a plane such as (001). The shear wave TV propagates with a velocity $V_{TV} = \sqrt{c_{66}/\rho}$ whatever the direction. The quasi-shear wave TH propagates with the velocity $V_{TH}(\theta)$, (θ : angle from direction [100]) and $V_{TH}(0) = V_{TV}$, its polarization lying in the (001) plane. However as the planes (010) and $(1\bar{1}0)$ are symmetry planes from the elastic point of view whatever the cubic class considered, the polarization of the wave TH is perpendicular to the wave vector for directions [100] and [110]. The two shear modes are pure along these directions. Consequently, a Rayleigh wave propagating along either direction consists of only two components in the sagittal plane and its propagation velocity V_R is independent of the velocity of the wave TH as the mechanical displacements are not coupled. It is not out of the question that V_R be greater than V_{TH}.

Two situations are to be distinguished according as the square of the ratio of the shear wave velocities along [110] : $A = (V_{TV} / V_{TH})^2 = 2\, c_{66}/(c_{11} - c_{12})$, which is the anisotropy factor of the solid, is less or greater than unity.

The first situation ($A < 1 \Rightarrow V_{TV} < V_{TH}$) is illustrated in figure 1a. The velocity V_{TH} of the wave TH increases with θ from its value along [100] and reached a maximum along [110]. The coupling between this wave TH and the Rayleigh wave is weak. Therefore the velocity of the Rayleigh wave $V_R(\theta)$ remains almost constant, slightly less than V_{TV}.

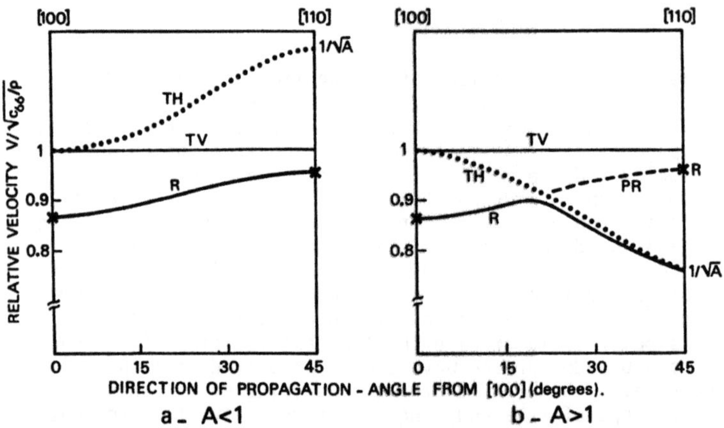

FIG .1. Velocity of surface waves (R: Rayleigh, PR: pseudo-Rayleigh) in the (001)-plane of a cubic crystal for anisotropy factor A less (a) or greater (b) than unity. The Rayleigh waves denoted (✕) are not coupled with TH bulk wave so that their velocity V_R can be larger than V_{TH}.

The second situation ($A > 1 \Rightarrow V_{TV} > V_{TH}$) is depicted in figure 1b. The velocity $V_{TH}(\theta)$ of the wave TH decreases continuously with θ from its value along [100] and reaches, along the direction [110], a minimum value less than the Rayleigh wave velocity $V_R(\pi / 4)$. Progressively as the angle θ increases, the coupling between the wave TH and the Rayleigh wave also increases. Eventually the Rayleigh wave degenerates into wave TH, the curve $V_R(\theta)$ mixing with the characteristic $V_{TH}(\theta)$. Now if we start from the representative point $V_R(\pi / 4)$ along [110] the coupling between the two waves TH and R also appears. The surface wave having a velocity $V_R > V_{TH}$ excites a growing amount of wave TH. This is no longer confined near the surface. It radiates towards the bulk of the material. It is then called *pseudo (or leaky) surface wave.*

Obviously, the foregoing reflections for $A > 1$ and plane (001) can also be valid for $A < 1$ and a plane other than (001) if the position of the velocity characteristics is similar. FARNELL [7] has shown that the results obtained for $A > 1$ and plane (001) are found qualitatively for $A < 1$ and plane (110). The preceding results incite to explore, in a methodical way, crystals less symmetric than cubic crystals and configurations without coupling similar to those examined and to determine analytically the equations for the mechanical displacement which consists of only two components.

3 - CONDITIONS FOR WAVE DECOUPLING

In a semi-infinite medium, unbounded in the directions x_1 and x_3 a surface wave propagating along x_1 is described by a particle displacement of the form

$$u_i = {}^\circ u_i \exp(-\,iqkx_2)\exp i(\omega t - kx_1) \qquad \text{with } \mathrm{Im}[q] < 0 \tag{1}$$

which expresses a decrease with depth below the free surface $x_2 = 0$. Taking into account the generalized Hooke's law which relates the stress T_{ij} to the displacement u_i by the stiffness tensor c_{ijkl} referred to the Cartesian axes x_i

$$T_{ij} = c_{ijkl}\, u_{l,k} \qquad (i,\, j,\, k,\, l = 1,\, 2,\, 3) \tag{2}$$

and substituting u_i and T_{ij} into the equation of motion

$$\rho\ddot{u}_i = T_{ij,j} = c_{ijkl}\, u_{l,jk} \qquad \text{for } x_2 > 0 \tag{3}$$

leads to the Christoffel equation.

$$(\Gamma_{il} - \zeta\delta_{il})\, {}^\circ u_l = 0 \qquad \text{with } \zeta = \rho V^2 \tag{4}$$

and

$$\Gamma_{il} = c_{i1l1} + (c_{i12l} + c_{i2l1})\, q + c_{i2l2}\, q^2 \tag{5}$$

System (4) splits into two parts if Γ_{13} and Γ_{23} vanish giving a Rayleigh wave polarized in the sagittal plane x_1x_2 and a shear horizontal wave polarized along x_3 that are uncoupled. This situation occurs when the six elastic constants with a single index of 3 are zero :

$$c_{14} = c_{24} = c_{15} = c_{25} = c_{64} = c_{65} = 0 \tag{6}$$

For the Rayleigh wave being considered $u_3 = 0$ and the mechanical boundary conditions on the free surface $x_2 = 0$

$$T_{i2} = -\,ik\, c_{i211}\, u_1 + c_{i212}(u_{1,2} - iku_2) + c_{i222}\, u_{2,2} = 0 \tag{7}$$

Crystal symmetry	ORTHORHOMBIC	TETRAGONAL (4, $\bar{4}$, 4/m)					
Propagation direction	A_2	⊥Z		//Z		X+ϕ_0	X+ϕ_0+π/4
Normal to the free surface	A_2'	X+ϕ_0	X+ϕ_0+π/4	X+ϕ_0	X+ϕ_0+π/4	//Z	

Crystal Symmetry	HEXAGONAL			TETRAGONAL (422, 4mm, $\bar{4}$2m, 4/mmm) or CUBIC					
Propagation direction	⊥Z	//Z	⊥Z	⊥Z		//Z		X	X+$\frac{\pi}{4}$
Normal to the free surface	⊥Z	⊥Z	//Z	X	X+$\frac{\pi}{4}$	X	X+$\frac{\pi}{4}$	//Z	
Cases solved by Stoneley			[2]					[1]	[1]

Table 1. The sixteen configurations for which the Rayleigh wave is decoupled from TH wave and the boundary conditions simplify. The angle ϕ_0 is given by $\tan 4\phi_0 = 4c_{16}/(c_{11} - c_{12} - 2c_{66})$.

is always fulfilled when i = 3. The two other equations (i = 1,2) simplify if

$$c_{1211} \equiv c_{61} = 0 \qquad \text{and} \qquad c_{1222} \equiv c_{62} = 0 \tag{8}$$

Symmetry considerations show that conditions (6) and (8) may be satisfied only in crystals belonging to orthorhombic, tetragonal, cubic and hexagonal systems. Sixteen suitable configurations, i.e. dispositions of the propagation direction (x_1) and of the normal (x_2) to the free surface with respect to the crystallographic axes XYZ, have been numbered [8]. Table 1 shows that they include the three particular cases (propagation in the (001) plane along directions [100] and [110] of cubic crystals considered in the foregoing section and in the basal plane of hexagonal crystal) previously studied by Stoneley [1,2].

4 - MECHANICAL DISPLACEMENT

In the following, conditions (6) and (8) are assumed to be satisfied; then the secular equation of system (4) reduces to

$$q^4 - Sq^2 + P = 0 \tag{9}$$

where P and S are the product and the sum of the roots q_1^2 and q_2^2 of this quadratic equation for q^2

$$P = q_1^2 q_2^2 = (c_{11} - \zeta)(c_{66} - \zeta) / c_{22}\, c_{66} \tag{10}$$

$$S = q_1^2 + q_2^2 = [2\, c_{12}c_{66} + c_{66}\zeta - c_{22}(c - \zeta)] / c_{22}c_{66} \tag{11}$$

where

$$c = c_{11} - c_{12}^2 / c_{22} \tag{12}$$

For the two valid solutions q_r (r = 1, 2), i.e. with a negative imaginary part, the components $°u_i^{(r)}$ of the eigenvector are given in Eq. (4); with $°u_1^{(r)} = 1$ and $°u_3^{(r)} = 0$:

$$p_r = °u_2^{(r)} = - \frac{c_{11} - \zeta + c_{66} q_r^2}{(c_{12} + c_{66})\, q_r} \qquad r = 1, 2 \tag{13}$$

The general solution is a linear combination of these two displacements propagating at the same velocity $V = \omega / k$:

$$\begin{cases} U_1(x_2) = A_1 \exp(- iq_1kx_2) + A_2 \exp(- iq_2kx_2) \\ U_2(x_2) = A_1p_1 \exp(- iq_1kx_2) + A_2p_2 \exp(- iq_2kx_2) \end{cases} \tag{14}$$

The weighting factors A_1, A_2 and the velocity V are determined by the boundary conditions (7) that, for i = 1 and 2, are expressed as

$$(q_1 + p_1)\, A_1 + (q_2 + p_2)\, A_2 = 0 \tag{15}$$

$$(c_{21} + c_{22}q_1p_1)\, A_1 + (c_{21} + c_{22}q_2p_2)\, A_2 = 0 \tag{16}$$

These two equations are compatible providing the determinant is zero

$$(p_1 - p_2)(c_{12} - c_{22}q_1q_2) + (q_1 - q_2)(c_{12} - c_{22}p_1p_2) = 0 \tag{17}$$

Substituting the value of $p_1 - p_2$ deduced from (13) and noting that the factor $c_{22}\,p_1p_2q_1q_2$ expresses, with the aid of the sum S and of the product P of the roots, as

$$c_{22}p_1p_2q_1q_2 = c_{11} - \zeta \tag{18}$$

Equation (17) reduces to

$$q_1q_2 = -\zeta(c_{11} - \zeta) / c_{22}(c - \zeta) \quad (19)$$

Substituting this expression into (10) leads to a wave velocity equation

$$f(\zeta) \equiv c_{66}(c_{11} - \zeta)\,\zeta^2 - c_{22}(c_{66} - \zeta)(c - \zeta)^2 = 0 \quad (20)$$

similar to that established by Stoneley [2] for the particular case of Rayleigh waves propagating in the basal plane of hexagonal crystals. However this equation is more general : it applies for the sixteen configurations given in Table 1.

When $\zeta = 0$, $f(\zeta)$ is negative and when $\zeta = \min[c_{66}, c]$, $f(\zeta)$ is positive since $c < c_{11}$, so that there is one and only one root ζ_R such that

$$0 < \zeta_R < \min[c_{66}, c] \quad (21)$$

The velocity V_R of the Rayleigh wave being now known ($\zeta_R = \rho V_R^2$), it is possible to derive the components of the mechanical displacement. Substituting $c_{11} - \zeta$ by Eq. (18) in Eq. (13) gives

$$p_r(c_{12} + c_{22}p_sq_s) = -c_{66}(p_r + q_r) \qquad r \neq s = 1,2$$

and taking Eqs. (15) and (16) into account yields

$$p_1 / p_2 = (A_2 / A_1)^2 \quad (22)$$

When the roots q_1^2 and q_2^2 of Eq. (9) are real and negative, i.e. $S < -2\sqrt{P}$, the two physically acceptable solutions are $q_r = -i\chi_r$, with $\chi_r > 0$. From Eq. (13), the coefficients p_r are purely imaginary : $p_r = -ia_r$, where the factor

$$a_r = (c_{11} - \zeta - c_{66}\chi_r^2) / (c_{12} + c_{66})\,\chi_r \quad (23)$$

is positive from Eq. (11). Equation (15) shows that the ratio A_2 / A_1 is real and negative, hence taking Eq. (22) into account,

$$A_2 / A_1 = -(a_1 / a_2)^{1/2} \text{ and } p_2A_2 = iaA_1 ,$$

with, from Eqs. (18) and (19)

$$a \equiv (-p_1p_2)^{1/2} = (a_1a_2)^{1/2} = (c/\zeta_R - 1)^{1/2} \quad (24)$$

Therefore the longitudinal and transverse components of the mechanical displacement, given by (14), are $\pi/2$ phase shifted :

$$\begin{cases} U_1(x_2) = A_1[e^{-\chi_1kx_2} - (a_1 / a_2)^{1/2}\, e^{-\chi_2kx_2}] \\ U_2(x_2) = iaA_1[e^{-\chi_2kx_2} - (a_1 / a_2)^{1/2}\, e^{-\chi_1kx_2}]. \end{cases} \quad (25)$$

Their decrease versus depth is identical to that of the components of the Rayleigh wave propagating in an isotropic solid for which $c_{66} = (c_{11} - c_{12})/2$. In particular, as a_r is a decreasing function of χ_r, the transverse component does not change sign whereas the longitudinal passes through zero.

When the roots of Eq. (9) are complex and then conjugate, i.e. $|S| < 2\sqrt{P}$, the two valid solutions are

$$q_r = (-1)^r g - ih \qquad \text{with} \qquad \begin{Bmatrix} g \\ h \end{Bmatrix} = [2\sqrt{P} \pm S]^{1/2} / 2 \quad (26)$$

Equation (13) shows that $p_1 = -p_2^* = ia\,e^{-i2\alpha}$ with, from Eqs. (22) and (15),

$$e^{-i2\alpha} = A_2 / A_1 = (q_2 + p_2)^* / (q_2 + p_2)$$

The two components (14) of the Rayleigh wave are again $\pi/2$ phase shifted, however, their amplitudes,

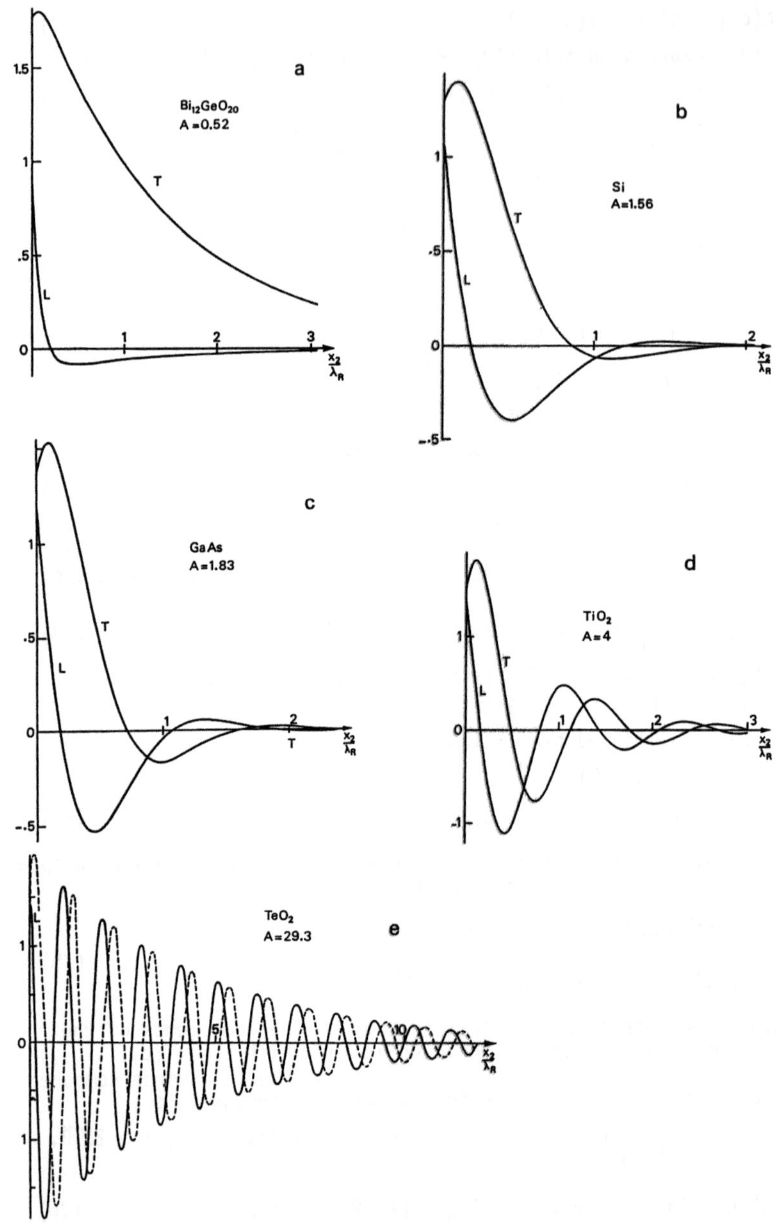

FIG. 2. Variation versus depth of the longitudinal (L) and transverse (T) components of the mechanical displacement of Rayleigh waves propagating along the [100] axis in the plane (010). For crystals (Si, GaAs, TiO_2, TeO_2) with A > 1, the larger the anisotropy factor A, the less the damping of the oscillations.

$$\begin{cases} U_1(x_2) = 2A_1 e^{-hkx_2}\cos(gkx_2 + \alpha) \\ U_2(x_2) = 2iaA_1 e^{-hkx_2}\cos(gkx_2 - \alpha) \end{cases} \qquad (27)$$

oscillate versus depth with a spatial period λ_R / g equal to the ratio of the wavelength λ_R and of the real part g of the roots ($q_2 = - q_1^*$). On the other hand, the decrease versus depth of the mechanical displacement, related to the imaginary part h of these roots, can be much slower than in the foregoing case. In fact, the difference stems from the anisotropy of the medium in the sagittal plane.

5 - INFLUENCE OF THE ANISOTROPY FACTOR

For crystals belonging to the cubic or tetragonal system, $c_{22} = c_{11}$ if the fourfold axis of symmetry is parallel to axis x_3. When the anisotropy factor in the sagittal plane : $A = 2c_{66}/(c_{11} - c_{12})$ is large with respect to unity, i.e., c_{12} slightly less than c_{11}, the solution ζ_R of Eq. (20) necessarily less than $c = c_{11} - c_{12}^2/c_{11}$, is small compared to c_{11} and c_{66} : $\zeta_R \simeq c/2$. Developing the sum S and the product P of the roots to the second order in ζ_R yields the imaginary part [formula (26)] in the form

$$h \simeq (1 + c_{66}/c_{11})/2A, \qquad A \gg 1. \qquad (28)$$

Consequently, *the larger the anisotropy factor, the less the damping of the oscillations*. This result is illustrated by the curves in Fig. 2 referring to propagation along the binary axis of symmetry [100] in crystals of $Bi_{12}GeO_{20}$, silicon (Si), gallium arsenide (GaAs), rutile (TiO_2), and tellurium dioxide (TeO_2). The numerical values of the stiffness constants used were from Refs. 9 - 13. Three of these crystals : $Bi_{12}GeO_{20}$, GaAs and TeO_2 are piezoelectric, however for the case considered, this effect does not have to be taken into account because the sagittal plane is perpendicular to a direct binary axis of symmetry [14]. It is worthwhile noting that as A increases, the ratio of the Rayleigh wave velocity V_R to the shear bulk wave velocity $V_T = \sqrt{c_{66}/\rho}$ decreases, as expected for $A \gg 1$:

$$\rho V_R^2 \simeq (c_{11}^2 - c_{12}^2)/2c_{11} \quad \Rightarrow \quad \frac{V_R}{V_T} \simeq \left[\frac{2}{A}\left(1 - \frac{c_{66}}{c_{11}A}\right)\right]^{1/2} \qquad (29)$$

For example for tellurium dioxide V_R is nearly four times less than V_T. In an isotropic solid V_R / V_T is between 0.874 and 0.955. Practically, Eqs. (28) and (29) are good approximations for $A > 4$.

6 - CONCLUSION

The propagation of Rayleigh waves in a crystal having a cubic symmetry has been examined from a physical point of view in considering in the (001)-plane the coupling between the Rayleigh wave and the horizontally polarized shear wave whose velocity varies with direction. The coupling explains the existence of the pseudo-surface wave when the anisotropy factor is greater than unity. Then the conditions for wave decoupling have been expressed analytically for any crystal. They lead to sixteen configurations for crystals belonging to orthorhombic, tetragonal, cubic and hexagonal symmetry classes. On the other hand, equations of the two mechanical displacement components have been established. For a sagittal plane perpendicular to a fourfold axis of symmetry it has been shown that the larger the anisotropy factor, the slower the decrease versus depth of the two mechanical components. This observation has been illustrated by examples ($Bi_{12}GeO_{20}$, Si, GaAs, TiO_2, TeO_2).

REFERENCES

1 - R. Stoneley, Proc. Roy. Soc. London, Ser. A232, 447-458 (1955).
2 - R. Stoneley, Mon. Not. Roy. Astron. Soc. Geophys. Suppl. 5,343-353 (1949).
3 - J. Synge, J. Math. Phys. 35, 323 (1957).
4 - H. Deresiewicz and R.D. Mindlin, J. Appl. Phys. 28, 669-671 (1957).
5 - D.C. Gazis, R. Herman and R.F. Wallis, Phys. Rev. 119, 533-544 (1960).
6 - J.J. Campbell and W.R. Jones, IEEE Trans. Sonics and Ultrasonics. SU.15, 209 (1968).
7 - G.W. Farnell : "Properties of Elastic Surface Waves" In Physical Acoustics (W.P. Mason and R.N. Thurston, eds.) Vol. 6, p. 109-166. New-York : Academic Press (1970).
8 - D. Royer and E. Dieulesaint, J. Acoust. Soc. Am. 76, 1438-1444 (1984).
9 - E.A. Kraut, B.R. Tittman, L.J. Graham and T.C. Lim, Appl. Phys. Lett. 17, 271 (1970).
10 - J.J. Hall, Phys. Rev. 161, 756 (1967).
11 - T.B. Bateman, H.J. Mc Skimin and J.M. Whelan, J. Appl. Phys. 30, 554 (1959 (1959).
12 - R.K. Verma, J. Geophys. Res. 65, 757 (1960).
13 - Y. Ohmachi and N. Uchida, J. Appl. Phys. 41, 2307 (1970).
14 - E. Dieulesaint and D. Royer : "Elastic waves in Solids" (Wiley, New-York, 1980), p. 280-281.

On Rayleigh Waves and Related Propagating Acoustic Waves

M.F. Lewis

Royal Signals and Radar Establishment, St Andrews Road, Malvern, Worcestershire, WR14 3PS, U.K.

1. INTRODUCTION

The propagation of mechanical disturbances in gases, fluids and solids has been studied for many centuries, and has attracted the attention of such celebrated physicists as Newton, Poisson, Hooke, Stokes and Maxwell. Although never enjoying the significance or popularity of electromagnetic wave propagation, it received increased interest in the late eighteenth and early nineteenth centuries with the investigations of RAYLEIGH [1], LOVE [2], STONELEY [3] and others into seismic wave propagation. It was during the course of such studies that exactly one hundred years ago Lord Rayleigh showed theoretically that a mechanical disturbance can propagate unperturbed on the surface of a semi-infinite isotropic solid, in many respects like waves on the sea. Rayleigh conjectured that such surface waves would be especially significant in seismic wave propagation because, unlike bulk acoustic waves, they would only diffract on the surface, not into the volume of the earth. In this conjecture he was partly correct, but it was shown later by LOVE [2] that surface waves arising from the layered structure of the earth's surface (and having a different particle motion, as discussed later) are probably more important in this respect. A number of other studies have been undertaken over the years in an attempt to define the modes of propagation of specific mechanical structures, such as the work of Pochhammer and Chree on cylindrical bars, and of STONELEY [3] on the waves that can propagate at a plane boundary between two solids. Unfortunately for many years this work remained largely of academic interest only, for two principal reasons. The first concerned the great mathematical complexity of calculating the properties of mechanical waves, eg compared to optical waves, while the second concerned the practical difficulty of exciting the desired mode efficiently and selectively. Two important exceptions are ultrasonic delay lines, and the ubiquitous quartz crystal oscillator, which have each played important roles in electronics and radio wave transmission for many decades. This situation has changed dramatically in recent years. In 1962 MORTLEY [4] invented the interdigital transducer, IDT, and successfully employed it in bulk acoustic wave dispersive delay lines, which were desperately needed to implement the new techniques of pulse compression radar. In 1965 WHITE and VOLTMER [5] demonstrated that the same transducer could be used to excite surface acoustic waves, SAW, efficiently and selectively on quartz. Such waves are a generalisation of Rayleigh's waves, since they propagate on substrates that are anisotropic and piezoelectric. This discovery has in turn led to the development of a whole new generation of electronic signal processing components operating in the frequency range from c. 10 MHz to 1 GHz and which are small, cheap, rugged, planar and reproducible. Many of these attributes follow directly from the basic properties of the SAW, surface confinement, low propagation loss on dielectric crystals, and low velocity, ie about 10^{-5} x the velocity of light, making SAW components smaller than their direct electromagnetic counterparts by a similar factor. The principal SAW devices developed to date are shown schematically in Figure 1 and are discussed

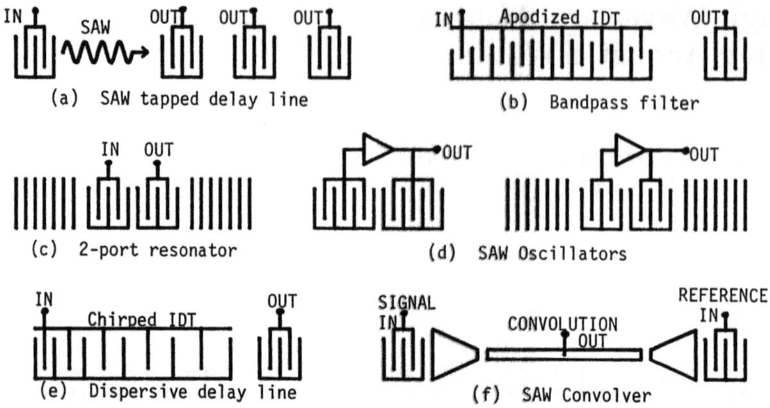

FIGURE 1. Schematic diagrams of six important SAW devices employing IDTs

in greater detail in other papers at this conference. The second vital factor in these developments is much less widely recognised and publicised, and concerns the advances in recent years in the performance of high speed digital computers. These advances have allowed us to compute the SAW and bulk wave properties of whole families of planes through crystals of arbitrary symmetry, SLOBODNIK [6], and indeed of much more complex structures, such as those composed of layers of different materials. In addition, digital techniques have proved invaluable in optimising the design of these analogue components, and in correcting such analogue device problems as acoustic diffraction.

During the course of these developments various workers have been led to consider the possiblity of employing waves other than SAW in such devices. Sometimes this has come about from deliberate attempts to seek alternative waves with new or improved characteristics, and on other occasions it has arisen through the accidental discovery of the excitation of these waves in SAW devices. In many cases these waves are similar to, or identical to those mentioned earlier, eg in connection with seismic studies. To provide an overall theme for this paper, we shall concentrate on those waves, and those aspects of their propagation, that are generally relevant to the design of devices employing interdigital transducers. (It is in this sense that the waves discussed here are related to Rayleigh waves, as implied in the title of this paper.) To this end we show in Table 1 the principal factors that influence the choice of acoustic substrate for such a device. It is interesting to note that the nature of the acoustic disturbance itself, eg the particle motion does not enter this table. Indeed the functions of most SAW devices can be performed using other waves altogether, eg em waves in a coaxial cable, fibre optics, magnetostatic waves, etc. In the following sections we shall provide a classification and brief description of the various alternative acoustic waves studied to date, and shall concentrate on those aspects pertinent to device operation. We also aim to show that between them the related waves do indeed offer significant individual practical advantages over SAW. Whether these advantages, singly or in combination, prove to be sufficiently important for such waves/substrates to replace the existing workhorse SAW substrates (YZ and 128° $LiNbO_3$, and ST-quartz) remains to be seen.

TABLE 1. Some Factors influencing the Choice of Substrate for Acoustic Devices

FACTOR	COMMENT
1. Availability of substrates of the required quality, size and orientation	Vital in mass production where low-cost 3" diameter substrates are preferred.
2. Piezoelectric coupling constant, k^2	This determines the efficiency-bandwidth product and is important in broadband devices. A high value is also necessary when using a multistrip coupler (MSC) to improve the performance in various respects.
3. Temperature coefficient of delay	Of primary importance in narrow band filters, resonators and oscillators.
4. Spurious wave excitation, and ease of eliminating same	Important in many specifications, especially IF bandpass filters.
5. Acoustic velocity	High velocities minimise propagation loss and facilitate photolithography at high frequencies. Low velocities provide smaller devices, and are useful for long delays, and for cheapness at low frequencies.
6. Acoustic attenuation	In single crystals this is usually only significant for devices above ~ 1 GHz, when the transducer metallisation may also contribute.
7. Sensitivity to substrate strain and contamination	Relevant to packaging, and to long term ageing. Can be exploited in sensors.

2. PLANE WAVES OF THE INFINITE MEDIUM

The first waves discussed here are the plane wave solutions of the infinite medium. This is quite appropriate for two reasons. Firstly, such waves form an invaluable starting point for the discussion of waves in bounded media, and secondly because the earliest reported use of the IDT was the original work of MORTLEY [4] on the bulk acoustic wave dispersive delay line shown schematically in Fig 2. The subject of such plane wave propagation has been treated by several authors in recent years, eg DIEULESAINT and ROYER [7], AULD[8] and FARNELL [9], whose general approach we shall follow. Although the problem is now well understood, the complications of piezoelectricity caused many conceptual and computational difficulties in the past, and much of the early work on acoustic waves in quartz involved simplified calculations by hand. Indeed today's most common frequency standard is based on the AT-cut quartz crystal which was identified in this way in 1935.

Choosing a rectangular co-ordinate system with axes x_i (i = 1,2,3) the equations governing the behaviour of an anisotropic piezoelectric crystal take the form:

$$S_{k\ell} = \tfrac{1}{2}\left(\frac{\partial u_k}{\partial x_\ell} + \frac{\partial u_\ell}{\partial x_k}\right) \qquad \text{infinitesimal strain components} \qquad (1)$$

$$\rho\,\frac{\partial^2 u_j}{\partial t^2} = \frac{\partial T_{ij}}{\partial x_i} \qquad \text{Newton's 2nd law} \qquad (2)$$

$$\left.\begin{aligned} \frac{\partial D_i}{\partial x_i} &= 0 \\ E_i &= -\frac{\partial \phi}{\partial x_i} \end{aligned}\right\} \quad \text{Maxwell's equations in the quasi-static approximation} \qquad (3)$$

$$\left.\begin{aligned} T_{ij} &= c^E_{ijk\ell}\, S_{k\ell} - e_{kij} E_k \\ D_i &= e_{ik\ell}\, S_{k\ell} + \varepsilon^S_{ik} E_k \end{aligned}\right\} \quad \text{linear piezoelectric constitutive relationships} \qquad (4)$$

$$(i,j,k,\ell = 1,2,3)$$

where u_i are the components of the displacement, S_{kl} components of the symmetricised infinitesimal strain tensor, T_{ij} components of the stress tensor, and ρ the density. D_i and E_i are components of the electric displacement and electric field, and ϕ the electrostatic potential which, as discussed later, is a valid approximation in view of the smallness of the velocity of sound relative to that of light. The quantities c^E_{ijkl}, e_{kij} and ε^S_{ik} are the elastic, piezoelectric and dielectric constants respectively and henceforth we shall drop their thermodynamic superscripts for simplicity. Summation over repeated indices is assumed throughout. Equations (1) to (4) lead directly to the following coupled equations of motion:

$$\rho \frac{\partial^2 u_j}{\partial t^2} = c_{ijk\ell} \frac{\partial^2 u_k}{\partial x_i \partial x_j} + e_{kij} \frac{\partial^2 \phi}{\partial x_k \partial x_i} \qquad (5)$$

$$\varepsilon_{ik} \frac{\partial^2 \phi}{\partial x_i \partial x_k} = e_{ik\ell} \frac{\partial^2 u_k}{\partial x_i \partial x_\ell} \qquad (6)$$

All the problems of interest in this paper are solutions of such equations in the acoustic medium, although in general they also involve solutions of Laplace's equation, $\nabla^2\phi = 0$, in the free space surrounding the medium, and must satisfy mechanical and electrical boundary conditions at the interfaces. To establish the plane wave solutions of the infinite medium, assume disturbances of the form

$$\left.\begin{aligned} u_j &= \alpha_j \exp jk(b_i x_i - vt) \\ \phi &= \alpha_4 \exp jk(b_i x_i - vt) \end{aligned}\right\} \qquad (7)$$

where v is the phase velocity and b_i the direction cosines of the wavevector whose magnitude is $k = \omega/v$. Substitution of (7) into (6) leads to

$$\alpha_4 = \frac{\alpha_k (e_{ik\ell}\, b_i\, b_\ell)}{(\varepsilon_{ik}\, b_i\, b_k)} \qquad (8)$$

If α_4 is now substituted back into (5) we obtain

$$\rho v^2 \alpha_j = c'_{ijk\ell}\, b_i\, b_\ell\, \alpha_k \qquad (9)$$

where

$$c'_{ijk\ell} = c_{ijk\ell}(1 + k^2_{ijk\ell}) \quad (10)$$

and the piezoelectric coupling constant, k^2_{ijkl}, is given by

$$k^2_{ijk\ell} = \frac{e_{mij}\, e_{nk\ell}\, b_m\, b_n}{c_{ijk\ell}(\varepsilon_{pq}\, b_p\, b_q)} \quad (11)$$

in which some dummy subscripts have been changed for clarity. The introduction of k^2 in (11) has allowed the equations of motion, (9), to be written in the same form as for a non-piezoelectric medium. The only difference is a 'stiffening' of the medium because the strain field is accompanied by an electric field, whose creation requires additional energy relative to a non-piezoelectric. The three equations of motion (9) can be written in the familiar form

$$(\Gamma_{jk} - \delta_{jk}\, \rho\, v^2)\, \alpha_j = 0 \quad (12)$$

where $\Gamma_{jk} = b_i b_l\, c'_{ijkl}$. The solution of (12) requires that the determinant $|(\Gamma_{jk} - \delta_{ij}\, \rho\, v^2)|$ vanish, and leads to 3 allowed values of ρv^2, with velocities v^n ($n = 1,2,3$). When the values v^n are substituted back into (12) they yield the 3 eigenvectors α_j^n, ie the polarisations (and hence ϕ). Thus in any direction in the crystal there are 3 plane wave solutions, and these waves have mutually orthogonal polarisations. In certain directions of high symmetry the polarisations may be exactly parallel or perpendicular to the wavevector, k, when the waves are pure-longitudinal or pure-transverse respectively, as in an isotropic medium. In other directions this will not be precisely true but it is still usually valid to describe them as quasi-longitudinal and quasi-transverse. Similarly the direction of energy flow, or group velocity, v_g, is not in general precisely parallel to k. As mentioned later this can lead to the possibility of more than 3 waves having their group velocities in a given direction. As is evident from (12) the phase velocities, v, are of the form $(c''/\rho)^{1/2}$ where c" is an appropriate linear combination of the stiffened elastic constants, c'. The difference, Δv, between v and its value in the absence of piezoelectricity is determined by k^2 in (10), and since k^2 is invariably much smaller than unity it is usual to make the approximation

$$k^2 = \frac{2\Delta v}{v} \quad (13)$$

The quantity, k^2, is of prime importance in determining the impedance of a transducer and the efficiency of acoustic excitation by electrical means, and the relationship in (13) is especially useful as it is often valid for other waves.

The problem outlined above is a relatively simple one for a number of reasons. Firstly it does not involve the boundaries or boundary conditions. Secondly, it involves plane wavefronts so that, for example, the wavefronts become equipotential surfaces, and the electric field is parallel to the wavevector. These simplifications enable the introduction of the stiffened elastic constants (10) which are, however, functions of the direction of propagation, b_i. Such simplifications do not apply to the more general type of wave propagation discussed later. Even so, to characterise a material fully using the procedure outlined in (1) to (12) usually requires a digital computer. A number of practical devices employ acoustic waves which are a close approximation to the plane waves of the infinite medium discussed above. These include (a) the dispersive delay line of Figure 2, (b) the quartz crystal resonator in which a plane wave is multiply reflected between two parallel faces of the crystal, (c) various acoustooptic devices, such as spectrum analysers, and (d) simple delay lines. The latter operate from a few MHz, EVELETH [10] to microwave frequencies, LEWIS [11]. The velocities of the waves employed range from c. 1 to 10 km/s, and provide delays from < 1 μs to > 1 ms, attenuation permitting. Most such devices employ bonded or evaporated resonant piezoelectric transducers, but the device of Figure 2 employs the IDT to launch waves at an angle, θ, to the surface when the phase-matching condition is satisfied, viz

$$p \cos\theta = m\lambda \tag{14}$$

where p is the spatial period of the IDT and m an integer, usually unity. As discussed later the launching of such waves often causes problems in SAW devices, but is exploited in other cases, principally in SSBW devices (section 3e). One advantage of this type of device over the planar SAW devices of Fig 1 is that once launched the waves do not encounter and interact with the rest of the transducer, or indeed with any surface damage or contamination.

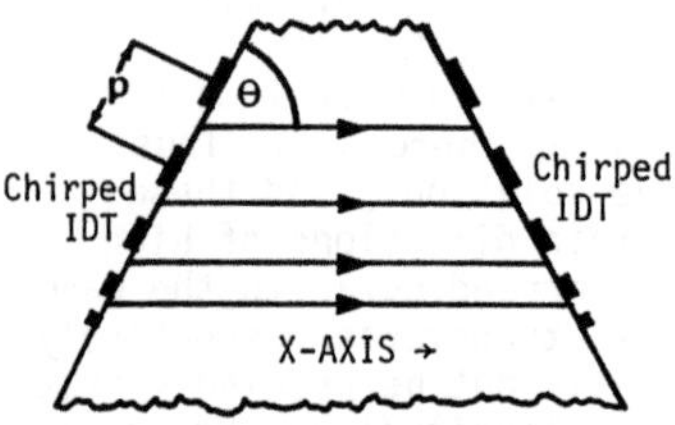

Figure 2. A dispersive bulk acoustic wave delay line employing IDTs. The chirped IDT is shown schematically in Figure 1(e)

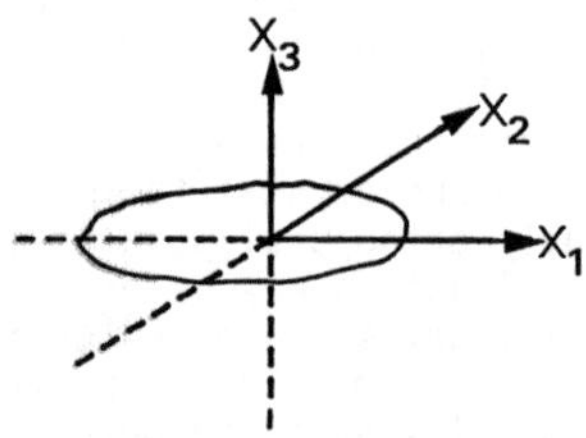

Figure 3. Co-ordinate system employed in section 3

3. WAVES ON A HALF-SPACE

(a) Rayleigh waves, SAW, and SH plane waves

In this section it is recognised that the operation of many devices, especially SAW devices, is influenced by or dominated by the substrate boundaries. As the topic of SAW propagation is the subject of the previous paper, the brief discussion included here is primarily for completeness, although it does lead naturally to a number of new 'related' waves. Consider the geometry of Fig 3 with propagation along the x_1-axis of the surface whose normal is parallel to x_3, and assume straight-crested waves of large aperture in the x_2-direction, so that derivatives with respect to x_2 vanish. Within the substrate (5) and (6) apply, but of course for a general substrate orientation the tensor quantities, eg c_{ijkl}, must be rotated to the co-ordinate system of Fig 3 using the standard rules for tensor transformation. As in section 2 the first objective is to determine the velocity of the SAW, from which the displacement and potential fields can be derived. In the present case however, it is necessary to employ a search procedure, ie a (phase) velocity, v, is guessed, and using this value an attempt is made to devise combinations of partial waves that satisfy the mechanical and electrical boundary conditions; if unsuccessful a new velocity is tried. One such procedure that has been used by many workers has been described in detail by FARNELL [9]. It is considerably more complicated than that described in section 2 for various reasons: (a) the potential field outside the substrate must be included, (b) electrical and mechanical boundary conditions must be satisfied at $x_3 = 0$, and (c) the elastic constants cannot be simplified as in (10), because the propagation/decay constants in the x_3-direction are not known beforehand. The procedure commonly employed is to seek linear combinations of the u_j and ϕ that propagate in the x_1-direction as $\exp j\omega(x_1/v-t)$, and in X_3 as $\exp jk_3x_3$ (where k_3 is in general complex) and which collectively satisfy (5) and (6). (As discussed later, one electrical boundary condition requires the continuity of ϕ at $x_3 = 0$, and ensures that the potential above the substrate is also of this form.) In general there are 8 such combinations, with values of k_3 in 4 complex-conjugate pairs. For an acceptable SAW solution the only relevant terms are the 4 combinations that decay into the substrate. It is then necessary to seek a linear combination (of these combinations) that satisfies the 4 additional boundary conditions at $x_3 = 0$, viz $T_{13} = T_{23} = T_{33} = 0$, and continuity of D_3. This leads to a complex determinental equation that is only amenable to solution with a digital computer. Even then, the determinant cannot be made to vanish exactly, but (through the iterations in v) is reduced to below some arbitrarily chosen small value. In general the SAW solution involves the potential, ϕ, and all 3 particle displacements, u_j, each with 4 complex components of propagation/decay into the substrate. Extensive tables of solutions for various substrate materials of practical interest have been provided by SLOBODNIK [6]. Despite the complexity of these results, a number of useful points may be noted:

(i) All substrate material/orientations investigated to date show a non-dispersive SAW solution with a velocity below that of the slowest bulk wave propagating in the same direction, typically by 10%. Thus no bulk wave can phase match to the SAW and draw energy from it. As in the case of the isotropic materials investigated by Lord Rayleigh, the particle motion at the surface is often retrograde elliptical, and predominantly confined to the sagittal plane (ie the plane containing x_1 and x_3) but it is only in specific cases that this is precisely true. One such case arises when the sagittal plane is a plane of mirror symmetry, as in the important practical case of SAW on Y-cut z-propagating $LiNbO_3$.

(ii) In the free space above the substrate the potential would ordinarily propagate as a component of an em wave satisfying the equation

$$k_1^2 + k_2^2 + k_3^2 = k_0^2 = (\omega/c)^2 \tag{15}$$

where c is the velocity of light. Since $c \gg v$, (15) essentially reduces to Laplace's equation, $\nabla^2\phi = 0$. Further, since $k_1 = k = \omega/v$, and $k_2 = 0$, it follows that $k_3 = \pm jk$. The SAW solution requires that $\phi \to 0$ as $x_3 \to \infty$, so that $k_3 = -jk$, when ϕ varies as

$$\phi = \phi_0 \exp j(\omega t - kx_1) \exp -kx_3 \tag{16}$$

Thus ϕ extends outside the substrate to a distance of a fraction of an acoustic wavelength, λ ($= 2\pi v/\omega$). This is common to other forms of acoustic wave propagation in/on piezoelectric substrates, and can be exploited in various ways, eg to isolate the transducer mechanically from the substrate, or to cause the waves to interact with a neighbouring semiconductor.

(iii) The velocity calculation outlined above can be repeated with the substrate surface covered by an infinitely thin perfect electrical conductor. For the same reason as in section 2 this leads to a small reduction in the SAW velocity, from which a piezoelectric coupling constant, k^2, can be derived using (13) and has the same significance for SAW as for the bulk waves considered earlier, ie it determines the impedance of an interdigital transducer, and is a measure of the maximum fractional bandwidth attainable without incurring an insertion loss penalty. Thus two of the most important practical properties of SAW propagation on a given substrate are summarised in the velocity, v, and the piezoelectric coupling constant, k^2. Two other properties of practical importance are the temperature coefficient of delay, and the beam-steering angle, β. The latter is the angle between the phase velocity $v = v_p$ and the group velocity, v_g, and is given by

$$\tan\beta = \frac{1}{v_p}\frac{\partial v_p}{\partial\theta} \qquad (17)$$

where θ is the angle of propagation (ie the k-vector) on the substrate. Again, tables of values of these quantities are given by SLOBODNIK [6].

It is evident from (16) that the decay of the potential field outside the substrate is a consequence of the slowing of the em wave from its free-space velocity, c, to the SAW velocity, v. Such behaviour is a recurring feature in this paper, and applies to both the potential and acoustic fields. Thus if a wave of interest is composed of partial waves (specifically plane waves of the infinite medium, see for example, section 4) that propagate with a higher velocity along x_1 than their natural velocity they acquire a real component of k normal to the substrate, leading to a sinusoidal variation of the disturbance into the medium (parallel to x_3 in Fig 3). By contrast if the partial wave travels along x_1 more slowly than its natural velocity it acquires an imaginary (or complex) component of k along x_3, and the disturbance decays into the medium. This behaviour is typified by Rayleigh waves on an isotropic substrate, in which each component of the displacement has two decay coefficients in the x_3-direction given by

$$\alpha^2 = k^2 - k_o^2 \qquad (18)$$

where $k_o = \omega/v_o$, and v_o is the velocity of shear or longitudinal waves in the infinite isotropic medium. Since the Rayleigh wave velocity is lower than the shear and longitudinal v_o by typically 10% and 50% respectively, the disturbances decay within a fraction of a wavelength of the surface.

The discussion up to this point has summarised the behaviour of many 'well-behaved' substrates. In other real substrates a number of more complicated phenomena occur, and lead to some interesting and potentially useful related propagating waves known as (i) Bluestein-Gulyaev waves, (ii) Pseudo-SAW, (iii) Tilted Bulk Waves, and (iv) surface skimming bulk waves, SSBW. It is useful to begin by recalling that even in the case of an isotropic substrate there is a second solution (ie in addition to the Rayleigh wave) that satisfies the stress-free boundary conditions, namely the shear horizontal (SH) bulk wave, ie a transversely polarised plane wave with particle motion along x_2 in Fig 3, and having no decay into the substrate. Such a wave has only one independent strain component, S_{12}, and this is totally decoupled from the surface stress components T_{13}, T_{23}, and T_{33} since the corresponding elastic constants (c_{56}, c_{46} and c_{36} in matrix notation) all vanish in an isotropic medium. In many earlier discussions this SH solution was ignored because it is not a true surface wave, yet it is directly or indirectly responsible for all the related waves mentioned above, and for others discussed later. Indeed throughout this paper the waves discussed divide rather naturally into two categories, those with particle motion predominantly in the sagittal plane, and those with pre-

TABLE 2. A Classification of Some Important Acoustic Waves. () indicates that the wave is not always present. A fuller account is given by AULD [8]

Substrate configuration	Particle motion predominantly in the saggital plane		Particle motion predominantly shear-horizontal, SH
Isotropic half-space	Rayleigh wave		SH plane wave
Anisotropic half-space	SAW (pseudo-SAW)		(SH plane wave) (tilted bulk wave) (pseudo-SAW)
Piezoelectric half-space, with surface electrically free or shorted	SAW (pseudo-SAW)		(Bluestein-Gulyaev) (SSBW) (tilted bulk wave) (pseudo-SAW)
Isotropic plate	Lamb Waves Symmetric	Antisymmetric	Sh plate modes
Isotropic half-space with acoustically slower isotropic layer	Rayleigh wave on the layer at high frequencies, on the substrate at low frequencies. Higher order modes, M_1	Sezawa wave (Stoneley interface waves) Higher order modes, M_2	Love waves

dominantly SH particle motion. This division is summarised in table 2. Only occasionally do these solutions couple appreciably so as to make the distinction meaningless.

(b) Bluestein-Gulyaev Waves

In the event that the sagittal plane is perpendicular to a crystallographic axis of two-fold rotational symmetry, the general SAW problem described above simplifies considerably. Thus the displacement components, u_1 and u_3 are decoupled from u_2 and the potential, ϕ. The former components combine to form a Rayleigh wave which is not coupled piezoelectrically and so cannot be excited with an IDT. The coupling between u_2 and ϕ suggests that an IDT will radiate horizontally-polarised bulk waves at an angle, θ, to the surface determined by (14). In particular, as $\theta \to 0$, the wave launched resembles the SH plane wave satisfying the boundary conditions in an isotropic medium. This leads to the SSBW devices of section 3(e), while BLUESTEIN [12] and GULYAEV [13] have shown that the piezoelectric coupling can also cause a true SH surface wave to propagate along the x-axis of Y-cut substrates of materials of 6 mm symmetry, eg ZnO and CdS. This is a rare case in which simple analytical expressions for the fields obtain. For an electrically-shorted surface the solution takes the form

$$\left.\begin{aligned} u_2 &= A \exp(k^2 \omega x_3/v) \exp j\omega(t-x_1/v) \\ \phi &= \frac{Ae_{15}}{\varepsilon_{11}} [\exp k^2 \omega x_3/v - \exp \omega x_3/v] \exp j\omega(t-x_1/v) \\ v &= [(1-k^4)\, \bar{c}_{44}/\rho]^{\frac{1}{2}} \end{aligned}\right\} \quad (19)$$

where $k^2 = e_{15}^2/(\varepsilon_{11}\bar{c}_{44})$, and $\bar{c}_{44} = c_{44} + e_{15}^2/\varepsilon_{11}$ is the piezoelectrically stiffened elastic constant controlling the shear wave velocity in the infinite medium. Since k^2 is invariably much small than unity it can be seen from (19) that the velocity is only marginally lower than that of the bulk shear wave, and that the penetration depth into the medium is much greater than the wavelength, as discussed in section 3(a). For an electrically free surface, the velocity is even closer to that of the bulk shear wave, and the penetration into the medium correspondingly deeper, (18).

It was also shown by BLUESTEIN [12] that such waves can propagate on a poled ferroelectric ceramic, and at one time this led to interest in the possibility of using such substrates to reduce the cost of SAW devices, principally the TV IF filter. However, this interest declined due to problems of reproducibility, and acoustic attenuation caused by scattering processes in the ceramic, and to the best of the author's knowledge no devices are yet in use that exploit Bluestein-Gulyaev waves.

(c) Pseudo-SAW

The modes of propagation in a bounded substrate comprise combinations of the partial waves of the infinite medium, in general some being cut-off as in the case of the potential, ϕ, discussed in section 3(a). For example, in a semi-infinite isotropic substrate, one solution that satisfies the boundary conditions is the SH-plane wave alone, while the Rayleigh wave comprises (cut-off) contributions from a longitudinal wave and a shear wave, each polarised in and propagating in the sagittal plane. Such a shear wave is often called shear vertical, SV, to distinguish it from the SH wave, even though its displacement may only contain a component in the vertical direction. Of course in an isotropic medium the shear waves are degenerate, so any direction may be deemed to support one SH and one SV shear wave. By contrast, in an anisotropic medium the velocities and polarisations of all the plane waves of the infinite medium are determined by the crystallography (section 2), and in particular the shear waves may not even approximately conform to the descriptions SH and SV. In a slightly anisotropic crystal like sapphire this has a number of relatively minor consequences, for example, in general (i) neither shear wave exactly satisfies the boundary conditions, (ii) the Rayleigh wave particle motion is not exactly in the sagittal plane, and (iii) the Rayleigh wave k_3 may contain small real components so that the decay of the disturbance into the substrate is slightly oscillatory. In more strongly anisotropic materials pseudo-SAW (or leaky SAW) solutions can arise; these are faster waves that do not usually exactly satisfy the boundary conditions, but which leak energy slowly into the substrate through a slower bulk wave, to which they are phase-matched along the surface, FARNELL [14]. In the author's own work such waves have been identified theoretically by relaxing the conditions on the boundary condition determinant described earlier. A more rigorous approach is to extend the velocity search procedure to include complex values of k_1, LIM and FARNELL[15]. Experimentally such waves are readily detected in devices employing the IDT structures of Fig

1; indeed the SAW substrate is often chosen to avoid such waves, which can interfere with the SAW response of the device. Although the occurrence of such pseudo-SAW is often considered to be erratic, the description below provides a plausible basis for expecting such waves. It is suggested that the appearance of pseudo-SAW in addition to SAW solutions reflects the two solutions in an isotropic medium, ie the Rayleigh and the SH plane wave. Pseudo-SAW solutions may be anticipated when the relevant quasishear waves of the infinite medium (ie those with wavevector in the sagittal plane and propagating close to x_1 in Fig 3) happen to be essentially horizontally and vertically polarised, and especially when the latter are the faster. This is illustrated by comparing the behaviour of waves in the specific case of the (001) plane of cubic crystals with different anisotropy constants, A, where

$$A = \frac{2c_{44}}{c_{11} - c_{12}} \tag{20}$$

If $A < 1$, the purely vertically-polarised shear wave, T_1 in Fig 4(a), is slower than the (quasi-) SH wave T_2, and a normal SAW solution obtains. In this case the second (SH) solution corresponds to the faster quasi-SH branch, T_2, which only satisfies the boundary conditions exactly when propagation is along the [100] or [110] axes. Following the notation of FARNELL [14], such isolated solutions are indicated by small squares in Fig 4. If $A > 1$, the vertically-polarised shear wave, T_1, is faster than the (quasi-) SH wave, T_2, except for propagation along [100], Fig 4(b). In this latter neighbourhood a normal behaviour occurs with a (slow) Rayleigh wave solution indicated by a cross in Fig 4(b). However as the [110] direction is approached, this Rayleigh wave continuously transforms into the SH plane wave solution. For propagation precisely along [110] a second surface wave solution obtains in the form of a pure Rayleigh wave (ie with zero leakiness and particle motion confined to the sagittal plane) and with a lower velocity than that of T_1, but higher than that of T_2; this evidently derives from the longitudinal and fast SV branches. For directions close to [110] a pseudo-SAW propagates, with energy being gradually leaked into the substrate through a phase-matched bulk wave. FARNELL [14] has examined the behaviour of such pseudo-SAW on the (001) plane of copper, and verified this description in considerable detail. Pseudo-SAW of this kind occur in many other cases for which the vertically-polarised shear wave is faster than the horizontally-polarised wave, eg for certain propagation directions on X-cut

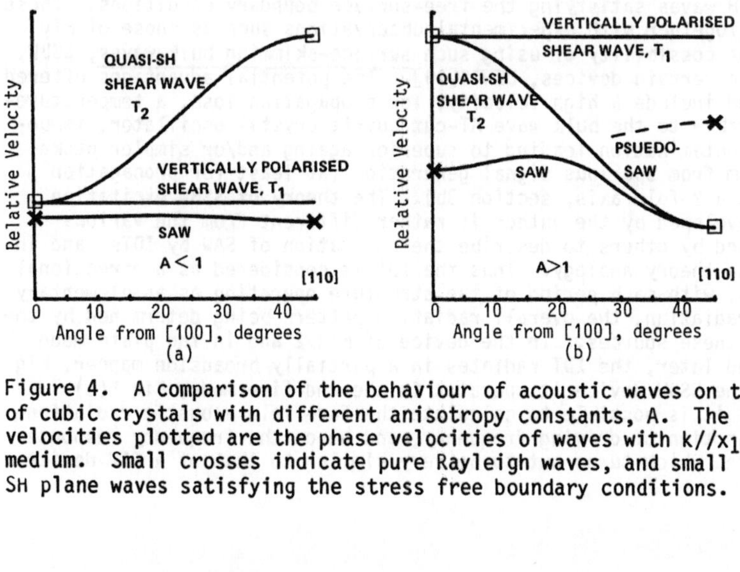

Figure 4. A comparison of the behaviour of acoustic waves on the (001) plane of cubic crystals with different anisotropy constants, A. The shear wave velocities plotted are the phase velocities of waves with $k // x_1$ in the infinite medium. Small crosses indicate pure Rayleigh waves, and small squares indicate SH plane waves satisfying the stress free boundary conditions.

quartz, ENGAN[16]. It would be of interest to examine critically the SAW and pseudo-SAW solutions at angles close to their crossover velocity to see if any coupling occurs at such points.

The practical interest in such waves hinges on the factors included in table I. For example, NEWTON [17] has shown that pseudo-SAW on 41° rotated Y-cut $LiNbO_3$ with x-axis propagation offer a very high value of k^2, while the same wave, and a newly discovered pseudo-SAW on quartz, offer superior temperature characteristics to conventional SAW on the respective materials, SHIMIZU and TANAKA [18]. The behaviour of this latter pseudo-SAW branch on rotated Y-cut quartz with x-axis propagation is consistent with the suggestions above, for example its leakiness is a minimum ($< 10^{-3}$ dB/λ) when the fast shear wave along the surface is approximately vertically polarised (θ ~ 30° in the notation of Shimizu and Tanaka, and close to the BC-cut), and also when the slow shear wave is vertically polarised (θ ~ 120°, and close to the AC-cut). A further potential advantage of such pseudo-SAW in some applications is their higher velocity relative to SAW, table I.

(d) Tilted Bulk Waves

In Fig 4(a) the crystal symmetry is such that the isolated SH solutions indicated by small squares suffer no beam steering, ie their phase velocity and group velocity are collinear. In more general cuts SH plane waves satisfying the boundary conditions can occur with their group velocity in the surface, but with their phase velocity pointing slightly into or out of the surface (17). Such waves occur, for example, on the (111) plane of silicon, and FARNELL [14] describes them as 'tilted bulk waves'. Emerging from such isolated points may be another pseudo-SAW branch which almost satisfies the boundary conditions, and which evidently derives from the SH plane wave solution of the isotropic medium. As discussed below, if such waves are launched by an IDT on the surface, they can be received by another IDT since the energy flow (group velocity) is in the surface, the direction of the phase velocity being immaterial.

(e) Surface Skimming Bulk Waves

The device of Fig 2 employs an IDT to launch bulk waves at an angle, θ, to the surface given by (14). When operated at such a frequency that λ = p, propagation essentially parallel to the surface (θ ~ 0) may be anticipated, especially in the case of SH waves satisfying the free-surface boundary conditions. These considerations, together with experimental observations such as those of Fig 5(a), suggest the possibility of using such surface-skimming bulk waves, SSBW, instead of SAW in certain devices, LEWIS[19]. The potential advantages offered are numerous, and include a high velocity, low propagation loss, a temperature-stability comparable to the bulk wave AT-cut quartz crystal oscillator, immunity to surface contamination leading to superior ageing and/or simpler packaging, and freedom from spurious signal generation (at least for propagation perpendicular to a 2-fold axis, section 3b). The theory of SSBW excitation and reception developed by the author is rather different from the various approaches adopted by others to describe the excitation of SAW by IDTs, and involves an antenna theory analogy. Thus the IDT is considered as a directional acoustic antenna, with each period of the structure operating as an elementary line source of radiation, the overall radiation pattern being determined by the interference of these sources. In the device of Fig 2 and in the plate mode devices described later, the IDT radiates in a partially broadside manner, Fig 5(d), while in the SSBW device it operates in the end-fire mode, Fig 5(c). Using this model it is possible to calculate the far-field acoustic radiation patterns as a function of driving frequency, and hence the frequency response of an SSBW device, which turns out to be very similar to that of a SAW device of the same structure but with the additional factor of (21). Further, by integrating the acoustic power launched by a single IDT over all angles, θ, as a function of frequency, see Figs 5(b) to (d), it is possible to explain the characteristic shape (λ) of the measured acoustic radiation resistance curves shown in Fig 5(a). An important feature of such directional antennas is their antenna gain, G. This is a measure of the concentration of the radiated energy in the required direction relative to that of a hypothetical isotropic radiator. Obviously the insertion loss of a device is reduced if the transmitting IDT has a large gain (which infers a large number of periods, N), and by reciprocity the gain operates in reception as well as transmission. Using these principles the author was able to predict and demonstrate an insertion loss of only 13 dB between two identical SSBW IDTs in close proximity, ie a loss little greater than that of a conventional SAW device. In this arrangement (which happens to be ideal for use as the feedback element of a single mode delay line oscillator) the combined antenna gains largely offset the diffraction (or spreading) loss from the radiating antenna. In the far-field the acoustic intensity at the substrate surface varies as x_1^{-1}, and the amplitude, a, as

$$a \propto (x_1)^{-\frac{1}{2}} \propto (vt)^{-\frac{1}{2}} \qquad (21)$$

where x_1 is measured from the input IDT, v is the velocity and t the time elapsed. This behaviour has been confirmed in all essential respects by experiment and by the more rigorous calculations of MILSOM [20] and LEE [21]. In addition LEE [21] has extended this quasi-isotropic theory to include the substrate anisotropy, and has emphasised that the IDTs communicate optimally when the group velocity lies along the surface, cf section 3(d). Inclusion of this factor has been shown by YEN [22] to give excellent agreement between the

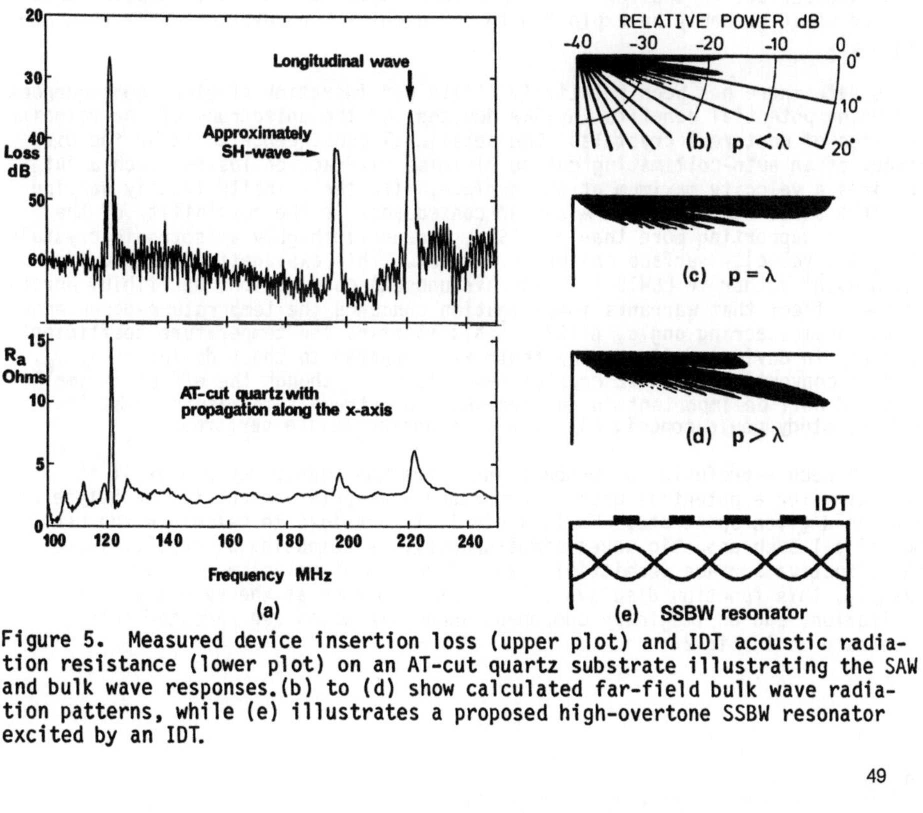

Figure 5. Measured device insertion loss (upper plot) and IDT acoustic radiation resistance (lower plot) on an AT-cut quartz substrate illustrating the SAW and bulk wave responses. (b) to (d) show calculated far-field bulk wave radiation patterns, while (e) illustrates a proposed high-overtone SSBW resonator excited by an IDT.

measured and calculated velocity and temperature coefficient by delay of SSBW on rotated Y-cut quartz. The principal use made of such devices to date is as the feedback elements of very high frequency oscillators, several groups reporting devices that operate well into the microwave bands.

As proposed by the author [19] it is also possible to make an SSBW substrate support a true surface wave by introducing various modifications to the propagating surface to reduce the velocity locally. (This mechanism is discussed in more detail in section 5.) The modifications proposed and subsequently demonstrated are (i) an electrically-conducting layer, when a shorted Bluestein-Gulyaev type of wave propagates, (ii) a mass-loading layer, when a Love wave (section 5(a)) propagates, LEE [23], and (iii) a corrugated surface when the slowing results from the dispersion introduced in the proximity of a stopband, YEN [22]. In this last arrangement the propagation is perpendicular to the corrugations, but a similar slowing effect could probably be effected for SH waves by implementing an array of finely-spaced corrugations parallel to the direction of propagation, and would operate by reducing the shear stiffness of the medium.

SSBW on quartz, $LiNbO_3$ and $LiTaO_3$ have also been used to demonstrate bandpass filters with various benefits over the corresponding SAW devices, including their ability to be sealed with an epoxy coating rather than a conventional package, BROWNING [24], YEN [22], BLOCH [25]. Some of these devices have been corrected for the $t^{-\frac{1}{2}}$ factor in their impulse response implicit in (21), and some have employed a multistrip coupler (MSC) to minimise spurious effects. The problem of reflections from the sample ends mentioned in section 2 is particularly severe in SSBW devices because the energy is largely below the surface in the form of a bulk shear wave, BLOCH [25]. However this feature could probably be exploited to implement a high-overtone SSBW resonator, Fig 5(e).

To date there has been relatively little consideration of other consequences, including potential benefits in SSBW devices, of the anisotropy of the velocity surfaces of most real crystals. One beneficial consequence could be the discovery of an auto-collimating cut to minimise diffraction losses; such a cut requires a velocity maximum at the surface, with the velocity ideally varying into the substrate as $\cos\theta$. A second consequence is the possibility of the substrate supporting more than one SSBW, because in highly anisotropic crystals the group velocity surface can be re-entrant. This explanation was provided by the present author to LEWIS [26] for his unusual observations on lithium borate. Another effect that warrants investigation concerns the temperature-dependence of the beam-steering angle, β (17). This modifies the temperature coefficient of delay in devices dominated by their v_g, compared to those dominated by v_p, such as conventional quartz crystal resonators. Although the effect is small, it could well be important in devices showing extreme temperature-stability, and its study could conceivably lead to superior device performance.

With such a profusion of propagating waves the reader may wonder if it is necessary for a potential user to undertake a computer search for each type of wave on a given substrate. Fortunately it is possible to calculate the behaviour of all such acoustic waves simultaneously by computing a quantity known as the effective surface permittivity as a function of velocity, MILSOM [20]. For example, this function displays a lossless resonance at the velocity of SAW excitation, and an imaginary component when bulk waves are radiated into the substrate as in Fig 2.

4. ACOUSTIC WAVES IN PLATES

The discussion in the previous section concerned waves that propagate on a semi-infinite substrate, and the properties of such waves from a useful basis for the first-order design of many acoustic devices. However, real devices must obviously employ a finite substrate, and so for a fuller understanding it is important to study the normal modes thereof. For practical reasons such substrates often take simple forms, eg bars, rods or plates; in this section the discussion will be confined to plates, since they form the basis of most modern planar acoustic components. To simplify the problem, it will be assumed initially that the plate is isotropic. As in the case of em waves in a metal waveguide, the propagation can be envisaged as comprising plane waves of the infinite medium zig-zagging along the waveguide at such an angle that between them they satisfy the boundary conditions, Fig 6(a). In the case of SH waves (polarised along x_2) this is a valid interpretation, with two such waves being multiply and specularly reflected at the boundaries $x_3 = \pm b/2$ without mode conversion. By contrast the orthogonally-polarised (SV) shear wave, and the longitudinal wave (L), do in general suffer mode conversion to each other at each reflection, Fig 6(c), and so form coupled-mode solutions known as a Lamb wave, AULD [8]. Thus the propagating modes of an infinite isotropic plate separate into the two families listed in table 2. Fortunately the SH problem is soluble analytically, and leads to a simple dispersion relationship and other useful conclusions, and so is included below.

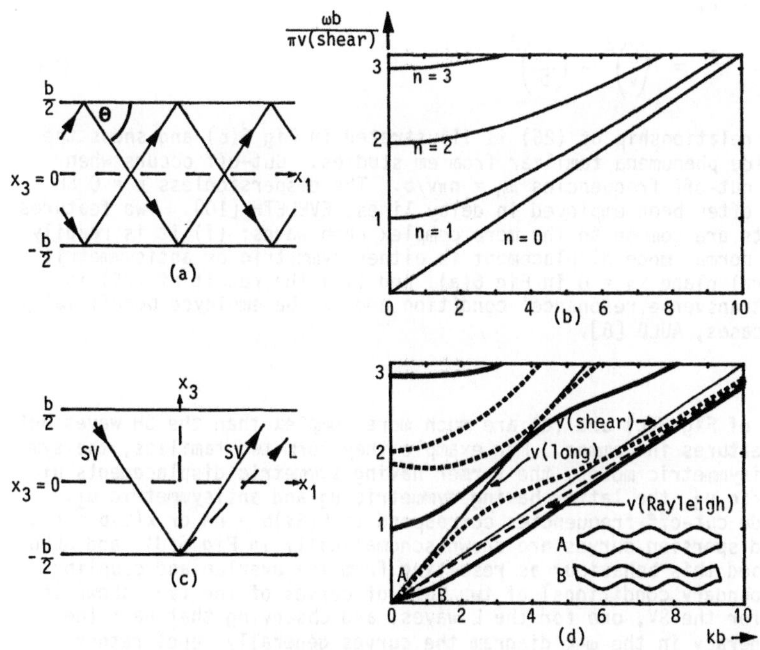

Figure 6. (a) and (b) show schematically an SH acoustic waveguide and its simple dispersion relationships. (c) illustrates the coupling between SV and L waves leading to the Lamb wave dispersion curves in (d) in which the symmetric modes appear as ··········, and the antisymmetric modes as ————

(a) SH-Plate Modes

Referring to Fig 6(a), assume plane wave solutions of the form

$$u_2 = A \exp j(\omega t - k_1 x_1 - k_3 x_3) + B \exp j(\omega t - k_1 x_1 + k_3 x_3) \tag{22}$$

The only non-vanishing strain components are S_{12} and S_{23}, while the stress-free boundary conditions require that $T_{13} = T_{23} = T_{33} = 0$ at $x_3 = \pm b/2$. In an isotropic medium the only non-vanishing elastic constant relating these components is c_{44}, which relates S_{23} to T_{23}. Thus S_{23} must vanish at $x_3 = \pm b/2$. Now S_{23} is given by (1) as

$$S_{23} = \tfrac{1}{2}\left(\frac{\partial u_2}{\partial x_3} + \frac{\partial u_3}{\partial x_2}\right) = \frac{jk_3}{2} \exp j(\omega t - k_1 x_1)\,[B \exp jk_3 x_3 - A \exp -jk_3 x_3] \tag{23}$$

so that

$$\left.\begin{aligned} A \exp -jk_3\, b/2 &= B \exp jk_3\, b/2 \\ A \exp jk_3\, b/2 &= B \exp -jk_3\, b/2 \end{aligned}\right\} \tag{24}$$

Equation (24) leads to $A = \pm B$, and $\exp jk_3 b = \pm 1 = \exp jn\pi$, where n is an integer and is even if $A = +B$, and odd if $A = -B$. Thus

$$k_3 b = n\pi \tag{25}$$

$$k_1^2 = k^2 = \left(\frac{\omega}{v}\right)^2 - \left(\frac{n\pi}{b}\right)^2 \tag{26}$$

The dispersion relationship of (26) is illustrated in Fig 6(b) and shows the cut-off waveguide phenomena familiar from em studies. Cut-off occurs when $kb = n\pi$ giving cut-off frequencies $\omega_n = n\pi v/b$. The dispersionless $n = 0$ SH plate mode has often been employed in delay lines, EVELETH [10]. Two features of these results are common to the more complex Lamb waves: (i) it is readily shown that the normal mode displacement is either symmetric or antisymmetric about the central plane $x_3 = 0$ in Fig 6(a), and (ii) the result of (25) is known as the 'transverse resonance' condition and can be employed beneficially in many other cases, AULD [8].

(b) Lamb Waves

The Lamb waves of Fig 6(c) and (d) are much more complex than the SH waves but have various features in common. For example they form two families, the symmetric and antisymmetric modes, the former having symmetric displacements u_1 and antisymmetric u_3, the latter having symmetric u_3 and antisymmetric u_1. Further the mode cut-off frequencies correspond to $k(SV)b = n\pi$ or $k(L)b = n\pi$. The resulting dispersion curves are shown schematically in Fig 6(d), and AULD [8] has described this behaviour as resulting from the overlap and coupling (through the boundary conditions) of two sets of curves of the type shown in Fig 6(b) (one for the SV, one for the L-waves) and observing that near the points of degeneracy in the ω-k diagram the curves generally repel rather than cross. The displacements of the lowest order symmetric and antisymmetric modes at low frequencies are shown schematically by the insets A and B in Fig 6(d). It is of interest to consider how these modes develop with increasing ω or kb. The first point to observe is that the phase velocities of all modes ultimately fall below the velocity of the longitudinal waves, so that this component of the motion becomes 'trapped' at the boundaries (section 3(a)) and the disturbance approaches that of the pure SV mode. In the case of the lowest order symmetric and antisymmetric modes only, the phase velocity ultimately falls below that of the shear wave, when both components become trapped at the surfaces. The limiting velocity, v_R, is the Rayleigh wave velocity, and it becomes clear that a Rayleigh wave on one surface can therefore be regarded as a linear combination of the lowest order symmetric and antisymmetric Lamb waves, and vice versa. Such an interpretation has a considerable appeal in various respects. The excitation of highly dispersive (ie small kb) Lamb waves in PZT plates using IDTs has been studied by TODA [27].

(c) Plate Modes

If the plate is anisotropic and piezoelectric the problem becomes immensely more complex, the general normal mode solution involving 6 acoustic partial waves (3 each with $\pm k_3$ components) and 4 (cut-off) em waves corresponding to the potential fields within and without the substrate. The resulting 10 by 10 determinant can only be solved numerically in a search procedure for the modal velocity otherwise similar to that described in section 3(a). In a tour de force WAGERS [28] has performed such an analysis on quartz and $LiNbO_3$ substrates in order to explain the spurious wave excitation observed in SAW devices. Excellent agreement between experiment and calculation was obtained by summing vectorially the contributions of the first 96 modes of each substrate. Obviously one could not contemplate using such a procedure to synthesise a given response. Some recent experiments have, however, demonstrated the potential use of such waves in devices, LEWIS [29], HELMICK [30]. The geometry employed is the family of rotated Y-cuts of quartz with propagation perpendicular to the x-axis, because these cuts all support an especially simple series of piezoelectrically-coupled SH plate modes similar to those illustrated in Figs 6(a) and (b). The principal attraction of such devices lies in their high fundamental frequency of operation, which can be derived from (14) by setting m = 1,

$$f(\theta) = v \sec \theta / p \qquad (27)$$

According to this relationship there is no limit to the upper frequency attainable by launching waves at a sufficiently large angle, θ, to the surface. In practice a response has been observed at six times the frequency of the corresponding SAW response, LEWIS [29] and suggests that such devices could be used well into the microwave regime. As in the case of the SAW devices of Fig 1, the operation of these devices is best described in the time-domain, in the present case in terms of 'bouncing' acoustic beams rather than the plate mode spectrum, Fig 7. Simple geometrical considerations show that the impulse response of such a device on an isotropic substrate comprises a series of delayed bounces, each of which is a downchirp, except for the zeroth (SSBW) response. The average path-length of the nth bounce is $P_n = L \sec \theta_n$, and the corresponding group delay is $\tau_n = L \sec \theta_n / v$. Similarly the centre frequency is given by (14) as $f_n = v\sec \theta_n / p$ where p is the IDT period. Thus the centre frequency should scale with the group delay. To approximate the behaviour of such an isotropic substrate the author has performed experiments on the orthogonal AC and BC cuts of quartz, because each of these has a velocity surface symmetrical about the plate normal (x_3-axis). In these circumstances the paths sketched in Fig 7(a) are correct, but should be interpreted as the group-velocity paths. Measured and calculated results on such cuts are shown in Fig 7(b), and show a high degree of agreement. For example it is observed that successive bounces are widely separated for the AC-cut, and closely spaced for the BC-cut; this is a direct consequence of the anisotropy of the (group) velo-

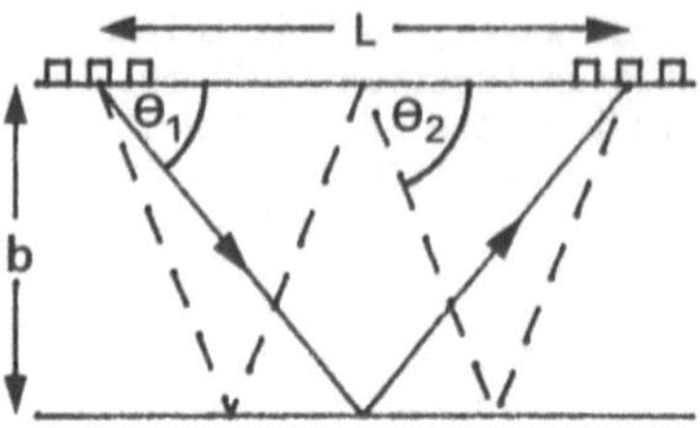

(a)

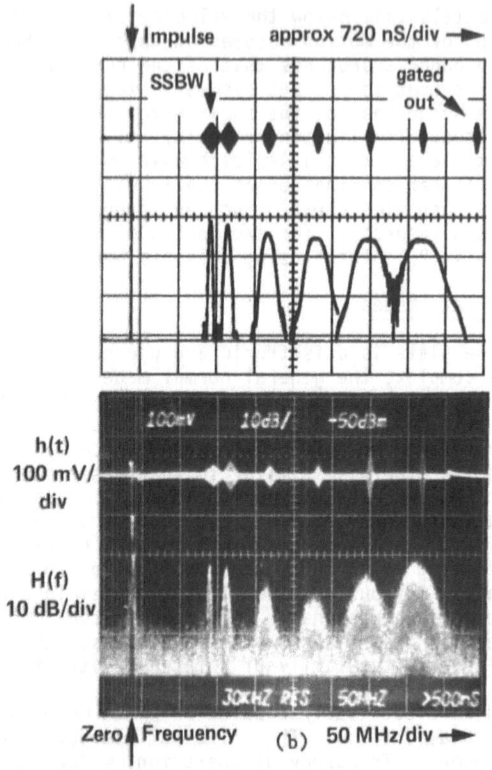

(b)

Figure 7. (a) A schematic plate mode device employing IDTS on AC- or BC-cut quartz.
(b) Experimental and calculated impulse response and frequency-response on a BC-cut substrate with L = 5 mm, B = 2.7 mm, and IDT period p = 32 μm. The timebase has been adjusted to show the proportionality of the centre frequency and group delay, and the 6th bounce has been gated out to eliminate a coincident spurious reflection from the substrate edge.

city surface, v_g being a maximum along the BC-axis (perpendicular to the AC-axis) and a minimum along the AC-axis.

Devices of this kind should not be too difficult to fabricate because the mechanical tolerances are similar to those regularly used in laser technology. The principal disadvantage at present is the high insertion loss on quartz, so an investigation of other materials with higher piezoelectric coupling could prove rewarding. In this context it is rather interesting to observe that while quartz and $LiNbO_3$ are both members of the trigonal system, they belong to different classes within this system, NYE [31]. As a consequence they have identical non-vanishing elastic constant components, but piezoelectric constants so different as to qualify as 'opposite', ie if a non-vanishing element is present in one class, it is absent in the other. This means that the rotated Y-cuts (with propagation perpendicular to the x-axis) that are so useful for exciting SH waves (SSBW and plate modes) in quartz, are not piezoelectrically coupled to the SH modes in $LiNbO_3$. Instead these cuts show coupling to the Rayleigh wave components, and they include one of the most important SAW substrates, viz the Y-cut with z-axis propagation.

5. WAVES ON A LAYERED HALF-SPACE

As the surface of the earth comprises a layered structure, numerous investigations of wave propagation in such a medium have been undertaken in the past by seismologists, EWING [32]. In the case of reasonably isotropic media, the propagating modes again separate into two basic classes, the SH modes, and modes whose displacements are confined to the sagittal plane.

(a) Love Waves

In the previous sections it has been shown that a semi-infinite isotropic medium supports a pure SH plane wave, and that similar results obtain for anisotropic substrates, although it is only in specific directions that such a wave satisfies the boundary conditions exactly. LOVE [2] has shown that if a layer is added to such a semi-infinite substrate, then SH surface waves can be supported providing the layer has a lower shear wave velocity than the substrate. Physically the reason for this can be described as follows: if it is assumed that a surface wave solution exists, then a finite proportion of the disturbance will be carried in the slower layer, and so the overall velocity is reduced relative to the substrate. As discussed in section 3a such a solution has a disturbance that decays into the substrate, a conclusion that is in accordance with the original assumption. Thus the surface wave solution is self-consistent. By contrast, an acoustically faster layer does not produce a surface wave solution. As the thickness, b, of the (slower) layer is increased relative to the acoustic wavelength, λ, higher order modes are introduced when the thickness supports integer numbers of half-wavelengths, the variation of the disturbance in the x_3-direction being oscillatory in the layer, and decaying in the substrate. This leads to a family of modes somewhat reminiscent of the SH plate modes of Fig 6(b), but in which the phase velocity can never exceed the velocity in the substrate, or the disturbance would radiate into the substrate. The resulting dispersion curves are shown schematically in Fig 8(a), and it will be noted that because the substrate thickness, b, has introduced a characteristic dimension into the problem, the propagation is dispersive. This even applies to the lowest order mode, in which the phase velocity makes a smooth transition from the velocity in the substrate for $\lambda \gg b$, to the velocity in the layer when $\lambda \ll b$.

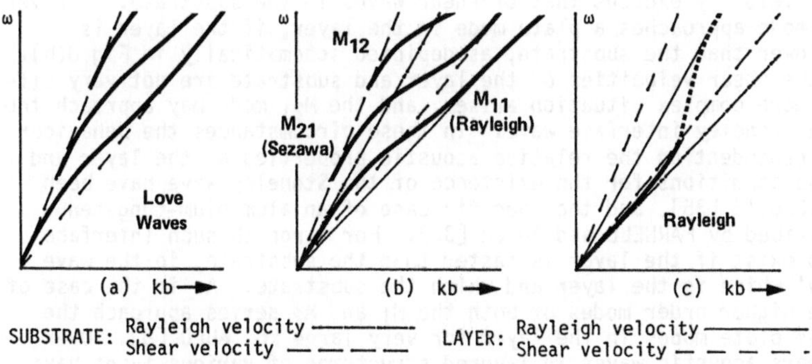

Figure 8. Schematic dispersion curves for waves propagating on a layered half-space. Only in (c) is the layer acoustically faster than the substrate.

(b) Rayleigh Modes

In contrast to the SH-waves discussed above, Rayleigh waves on a half-space form a tightly-bound surface wave solution even in the absence of a slowing mechanism or layer at the surface. Thus the effect of a thin layer (of thickness $b \ll \lambda$) is primarily to modify the properties of the existing Rayleigh mode, as illustrated in Fig 8(b). One effect of a slow layer is to reduce further the velocity of the Rayleigh wave below that of a shear wave in the substrate. It also introduces dispersion because, as is physically quite plausible, the Rayleigh wave velocity varies smoothly from that in the substrate for $b \ll \lambda$, to that in the layer for $b \gg \lambda$. In SAW devices this is an undesirable effect, and is minimised by choosing the metal of the electrode structure to match the substrate acoustically, eg Al on ST-quartz, and Au on BGO/BSO. If the layer is acoustically faster than the substrate a surface wave solution still obtains for small kb, while for larger kb it becomes leaky when the velocity rises to equal the shear wave velocity in the substrate, ASH [33]. This is illustrated in Fig 8(c). In most cases of practical interest it is possible to calculate the properties of the modified Rayleigh wave by perturbation techniques, TIERSTEN[34], AULD[8]. For completeness it should also be mentioned that it is possible to form a two dimensional Rayleigh wave waveguide by various means, eg by depositing a strip of slower material on a half-space, or by using a metallic strip to perform localised piezoelectric-shorting of SAW on a piezoelectric substrate. Such guided surface wave or microsound structures can be used to confine the acoustic energy not just vertically but also transversely (ie in the x_2-direction in Fig 3) in a width of order one wavelength, and could form the basis of a generation of microwave devices which are physically smaller than their em counterparts by a factor of 10^5, ASH [33], TIERSTEN [34].

(c) Higher Order Rayleigh Waves

If the velocity of shear waves in the layer is lower than in the substrate, new modes appear as ω (or kb) is increased, in a similar manner to the Love waves of section 5(a). These modes form two families called the M_1 and M_2 series, which degenerate into the symmetric and antisymmetric Lamb waves of the layer (section 4b) if the density or stiffness of the substrate are made to vanish. The lowest order M_1 mode, M_{11}, corresponds to the modified Rayleigh wave discussed in section (b) and exists for all ω or kb. The lowest order mode of the second series, M_{21}, is called a Sezawa wave and is leaky below a certain value of kb when its velocity exceeds that of shear waves in the substrate. For very large kb this mode approaches a plate mode in the layer, if the layer is appreciably slower than the substrate, as depicted schematically in Fig 8(b). If, however, the shear velocities of the layer and substrate are not very different a much more complex situation arises, and the M_{21} mode may approach the behaviour of a Stoneley interface wave. In these circumstances the behaviour is critically dependent on the relative acoustic properties of the layer and substrate. The conditions for the existence of the Stoneley wave have been discussed by SCHOLTE [35], and the specific case of an aluminium-tungsten interface described by FARNELL and ADLER [36]. For large kb such interface waves can also exist if the layer is faster than the substrate, ie the wave does not 'know' which is the layer and which the substrate. As in the case of Lamb waves the higher order modes of both the M_1 and M_2 series approach the behaviour of SV plate modes in the layer for very large kb, AULD [8]. Over the past 15 years acoustic waves in layered structures of various types have often been considered for use in planar electronic devices similar to those of Fig 1. For example, LARDAT [37] considered their potential use in dispersive delay lines, while many others have investigated deposited piezoelectric thin films, principally ZnO, for acoustic wave excitation on non-piezoelectric substrates. In particular, SHIOSAKI [38] has described the use of such ZnO films for SAW transduction on cheap glass substrates for TV IF filter production. Similarly the attractions of integrating acoustic and electronic devices on a single substrate led ARMSTRONG [39], BOWERS [40] and others to study the use of ZnO for efficient SAW and Sezawa wave excitation on a silicon substrate. Unfortunately little of this work has been exploited in production because of numerous difficulties, including multimode propagation (within and between the series of modes discussed above), the problems of controlling the reproducibility (of layer thickness, density, grain size, adhesion.....), and dispersion, although in principle this can be taken out via the transducers if it is reproducible. Concerning dispersive delay lines themselves, these now invariably employ non-dispersive propagation with dispersion introduced reproducibly through the photolithography of the dispersive transducers, Fig 1(e), or through reflector arrays.

6. CONCLUSION

The propagation of acoustic waves in bounded solid media is extremely complex, but is important in seismic studies, in mineral prospecting, and in many modern devices that employ ultrasonic waves to perform sophisticated electrical functions. In this review the author has attempted to build up a physical description of the modes of some of the more important structures encountered in practice, starting from the relatively simple case of plane wave propagation in the infinite medium. This process has been facilitated by dividing the propagating waves into two categories, those with shear-horizontal particle motion, and those with particle motion in the sagittal plane. In the text it may have appeared accidental that such a division was possible for both the semi-infinite isotropic substrate, and for the bounded planar substrates, but the underlying reasons are related. For example if θ is allowed to approach 0 in Fig 6(a), the incident and reflected SH waves coincide, and form the SH plane wave of section 3a. Throughout this review the practical consequences have been mentioned where appropriate, and various aspects warranting further attention have been noted. The field offers many more opportunities of both practical and academic interest. For example devices employing acoustic waves of the types described here are potentially useful in many kinds of sensor. One particular suggestion is to use an SSBW (or related device) as a strain gauge or accelerometer, since its immunity to its surroundings would allow it to be liquid-damped. The effect of such an inhomogeneous quasi-static strain on the leakiness of weakly-bound surface waves could also form an interesting study.

REFERENCES

1. Rayleigh, Lord, Proc London Math Soc 17 (1885), 4-11
2. A E H Love Some Problems of Geodynamics, University Press, Cambridge 1911
3. R Stoneley, Proc Roy Soc A, 106 (1924), 416-428
4. W S Mortley, Marconi Review (1965), 273-290
5. R M White and F W Voltmer, Appl Phys Lett 7 (1965) 314-6
6. A J Slobodnik, E D Conway and R T Delmonico, the Microwave Acoustics Handbook series, Air Force Cambridge Research Laboratories, 1973
7. E Dieulesaint and D Royer Elastic Waves in Solids, Wiley, 1980
8. B A Auld Acoustic Fields and Waves in Solids Vols I and II Wiley, 1973
9. G W Farnell in Topics in Applied Physics, Vol 24, Acoustic Surface Waves, Springer-Verlag, 1978
10. J H Eveleth, Proc IEEE 53 (1968), 1406-1428
11. M F Lewis, Electronics Letts 8 (1972), 131-2
12. J L Bluestein, Appl Phys Lett 13 (1968), 412-3
13. Y V Gulyaev, JETP Letters 9 (1969), 37-38
14. G W Farnell in Physical Acoustics, Vol 6, Academic Press, 1970
15. T C Lim and G W Farnell, J Acoust Soc Am 45 (1968) 845-851
16. H Engan, K A Ingebrigtsen, and A Tonning, Appl Phys Lett 10 (1967) 311-313
17. C O Newton, Electronics Lett 9 (1973) 481
18. Y Shimizu and M Tanaka, Electronics Letts 21 (1985) 225-6
19. M F Lewis, Proc 1977 IEEE Ultrasonics Symposium, 744-752
20. R F Milsom, N H C Reilly and M Redwood, IEEE Trans SU-24 (1977) 147-166
21. D L Lee, IEEE Trans SU-27 (1980) 22-30

22. K H Yen, K F Lau and R S Kagiwada Proc 1979 IEEE Ultrasonics Symposium 776-785
23. D L Lee, IEEE Trans SU-28 (1981) 330-341
24. T I Browning, D J Gunton, M F Lewis and C O Newton, Proc 1977 IEEE Ultrasonics Symposium, 753-6
25. P D Bloch, E G S Paige and M G Wagg IEEE Trans SU-31 (1984) 77-83
26. B Lewis, N M Shorrocks and R W Whatmore, Proc 1982 IEEE Ultrasonics Symposium 389-393
27. K Toda, J Appl Phys 44 (1973) 56-62
28. R S Wagers, IEEE Trans SU-23 (1976) 113-127
29. M F Lewis, Electronics Letts 17 (1981) 819-22
30. C N Helmick Jr., D J White and K M Lakin, Proc 1981 IEEE Ultrasonics Symp, 280-285
31. J F Nye, Physical Properties of Crystals, Oxford University Press, 1976
32. M W Ewing, W S Jardetzky and F Press, Elastic Waves in Layered Media, McGraw-Hill, 1957
33. E A Ash, R M De La Rue and R F Humphryes, IEEE Trans MTT-17 (1969) 882-892
34. H F Tiersten, J Appl Phys 40 (1969) 770-789
35. J G Scholte, Roy Astron Soc London, Monthly notices, Geophysics suppl 5 (1947) 120-126.
36. G W Farnell and E L Adler in Physical Acoustics Vol 9, Academic Press, 1972
37. C Lardat, C Maerfeld and P Tournois, Proc IEEE 59 (1971) 355-368
38. T Shiosaki, Proc 1978 IEEE Ultrasonics Symposium, 100-110
39. G A Armstrong and S Crampin, Electronics Letts 9 (1973) 322-3
40. J E Bowers, B T Khuri-Yakub and G S Kino, Appl Phys Lett 36 (1980) 806-7

Part III

Generation

Rayleigh Wave Transducers

D.P. Morgan

Plessey Research (Caswell) Ltd, Allen Clark Research Centre,
Caswell, Towcester, Northants, NN12 8EQ, England

1. Introduction

In modern Rayleigh-wave devices, transducers are needed to generate Rayleigh waves in response to an applied electrical signal and to sense the waves so as to give an electrical output waveform. These are clearly key requirements for electronic devices, and are usually met by using interdigital transducers on piezoelectric substrates. The flexibility of modern fabrication techniques has enabled a very wide variety of transducer types to be developed, including for example apodised or chirped transeducers which perform signal processing operations in addition to transduction. In many cases the device response is determined mainly by the transducer behaviour, and a detailed understanding of the latter is therefore essential for both analysis and design of devices.

This paper concentrates on some fundamental aspects of interdigital transducers. Section 2 below discusses the main types of transducer. Section 3 considers transducer analysis, particularly the approximate "quasi-static" method which yields simple results applicable to a wide variety of transducer problems. Section 4 is concerned with electrode interactions in transducers.

2. Types of Rayleigh-wave Transducer

The wide variety of transducers for Rayleigh waves is well illustrated by White's extensive review of 1970[1], which mentioned 24 different methods. In addition to transducers for electronic devices, White included many methods developed earlier for use in non-destructive testing applications, examples being the wedge and the comb, shown in Figure 1. Both of these operate by converting a bulk wave, generated by a piezoelectric plate transducer, into a surface wave. Also noteworthy is the variety of methods for probing Rayleigh waves, that is, for measuring the amplitude and phase distribution of the waves over a specified area. For a piezoelectric substrate, this can be done by using a metal stylus to sense the potential associated with the wave. More generally, a variety of optical techniques have been used[2] - for example the wave may be made to diffract the light in a manner very similar to a conventional diffraction grating.

Rather than attempting to survey the entire field, this paper will concentrate on transducers used in present-day electronic devices. Even here, the range of types is very extensive, with many new types emerging since White's review, and some selectivity is necessary. However, since transducers are of fundamental importance in Rayleigh-wave devices, the reader will find more information in other papers in this Symposium.

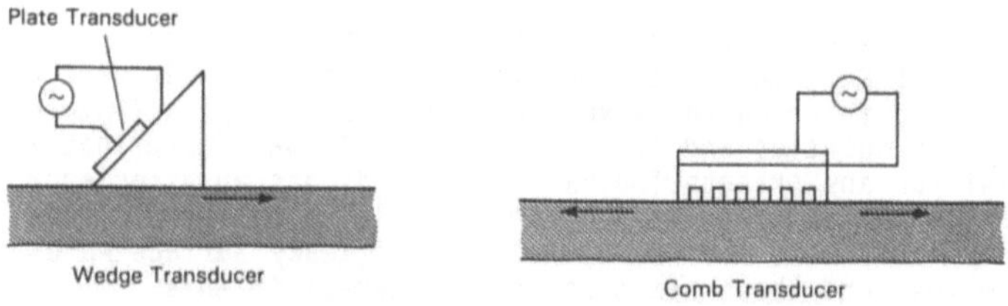

Fig. 1 Wedge and comb transducers

The basic interdigital transducer

A key factor in most electronic devices is the use of a piezoelectric crystalline substrate material, usually quartz or lithium niobate. For such a material the Rayleigh wave can be generated or sensed by means of an interdigital transducer consisting of interleaved metal electrodes on the surface, as shown in Figure 2(a).

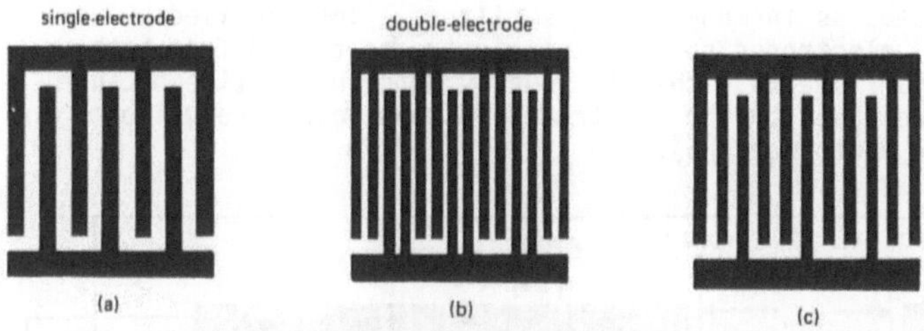

Fig. 2 Unapodised interdigital transducers (from [17])

This type of transducer was first demonstrated for Rayleigh wave excitation by White and Voltmer[3] in 1965, though it is also the subject of earlier patents[4] and was also demonstrated for generation of acoustic bulk waves[5]. The advent of this transducer had some very significant practical consequences for Rayleigh-wave devices:

a) The transducer is planar and may be fabricated by techniques already established for integrated circuits (film deposition and lithography), and for many devices these techniques are economically effective and suitable for large-scale production.

b) Fabrication techniques enable patterns of almost arbitrary geometry to be produced, giving very substantial versatility. This has led to a wide variety of transducer types, for example apodised, chirped, withdrawal-weighted, though even so the geometrical versatility has not been fully exploited.

c) The relevant transducer properties, such as the frequency response, can be remarkably accurate and reproducible. This is due to the very high accuracy of modern pattern generation machines used to make photo-masks, and the highly reproducible acoustic properties of the crystalline substrate materials. For example, the Rayleigh-wave velocity is often controlled to an accuracy of a few parts per million and for some materials, notably quartz, it can also be highly temperature stable.

Owing to these attractions, the introduction of the interdigital transducer led to extensive studies of Rayleigh waves and other acoustic waves in piezoelectric crystalline media, in which the wave properties have complex features depending on the crystal properties and its orientation. It was found that a Rayleigh wave can exist for any orientation, and usually it has an associated electric field and can therefore be excited by an interdigital transducer. For some cases other types of wave are found, for example leaky surface waves or Bleustein-Gulyaev waves, and these are discussed elsewhere in this Symposium. For practical devices the substrate material and orientation are usually chosen such that the only surface wave present is a non-leaky surface wave, and in this paper it will be assumed that such a choice has been made. In addition, some bulk wave excitation is nearly always present, though in most cases this does not significantly degrade the device performance.

Apodised transducers

The basic transducer structure of Figure 2(a) can be modified in many ways. One of the commonest modifications is apodisation, in which the lengths of the electrodes are varied in order to enable a variety of responses to be obtained. Figure 3 shows a Rayleigh-wave device using one apodised transducer and one unapodised, as in many bandpass filters. The apodised transducer has "dummy" electrodes so that the electrodes extend from both bus-bars, with only small breaks - this is usually done in practice because the perturbation of the waves by the electrodes is thus made more uniform, and this leads to improved performance[6].

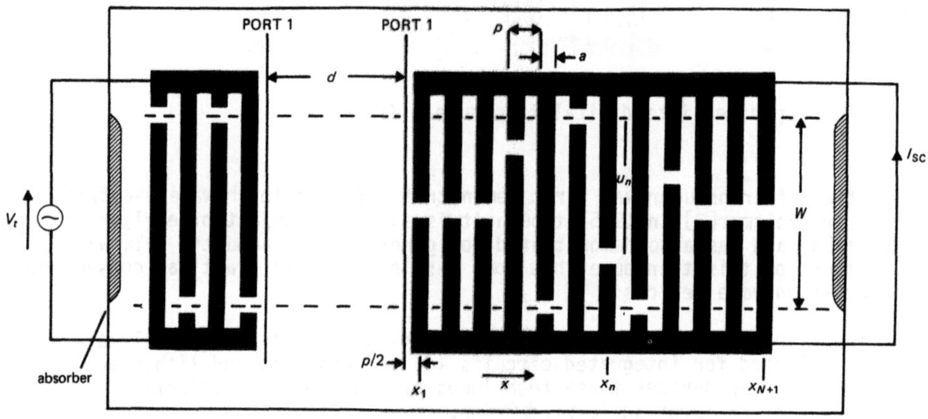

Fig. 3 Bandpass filter using apodised transducer (from [17])

The frequency response of the device can be obtained approximately by using a simple "delta-function model", similar to that introduced by Tancrell and Holland[7]. In the apodised transducer, assumed here to be receiving surface waves, each inter-electrode gap is regarded as a Rayleigh-wave detector, contributing an output current proportional to the overlap of the adjacent electrodes with different polarities. Taking the electrode locations to be x_n and the electrode pitch to be p, the location of the nth gap is $x_n + p/2$, and this determines the phase of the current contribution. If u_n is the electrode break location, as in the figure, the electrode overlap is $u_{n+1} - u_n$. The sum of the contributions is essentially the transducer frequency response $H_t(\omega)$, but for more rigour we also include an element factor $E(\omega)$ which represents the frequency response of an individual gap. The element factor is evaluated later. The transducer response is therefore

$$H_t(\omega) = E(\omega) \sum_{n=1}^{N} \frac{u_{n+1} - u_n}{W} \exp\left[-jk_0 (x_n - x_1 + p)\right] \quad (1)$$

where k_0 is the Rayleigh-wave wavenumber at frequency ω, N is the number of gaps and W is the maximum aperture. The same expression may be used for the response of the unapodised transducer. The overall response of the device can be defined as the current I_{sc} produced by one transducer when it is shorted, divided by the voltage V_t applied to the other transducer. Taking the two transducer responses to be $H_t^a(\omega)$ and $H_t^b(\omega)$, the device response is easily shown to be

$$H(\omega) = I_{sc}/V_t = H_t^a(\omega)\, H_t^b(\omega) \exp(-jk_0 d) \quad (2)$$

where d is the separation of the two transducers. Thus the device response is given simply by the product of the transducer responses. Note that if both transducers are apodised this relation is no longer true, though it does remain true if the transducers are coupled via a multi-strip coupler.

The above equations show that, to first order, the device response is easily calculated. The ability to vary the electrode overlaps in an arbitrary fashion gives enormous flexibility. The element factor $E(\omega)$ in equn (1) varies slowly with frequency, and an estimate of the transducer response can be obtained by taking $E(\omega)$ to be constant. The response of equn(1) then has the same form as that of several other devices, for example digital finite-impulse-response filters[8], which are collectively known as <u>transversal filters</u>. The transducer "impulse response", obtained by Fourier transforming equn(1) to the time domain, is a sequence of delta functions at times corresponding to the gap locations and with amplitudes proportional to $u_{n+1} - u_n$. An important consequence is that established techniques used to design a transversal filter to give a specified frequency response[8] can also be applied to Rayleigh-wave transducer design. For accurate results the required frequency response of the transducer is first divided by $E(\omega)$.

Other types of interdigital transducer

In practice, a wide variety of other transducer types are used in practical devices, and here we mention the main types briefly. More details are given by Lewis et al[9]. Figures 2(b) and 2(c) show two types of "multi-electrode" transducer[10], which have more than two electrodes per period. These are often used in order to minimise electrode interaction effects, which are discussed in Section 4 below. For chirp filters, used in pulse-compression radar systems, the impulse response is required to be a pulse whose frequency varies with time, and this is usually obtained by varying the pitch of the electrodes along the length of the transducer. It is of course possible to combine these factors; thus a multi-electrode transducer may have both varied pitch and apodisation.

A withdrawal-weighted transducer is illustrated in Figure 4. Here a relatively long transducer is weighted by removing selected electrodes, enabling the frequency response to be modified so as to approximate a required response. This is an alternative to apodisation weighting, described above, and has the advantage that the response is less affected by Rayleigh-wave diffraction. However, withdrawal weighting is less flexible than apodisation, and is generally suitable only for transducers with relatively small bandwidths. Many other weighting techniques have been investigated, and these are reviewed by Engan[11].

Fig. 4 Withdrawal-weighted transducer (from [17])

Figure 5 shows a unidirectional transducer in which the electrodes are connected sequentially to three bus-bars, using a dielectric spacer to enable insulated cross-over connections to be made. A driving network consisting of a few reactive components is used to apply voltages phased such that Rayleigh waves are generated only in one direction, unlike the previously mentioned transducers which are all bidirectional. When electrically matched the unidirectional transducer gives low conversion loss and also suppresses acoustic reflections. It thus minimises spurious signals due to multiple acoustic transits of the device, and this is particularly important for devices required to have low insertion losses. There are several other types of unidirectional transducer, including some recent types in which some of the electrodes are used to deliberately introduce mechanical or electrical loading in an asymmetrical fashion[9,12]. The latter types avoid the inconvenience of fabricating cross-overs.

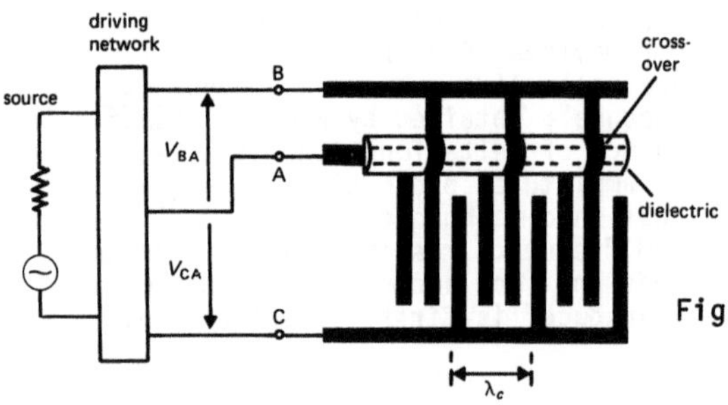

Fig. 5 Three-phase unidirectional transducer (from [17])

3. Analysis of Interdigital Transducers

Since all practical electronic devices rely on interdigital transducers, the analysis of the transducers is of crucial importance, not only for device analysis, but also for device design. The transducer analysis is however a very complex issue owing to a variety of factors:

a) Interdigital transducers have a wide variety of forms, as discussed in Section 2 above.

b) For each type a variety of parameters must be evaluated - the amplitude of launched surface waves, the transducer admittance, and the scattering coefficients for incident Rayleigh waves. Also, device analysis must allow for combinations of transducers, sometimes with a multi-strip coupler, and must evaluate the circuit effect and the unwanted triple-transit signal.

c) A variety of physical phenomena, many unwanted, can occur - electrostatic effects (determining harmonic strengths and end effects), electrode interactions, diffraction, electrode resistivity and unwanted acoustic modes (such as bulk waves or leaky surface waves, though in some devices one of these is the main mode).

For practical device development a convenient analysis applicable to a wide variety of transducers is desirable, even if this is obtained at the expense of excluding some of the phenomena mentioned in (c) above. For this reason the simple delta-function model (Section 2) is widely used, even though it excludes most of the physical phenomena and does not give most of the parameters in (b) above. The crossed-field network model of Smith et al[13] is more sophisticated, giving all the parameters needed, but does not allow for bulk waves. The model can be extended to allow for electrostatic effects [14]and electrode interactions[15], but at some cost in complexity. Milsom's Green's function analysis[16] gives all the required parameters and includes most of the phenomena in (c), but again this is too complex for device development in general.

Section 3.2 below describes some results obtained from the "quasi-static" analysis, which is a simplification of Milsom's method. These results enable most of the transducers in (a) above to be analysed quite simply, giving all the parameters in (b). The simplicity is a consequence of ignoring bulk waves and electrode interactions, which are not very significant in most practical devices, but an extended analysis allowing for interactions is described in Section 4. However, before embarking on these topics, Section 3.1 describes some fundamental relationships.

In the discussion here, most of the results are stated without proof. The derivations required are given in detail elsewhere[17].

3.1 Fundamental relationships

A fundamental theorem applying to any insulating piezoelectric material is the reciprocity relation given by Auld[18]. The relation itself is not given here, but we give one of its important consequences. Consider a device having two transducers on the surface of a piezoelectric solid, as in Figure 6. The transducers consist of metal electrodes with zero resistivity, and mechanical loading is neglected. Define ϕ_a and ρ_a as the surface potential and charge density when unit voltage at frequency ω is applied to transducer A, with transducer B absent. Similarly, ϕ_b and ρ_b are defined for transducer B. With

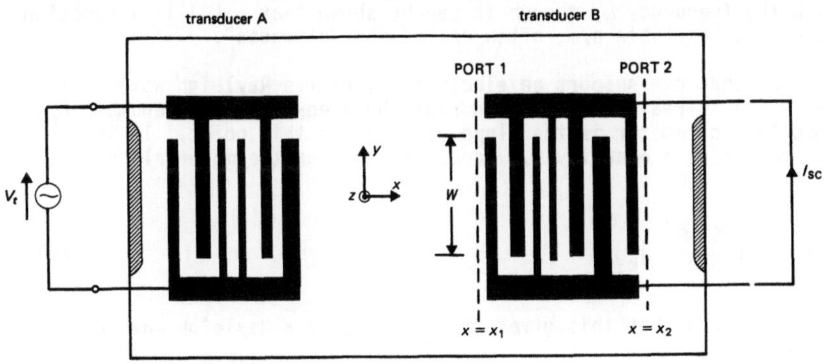

Fig. 6 General unapodised transducers (from [17])

both transducers present, we now apply a voltage V_t to transducer A and short circuit transducer B. Using Auld's reciprocity relation, the output current can be shown to be given by the surface integrals

$$I_{sc} = -j\omega V_t \int_S \phi_a \rho_b \, dS = -j\omega V_t \int_S \phi_b \rho_a \, dS \qquad (3)$$

This equation gives the output of a receiving transducer in terms of quantities obtained by analysing launching transducers. It is valid whatever types of acoustic waves are present - Rayleigh waves, leaky waves, bulk waves and so on - even if they are reflected from the substrate surfaces.
We now assume that only Rayleigh waves are incident on transducer B, which is the case if transducer A launches these waves and if the transducer separation is large. Assuming the transducers to be unapodised, the potential of the Rayleigh wave in the region of transducer B is written as $\phi_o \exp(-jk_o x)$, where k_o is the free-surface wavenumber. Equating this with $V_t \phi_a$ in equn (3) we have

$$I_{sc} = -j\omega W \phi_o \int_{-\infty}^{\infty} \rho_b (x) \exp(-jk_o x) \, dx$$

$$= -j \omega W \phi_o \bar{\rho}_b (k_o) \qquad (4)$$

where we define $\bar{\rho}_b(\beta)$ as the Fourier transform of $\rho_b(x)$, and W is the transducer aperture. Equation (4) is valid when electrode interaction effects are present, and also if transducer B can couple to bulk waves or other acoustic modes.

Effective Permittivity The analysis for a transducer launching Rayleigh waves is facilitated by defining the effective permittivity $\varepsilon_s(\beta)$, introduced by Ingebrigtsen[19]. We imagine a piezoelectric half-space with its surface free of mechanical forces. The potential and charge density at the surface are $\phi(x)$ and $\sigma(x)$, at frequency ω. The charge density can only exist if electrodes are present, but the electrodes need not be considered explicitly at this stage. If $\bar{\Phi}(\beta)$ and $\bar{\sigma}(\beta)$ are the Fourier transforms of $\phi(x)$ and $\sigma(x)$, we define $\varepsilon_s(\beta)$ by

$$\varepsilon_s(\beta) = \frac{\bar{\sigma}(\beta)}{|\beta|\bar{\Phi}(\beta)} \qquad (5)$$

This function depends on the half-space material and orientation. It also varies with the frequency ω, though it can be shown that $\varepsilon_s(\beta)$ is a function of the normalised variable β/ω, often called the "slowness".

For a material that can support an electrically-active Rayleigh wave it is found that $\varepsilon_s(\beta)$ is real in the region near the free-surface wavenumber k_o, and an important parameter is the slope of $\varepsilon_s(\beta)$ at this point. It is convenient to define a quantity Γ_s, dependent only on the material and orientation, by

$$1/\Gamma_s = -k_o \left[\frac{d\varepsilon_s(\beta)}{d\beta}\right]_{k_o} \qquad (6)$$

It can be shown [19] that this gives the power P_s of a Rayleigh wave beam using the formula

$$P_s = \tfrac{1}{4} \omega W |\phi_s(x)|^2 / \Gamma_s \qquad (7)$$

where W is the beam width and $\phi_s(x)$ the surface potential. The same parameter gives the potential of Rayleigh waves launched by an unapodised transducer. If $\sigma(x)$ is the charge density on the electrodes, with Fourier transform $\bar{\sigma}(\beta)$, the potential of the wave launched in the -x direction is [16,17]

$$\phi_s(x) = j\,\Gamma_s\,\bar{\sigma}\,(k_o)\,\exp(jk_o x) \tag{8}$$

This equation ignores mechanical loading and Rayleigh-wave diffraction, but is otherwise very general in that it allows for electrode interactions and is valid if the transducer generates bulk waves as well as Rayleigh waves.

Equations (8) and (4) can be combined to give a reciprocity relation, relating the receiving process to the launching process:

$$\left[\frac{I_{sc}}{\phi_i(x_1)}\right]_{\text{receive}} = -\frac{\omega W}{\Gamma_s}\left[\frac{\phi_s(x_1)}{V_t}\right]_{\text{launch}} \tag{9}$$

where V_t is the voltage on the launching transducer, $\phi_s(x)$ is the potential of the launched wave and, for reception, $\phi_i(x)$ is the potential of the incident wave. The wave potentials are evaluated at the point $x = x_1$, which can be taken to be close to the edge of the transducer and can be regarded as the acoustic port, though the precise value of x_1 is of no consequence.

<u>Ingebrigtsen's approximation</u> Generally, the effective permittivity $\varepsilon_s(\beta)$ must be found using an extensive numerical procedure, but for practical purposes a simple approximate form can be used. We define $k_o = \omega/v_o$ as the Rayleigh-wave wavenumber for a free surface, and $k_m = \omega/v_m$ as the wavenumber for a metallised surface (with the surface still assumed to be force-free). Equation (5) shows that $\varepsilon_s(\beta)$ must be zero at $\beta = k_o$ and infinite at $\beta = k_m$. By taking Taylor expansions in the vicinity of these points and bearing in mind that symmetry requires $\varepsilon_s(-\beta) = \varepsilon_s(\beta)$, we obtain[19]

$$\varepsilon_s(\beta) \approx (\varepsilon_o + \varepsilon_p^T)\,\frac{\beta^2 - k_o^2}{\beta^2 - k_m^2} \quad \text{where} \tag{10}$$

$$\varepsilon_p^T = (\varepsilon_{xx}\varepsilon_{zz} - \varepsilon_{xz}^2)^{1/2} \tag{11}$$

with the permittivity elements ε_{ij} evaluated at constant stress. The use of the coefficient $(\varepsilon_o + \varepsilon_p^T)$ here is a good approximation which becomes exact if the material is non-piezoelectric, in which case $k_o = k_m$ and $\varepsilon_s(\beta) = \varepsilon_o + \varepsilon_p$. Equation (10) does not allow for bulk-wave excitation, and it is not valid for Bleustein-Gulyaev waves.

Using equn (10) and the fact that $v_o \approx v_m$, the constant Γ_s defined by equn (6) is found to be

$$\Gamma_s \approx \frac{1}{\varepsilon_o + \varepsilon_p^T}\;\frac{v_o - v_m}{v_o} \tag{12}$$

This formula is very convenient to use since values of ε_p^T, v_o and v_m are known for a wide variety of materials and orientations. It also confirms the early conclusion of Campbell and Jones[20] that the piezoelectric coupling for

Rayleigh waves is characterised by the fractional velocity change $(v_o - v_m)/v_m$ which occurs when the surface is metallised.

3.2 Quasi-static Analysis

Milsom et al[16] have developed a method of transducer analysis based on the above equations, using a Green's function derived from the effective permittivity. This method is very versatile, including for example the effects of electrode interactions and bulk wave excitation, but its complexity makes it rather inconvenient to use. Here we describe the "quasi-static" analysis[21], which is a simplification of Milsom's method obtained by ignoring electrode interactions and bulk wave excitation. The present description is based on a revised formulation[17] which is more self-consistent and can also be applied to multi-strip couplers. A higher-order version of the quasi-static analysis can be used to analyse electrode interactions, as discussed in Section 4.

We consider a piezoelectric half-space with surface potential $\phi(x)$ and charge density $\sigma(x)$, assuming that these vary only in the x-direction. The Green's function $G(x)$ is defined such that

$$\phi(x) = G(x) * \sigma(x) \tag{13}$$

at frequency ω, where the asterisk indicates convolution. The Green's function can be approximately written as[16]

$$G(x) \approx G_e(x) + G_s(x) \tag{14}$$

where $G_e(x)$ and $G_s(x)$ are respectively contributions due to electrostatic effects and surface-wave excitation, given by

$$G_e(x) = - \frac{\ln|x|}{\pi(\varepsilon_o + \varepsilon_p^T)} \tag{15a}$$

and

$$G_s(x) = j\,\Gamma_s \exp(-j\,k_o|x|) \tag{15b}$$

with Γ_s defined by equn (6). Equation (14) neglects an additional contribution due to bulk-wave excitation. It can be shown that equns (14) and (15) lead to an expression for the effective permittivity almost the same as Ingebrigtsen's approximation, equn (10).

To analyse a transducer launching surface waves, it is assumed that the charge density is dominated by an electrostatic term $\sigma_e(x)$, defined as the solution obtained when surface-wave excitation is ignored. Thus $\sigma_e(x)$ is the solution of

$$\phi_e(x) = G_e(x) * \sigma_e(x) \tag{16}$$

where $\phi_e(x)$ is a surface potential equal to the electrode voltage whenever x is on an electrode. The total charge density is now written as

$$\sigma(x) \approx \sigma_e(x) + \sigma_a(x) \tag{17}$$

where $\sigma_a(x)$ is a small term due to the surface waves present. The surface potential, given by equns (13), (14) and (17), now has four terms, but the term $G_s(x) * \sigma_a(x)$ is small and is omitted. Thus, in the quasi-static approximation we have

$$\phi(x) = G_e(x)*(\sigma_e(x) + \sigma_a(x)) + G_s(x)*\sigma_e(x) \qquad (18)$$

This equation can be used to deduce $\sigma_a(x)$ from $\sigma_e(x)$. A second-order Green's function analysis can be obtained by writing $\sigma(x) = \sigma_e(x) + \sigma_a(x) + \sigma_a'(x)$ and omitting a term $G_s(x) * \sigma_a'(x)$ in the potential. This enables $\sigma_a'(x)$ to be deduced from $\sigma_a(x)$.

We now consider a two-terminal transducer, such as those in Figure 6, with each electrode connected to one of the two bus-bars. The transducer is taken to be unapodised but the electrode and gap widths are at this stage arbitrary - thus the analysis is very general and applies, for example, to withdrawal-weighted transducers. Diffraction is ignored. When a voltage is applied the surface potential $\phi(x)$ is given by equn (18), in the quasi-static approximation. At points remote from the transducer the potential due to the electrostatic Green's function $G_e(x)$ is small, and hence the potential $\phi_s(x)$ associated with the surface waves generated is given by the remaining term

$$\phi_s(x) = G_s(x) * \sigma_e(x) \qquad (19)$$

It is convenient to define $\rho_e(x)$ as the electrostatic charge density for unit voltage applied across the transducer, with Fourier transform $\bar{\rho}_e(\beta)$. Thus, if V_t is the transducer voltage we have $\sigma_e(x) = V_t\, \rho_e(x)$. Using equn (19) with equn (15b) for $G_s(x)$, the surface-wave potential for points outside the transducer is found to be

$$\phi_s(x) = j\, \Gamma_s\, V_t\, \bar{\rho}_e(\pm k_o)\, \exp(\pm j\, k_o\, x) \qquad (20)$$

where the upper sign is used for points to the left of the transducer, and the lower sign for points to the right. This result shows that the quasi-static analysis neglects electrode interactions, since it predicts that the surface-wave amplitude is proportional to the Fourier transform of the electrostatic charge density.

Some other results from the quasi-static analysis are now given, without derivation. The transducer admittance is written as

$$Y_t(\omega) = G_a(\omega) + j\, B_a\,(\omega) + j\, \omega\, C_t \qquad (21)$$

where the capacitance C_t is obtained by integrating the electrostatic charge density $\rho_e(x)$ over all the electrodes connected to one bus-bar. The acoustic conductance $G_a(\omega)$ and susceptance $B_a(\omega)$ arise from the acoustic charge density $\sigma_a(x)$, giving

$$G_a(\omega) = \omega\, W\, \Gamma_s\, |\bar{\rho}_e\,(k_o)|^2 \qquad (22)$$

The susceptance is found to be the Hilbert transform of $G_a(\omega)$, a result to be expected because the transducer must be causal. The conductance can also be obtained by evaluating the total surface-wave power generated by the transducer, and this agrees with equn (22). For a shorted transducer receiving surface waves, the output current given by the quasi-static analysis is found to agree with the reciprocity relation, equn (9), using equn (20) for the launching process.

For a surface wave incident on a transducer connected to a load with admittance Y_L, the power reflection coefficient is found to be

$$R(\omega) = G_a^2 \,/\, |Y_t + Y_L|^2 \qquad (23)$$

and the power transmission coefficient is found to be

$$T(\omega) = 1 - (G_a^2 + 2G_aG_L) \,/\, |Y_t + Y_L|^2 \tag{24}$$

where G_L is the real part of the load admittance Y_L. The conversion coefficient, giving the power dissipated in the load divided by the power of the incident wave, is

$$C(\omega) = 2G_aG_L \,/\, |Y_t + Y_L|^2 \tag{25}$$

By reciprocity, this equation also applies for a launching transducer, as can be shown using simple network analysis. The three power coefficients, equns (23)-(25), are valid at all frequencies and their sum equals unity, so that all of the power incident on the transducer is accounted for. Some experimental results confirming these equations were obtained by Smith et al[13,15], using uniform transducers operated at the fundamental centre frequency. Note in particular that the reflection coefficient $R(\omega)$ is zero if the transducer is shorted, that is, if $Y_L = \infty$. This is confirmed experimentally, though not for single-electrode transducers with large numbers of electrodes - in this case electrode interactions become significant.

Although the transducer has been taken to be unapodised, the analysis is readily generalised to the case of an apodised transducer by dividing the transducer into a number of parallel channels, as done by Tancrell and Holland[7].

3.3 Transducers with regular electrodes - element factor

The above analysis gives the quasi-static properties of an unapodised transducer in terms of the electrostatic charge density $\rho_e(x)$ and its Fourier transform $\bar{\rho}_e(\beta)$. For a general transducer such as those in Figure 6 it is necessary to evaluate $\rho_e(x)$ numerically, and a variety of techniques have been used[14,22]. However, many practical transducers have regular electrodes, that is, the electrode width and pitch are uniform throughout the transducer, as illustrated in Figure 7. In this case $\rho_e(x)$ can be obtained from an analytic formula, and this leads to a formulation corresponding closely to the simple delta-function model[23].

We first state the charge superposition principle, which applies for arbitrary electrode geometries. Suppose that the electrodes of a transducer are disconnected from each other and unit voltage is applied to electrode n, with zero voltage on all the other electrodes. In this situation the electrostatic

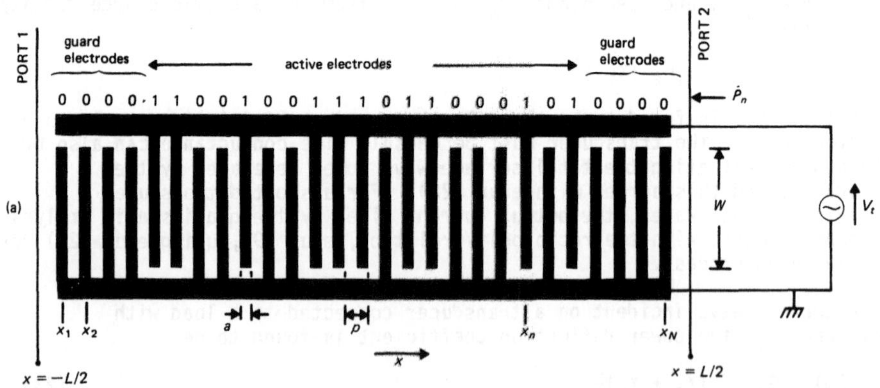

Fig. 7 Unapodised transducer with regular electrodes (from [17])

charge density on the electrodes is defined to be $\rho_{en}(x)$. If we now apply voltages V_n to the electrodes, the charge superposition principle states that the electrostatic charge density is given by [24]

$$\sigma_e(x) = \sum_n V_n \rho_{en}(x) \qquad (26)$$

and hence $\sigma_e(x)$ can be found for any set of voltages V_n once the functions $\rho_{en}(x)$ are known. For transducer analysis we define the polarity of electrode n as $P_n = 0$ or 1, depending on which bus-bar the electrode is connected to, as shown in Figure 7. The function $\rho_e(x)$ is then given by equn (26), with V_n replaced by P_n.

For a transducer with regular electrodes the functions $\rho_{en}(x)$ are all much the same except for lateral displacements. This is not in fact true for the electrodes near the ends, but it is assumed here that a few guard electrodes, with $P_n = 0$, are included at each end. For the guard electrodes, the corresponding charge densities $\rho_{en}(x)$ do not contribute to the total charge density $\rho_e(x)$, and this has the very useful consequence that end effects are practically eliminated. The total charge density $\rho_e(x)$ thus has the form

$$\rho_e(x) = \sum_n P_n \rho_f(x - x_n) \qquad (27)$$

where x_n is the centre location of electrode n. The function $\rho_f(x)$ is the elemental charge density, defined as the charge density on an infinite array of regular electrodes when one electrode has unit voltage and all the others have zero voltage. The Fourier transform of equn (27) is

$$\bar{\rho}_e(\beta) = \bar{\rho}_f(\beta) \sum_n P_n \exp(-j\,\beta\, x_n) \qquad (28)$$

This formulation simplifies the calculation of all the transducer properties mentioned earlier. For example, the amplitude of a wave launched by the transducer, equn (20), is seen to be given by the product of an element factor $\bar{\rho}_f(k_o)$ and an array factor $\sum P_n \exp(-j\, k_o\, x_n)$. In the x-domain, the array factor corresponds to a set of delta functions located at the electrode centres x_n, with amplitudes P_n. Thus the analysis is very similar to the simple delta-function model, but with the important distinctions that it correctly allows for electrostatic effects and it enables the transducer admittance and scattering parameters to be evaluated.

The function $\bar{\rho}_f(\beta)$ is given by[25]

$$\bar{\rho}_f(\beta) = (\varepsilon_o + \varepsilon_p^T) \frac{2 \sin \pi s}{P_{-s}(-\cos\Delta)} P_m(\cos\Delta) \qquad (29)$$

$$\text{for} \quad m < \frac{\beta p}{2\pi} < m + 1$$

where $\Delta = \pi a/p$, $s = \beta p/(2\pi) - m$, $P_{-s}(x)$ is a Legendre function and $P_m(x)$ is a Legendre polynomial. Also, a and p are respectively the electrode width and pitch. Figure 8 shows $\bar{\rho}_f(\beta)$ for several values of the metallisation ratio a/p. In most practical transducers the array factor varies rapidly with frequency, so that the frequency variation of the element factor has relatively little effect. The points where βp is an odd multiple of π correspond to the fundamental centre frequency and the harmonics of a single-electrode transducer. For these points $s = ½$, and the Legendre function in equn (29) is

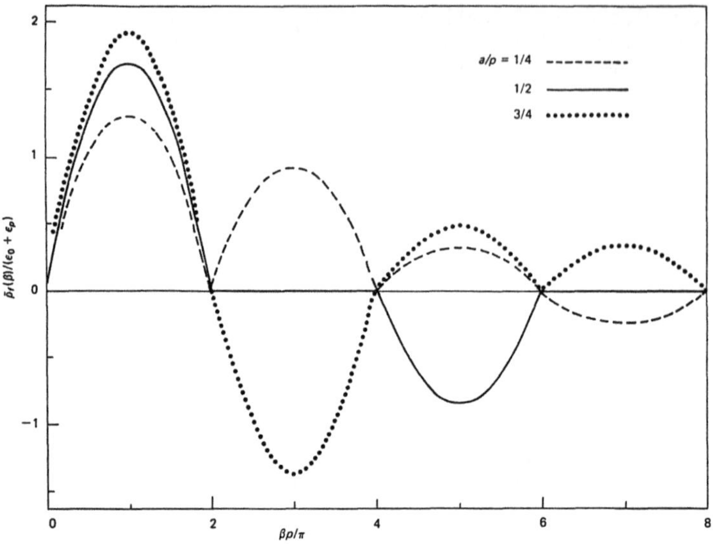

Fig. 8 Elemental charge density in the β domain, for three metallisation ratios (from [17])

equal to an elliptic integral. This gives Engan's formula [26] for the relative harmonic strengths.
We now note briefly some other applications of the quasi-static analysis[17].

a) The analysis is readily extended to apodised transducers such as that in Figure 3, and can be written in terms of elements associated with the inter-electrode gaps instead of with the electrodes. This enables the element factor $E(\omega)$ of equn(1) to be written as

$$E(\omega) = (\omega W \Gamma_s)^{\frac{1}{2}} \frac{\bar{\rho}_f(k_o)}{2j \sin(k_o p/2)} \tag{30}$$

b) For a chirp transducer the electrodes are not regular. However, a simple approximation can be used to express the response in terms of an element factor and an array factor, as above.

c) The charge superposition principle, equn(26) can be used to analyse a three-phase unidirectional transducer such as that in Figure 5. This relates the unidirectional transducer to three two-terminal "constituent" transducers, analysed by the above methods.

d) The quasi-static method can also be applied to multi-strip couplers. This is simpler than other coupler theories, but owing to the neglect of electrode interactions, does not predict the stop bands.

e) For multi-electrode transducers the quasi-static method gives admittance formulae agreeing with those of Engan[10].

4. Electrode Interactions in Transducers

The term "electrode interactions" refers to the perturbation of the surface wave due to the electrical or mechanical loading caused by the electrodes. The main practical consequences are that the effective velocity within the transducer is reduced, the transducer reflection coefficient is affected so that it reflects even when it is shorted, and the transducer frequency response and conductance are distorted. These effects are not predicted by the quasi-static analysis given above. In practice they are not usually very significant except for single-electrode transducers, in which constructive interference occurs at the centre frequency, where the electrode pitch equals the half-wavelength. Earlier methods of analysis[27-29] have proved effective, though they have not been widely used owing to their complexity. The present description[17] is, as for previous accounts, restricted to transducers with regular electrodes.

Interaction effects are best described in terms of a reflective array model in which all the electrodes are taken to scatter surface waves in an identical fashion - for a transducer, this implies that a shorted receiving transducer must be considered, but reciprocity can be invoked to apply the results to a launching transducer. Define r and t as the reflection and transmission coefficients for one electrode, with phase referred to the electrode centre. It is assumed that there is no power loss, and this is found to imply

$$|r|^2 + |t|^2 = 1 \tag{31}$$

and

$$r/t = \pm j|r/t| \tag{32}$$

Thus r and t are in phase quadrature. Apart from the sign ambiguity t can be deduced from r, and in many cases the ambiguity is removed because t is known to be close to unity. For a regular array of electrodes, with pitch p, we define a transmission matrix relating the wave amplitudes c_1 and b_1 to the left of one electrode to the amplitudes c_2 and b_2 to the right, where the c's refer to waves propagating to the right and the b's to waves propagating to the left. The relation is found to be

$$\begin{bmatrix} c_2 \\ b_2 \end{bmatrix} = \begin{bmatrix} 1/\tau^* & r/t \\ -r/t & 1/\tau \end{bmatrix} \begin{bmatrix} c_1 \\ b_1 \end{bmatrix} \tag{33}$$

where the c's and b's are evaluated midway between the electrodes, and $\tau = t\exp(-jk_0p)$, with $k_0 = \omega/v_0$.

For an infinite regular array, the wavenumber γ can be found by assuming that $c_2 = c_1 \exp(-j\gamma p)$ and $b_2 = b_1 \exp(-j\gamma p)$. Using equn (33), this gives

$$\cos\gamma p = \frac{\cos(k_0 p - \theta_t)}{|t|} \tag{34}$$

where θ_t is the phase of t. Since $|t| < 1$, this equation shows that γ is complex, giving a stop band, when k_0p is close to a multiple of π. If r is small the width of the stop band can be shown to be $\Delta\omega \approx 2v_0|r|/p$, where v_0 is the free-surface velocity.

For electrodes on a strongly piezoelectric substrate, such as lithium niobate, the electrode reflection coefficient r is primarily due to electrical loading, and can be evaluated[17] using the second-order Green's function analysis mentioned in Section 3.2. For low frequencies, where $\omega < 2\pi v_0/p$, the result is

$$r \approx -\tfrac{1}{2}j \frac{\Delta v}{v} k_0 p \left[P_{2s} (\cos \Delta) \pm \frac{P_s(\mp\cos \Delta)}{P_{-s}(\mp\cos \Delta)} P_{-2s} (\cos \Delta) \right] \tag{35}$$

where $s = k_0 p/(2\pi)$ and $\Delta v/v = (v_0 - v_m)/v_0$. The upper signs refer to shorted electrodes and the low signs to open-circuit electrodes (as in a multi-strip coupler). This agrees with the formula given by Datta[30] which is also valid for higher frequencies, that is, for $\omega > 2\pi v_0/p$. The reflection coefficient varies slowly with frequency. At $\omega = \pi v_0/p$, or $s = \tfrac{1}{2}$, the electrode pitch equals the half-wavelength, and for this frequency the value of r is shown as a function of the metallisation ratio a/p in Figure 9. For $s = \tfrac{1}{2}$, r is negative imaginary for shorted electrodes and positive imaginary for open-circuit electrodes. Note that, since r is purely imaginary, the transmission coefficient t must be real, as shown by equn (32). Thus the analysis leading to equn (35) is not accurate enough to predict the velocity slowing due to the presence of the electrodes. However, the velocity slowing can be evaluated by other methods[31]. Measurements on electrode arrays[32] have confirmed equn (35).

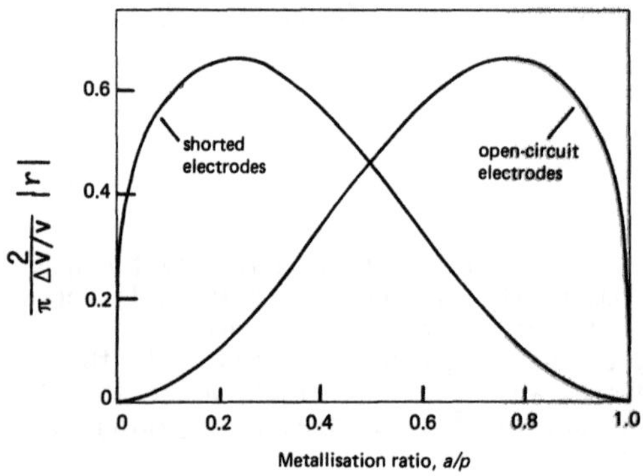

Fig. 9 Magnitude of reflection coefficient due to electrical loading, for one electrode in a regular array (from [17])

For a finite array of electrodes, such as a shorted interdigital transducer, equn (33) can be used recursively to find the surface-wave amplitudes at all points, and the latter can be expressed analytically. In particular, for a wave incident at one end the reflection coefficient of the array can be found. The power reflection coefficient R is found to be given by

$$1/R = 1 + |t/r|^2 \frac{\sin^2(\gamma p)}{\sin^2(N\gamma p)} \tag{36}$$

where N is the number of electrodes and γ is defined as the solution of equn (34). The amplitude reflection coefficient is shown in Figure 10 for two values of N, taking $|r| = 0.0154$, appropriate for electrical loading with a/p = ½ and a Y,Z lithium niobate substrate. For $N|r| \ll 1$, the second term on the right of equn (36) is dominant, and R has the form expected when multiple reflections are insignificant. For $N|r| \gg 1$ the reflection coefficient is close to unity when the frequency is in the range corresponding to the stop band of an infinite array. The curve for N = 40 corresponds quite well with De Vries' experimental result[33].

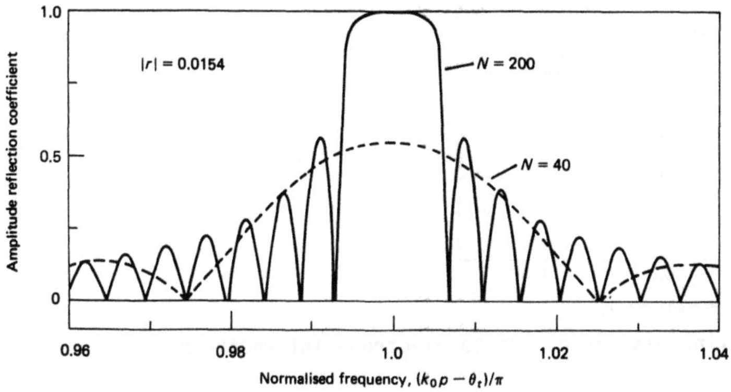

Fig. 10 Reflection coefficient for a regular array of N electrodes (from [17])

To evaluate the distortion in the frequency response of a transducer, we consider a shorted transducer receiving a uniform surface-wave beam. The surface-wave amplitudes throughout the transducer are obtained by cascading the transmission matrix, equn (33), and this is valid even if the transducer is apodised, provided dummy electrodes are included. The current entering electrode n from the upper bus-bar is found to be

$$I_n = -j\omega u_n \int_{-\infty}^{\infty} \rho_{en}(x)\ \phi_s(x)dx \tag{37}$$

where $\phi_s(x)$ is the surface-wave potential, including waves propagating in both directions, and u_n is the length of electrode n measured from the upper bus-bar (Figure 3). The currents are summed to give the total current I_{sc} flowing in the short circuit. For simplicity, it is assumed here that the transducer geometry is symmetrical and that the number of electrodes, N, is odd. This is found to give

$$I_{sc}/\phi_0 = H_0(\gamma)\ A(\gamma) \tag{38}$$

where γ is the solution of equn (34) and ϕ_0 is the potential of the incident surface wave, at the edge of the transducer. Equation (38) is essentially the frequency response of the transducer, and can also be applied to a launching transducer by using reciprocity, equn (9). The function $H_0(\gamma)$ is defined such that $H_0(k_0)$ is the response obtained from the quasi-static analysis, ignoring interactions. The complex function $A(\gamma)$, given by an analytic formula, is equal to unity when interactions are absent, that is, when $r = 0$ and $\gamma = k_0$. In practice, $H_0(\gamma)$ is usually very similar to $H_0(k_0)$, so the effect of interactions can be obtained simply by multiplying by $A(\gamma)$. Moreover, $A(\gamma)$ is independent of the form of the apodisation. Figure 11 illustrates the magnitude of $A(\gamma)$.

Figure 12 shows the effect of interactions in a bandpass filter on Y,Z lithium niobate, using an apodised single-electrode transducer with 51 electrodes and an unapodised double-electrode transducer. The continuous theoretical curve is the insertion loss calculated using the quasi-static analysis, and the broken curve is obtained by multiplying the response by $A(\gamma)$. Similar results have been found by Skeie and Engan[28].

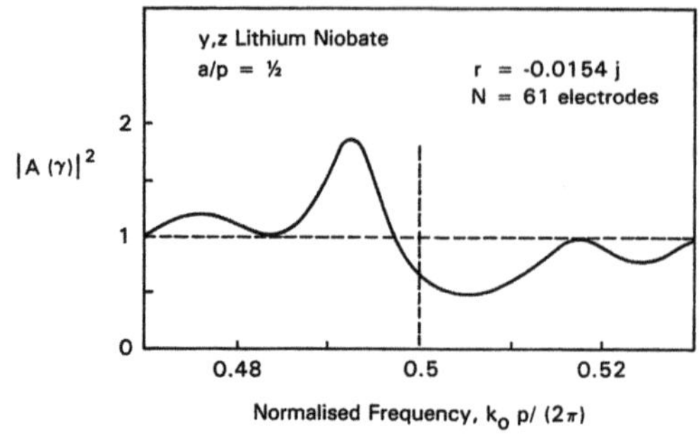

Fig. 11 Function giving distortion due to electrode interactions

The same analysis may also be used to evaluate the parallel conductance $G_a(\omega)$ of the transducer. Assuming the transducer to be unapodised, it is found that

$$G_a(\omega) \doteq \Gamma_s|H_0(\gamma)\ A(\gamma)|^2/(\omega W) \tag{39}$$

where W is the aperture. This is found to agree well with the measurements of Daniel and Emtage[34] on uniform single-electrode transducers using Y, Z lithium niobate. The distortion is greatest for a metallisation ratio a/p = 0.25 because the electrode reflection coefficient r is maximised at this point, as can be seen from Figure 9.

For a weakly-piezoelectric substrate, such as quartz, interaction effects arise primarily from mechanical, rather than electrical, loading. The analysis given above is still valid for the case of mechanical loading, provided an appropriate value for the electrode reflection coefficient r is used. Several authors, for example Datta[35], have analysed this. Measurements using aluminium electrodes on quartz have shown reasonable agreement, and in particular Marshall[36] showed that r is proportional to the film thickness, thus demonstrating that the perturbation is primarily mechanical rather than electrical.

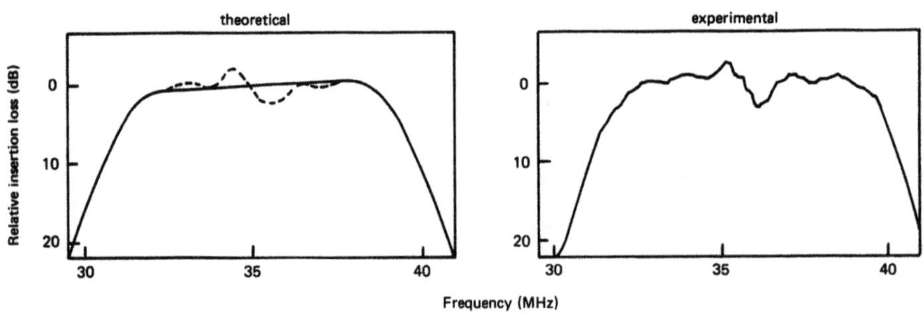

Fig. 12 Effect of interactions in interdigital bandpass filter (from [17])

Acknowledgements

It is a pleasure to acknowledge many helpful discussions with B Lewis, R F Milsom and E G S Paige. I am also indebted to Elsevier Science Publishers B V, for permission to reproduce figures from Reference [17].

References

1. R M White: Proc IEEE 58 1238-1276 (1970)
2. G I Stegeman: IEEE Trans SU-23 33-63 (1976)
3. R M White and F W Voltmer: Appl Phys Lett 7 314-316 (1965)
4. W S Mortley: British Patent 988,102 (1963). Also J H Rowen, US Patent 3,289,114 (1963)
5. W S Mortley: Marconi Rev No.159 273-290 (1965)
6. R H Tancrell and R C Williamson: Appl Phys Lett 19 456-459 (1971)
7. R H Tancrell and M G Holland: Proc IEEE 59 393-409 (1971)
8. L R Rabiner and B Gold: 'Theory and Application of Digital Signal Processing'. Prentice-Hall 1975.
9. M F Lewis, C L West, J M Deacon and R F Humphryes: Proc IEE 131A 186-215 (1985)
10. H Engan: IEEE Trans SU-22 395-401 (1975)
11. H Engan: Wave Electronics 2 133-154 (1976)
12. C S Hartmann, P V Wright, R J Kansy and E M Garber: IEEE Ultrasonics Symp, 1982 40-45
13. W R Smith, H M Gerard, J H Collins, T M Reeder and H J Shaw, IEEE Trans, MTT-17 856-864 (1969)
14. W R Smith and W F Pedler: IEEE Trans MTT-23 853-864 (1975)
15. W R Smith: IEEE Trans MTT-22 960-964 (1974)
16. R F Milsom, N H C Reilly and M Redwood: IEEE Trans SU-24 147-166 (1977)
17. D P Morgan: "Surface-wave Devices for Signal Procesing" Elsevier 1985
18. B A Auld: "Acoustic Fields and Waves in Solids" Wiley 1973 Vol.II
19. K A Ingebrigtsen: J Appl Phys, 40 2681-2686 (1969)
20. J J Campbell and W R Jones: IEEE Trans SU-15 209-217 (1968)
21. D P Morgan: IEEE Trans SU-27 111-123 (1980)
22. C S Hartmann and B G Secrest: IEEE Ultrasonics Symp 1972, 413-416. Also K R Laker, E Cohen and A J Slobodnik: IEEE Ultrasonics Symp 1976, 317-321.
23. B Lewis, P M Jordan, R F Milsom and D P Morgan: IEEE Ultrasonics Symp 1978 709-714. Also S Datta, B J Hunsinger and D C Malocha: IEEE Trans SU-26 235-242 (1979)
24. S Ramo, J R Whinnery and T van Duzer: "Fields and Waves in Communication Electronics" Wiley 1965, pp154,314.
25. R C Peach: IEEE Trans SU-28 96-105 (1981)
26. H Engan: IEEE Trans ED-16 1014-1017 (1969)
27. H Skeie: J Acoust Soc Am 48 1098-1109 (1970)
28. H Skeie and H Engan: Radio and Electron Eng 45 207-220 (1975)
29. P R Emtage: J Acoust Soc Am 51 1142-1155 (1972)
30. S Datta and B J Hunsinger: J Appl Phys 51 4817-4823 (1980)
31. K Blotekjaer, K A Ingebrigtsen and H Skeie: IEEE Trans ED-20 1139-1146 (1973)
32. G L Matthaei, F Barman and E B Savage: Electronics Lett 12 556-557 (1976). Also Wave Electronics 4 251-272 (1983).
33. A J De Vries, in H Matthews (ed) "Surface Wave Filters" Wiley 1977, pp263-305
34. M R Daniel and P R Emtage: Appl Phys Lett 20 320-322 (1972)
35. S Datta and B J Hunsinger: J Appl Phys 50 5661-5665 (1979)
36. F G Marshall: IEEE Ultrasonics Symp 1975 pp290-292.

Excitation, Detection and Reflexion of Rayleigh and Surface Skimming Bulk Waves

M. Planat

Laboratoire de Physique et Métrologie des Oscillateurs du C.N.R.S., associé à l'Université de Franche-Comté-Besançon, 32, avenue de l'Observatoire, F-25000 Besançon, France

1. Introduction

Since the pioneer work by RAYLEIGH [1] (1885) at the end of the last century and due to the progress of electronic techniques, Surface Wave Ultrasonics is now a subject of its own. Everybody knows that the Rayleigh wave can exist at the mechanical free surface of an elastic homogeneous half space. More mysterious are the origin and characteristics of the body waves also radiated by the excitation source. This last subject is of so strange appearance that even serious treatises of theoretical acoustics do not mention it. On the other hand physicists have encountered these radiated waves since it seems that the first interdigital transducer was introduced to transmit bulk waves at a chosen angle from the surface [2].

In fact these waves are not a specific attribute of spatially periodic sources since they exist also at a point load. It is of major importance to read again the famous LAMB's paper (1903) [3]. He found "that the surface disturbance produced by a single impulse of short direction is to be analysed roughly into two parts, which we may distinguish as the **minor tremor** and the **main shock** respectively ; the minor tremor sets in at any place after an interval equal to the time which a wave of longitudinal displacement would take to traverse the distance from the source. Except for certain marked features at the inception, and again at an epoch corresponding to that of direct arrival of transverse waves, it may be described as consisting of a long undulation leading up to the main shock, and dying out gradually after this has passed ... The main shock, on the other hand, is propagated unattenuated at the velocity of free Rayleigh waves".

Then the Rayleigh wave cannot exist alone. It is not a wave in the "usual sense" since the spectrum of acoustic waves emanating from an arbitrary transducer at the surface of an half plane contains other "tremors" than that corresponding to the classical Rayleigh wave ; this "tremor" contains the so-called surface skimming bulk waves (SSBWs) or shallow bulk acoustic waves (SBAWs) and have quasi longitudinal or quasi transversal displacements as plane bulk waves in an unbounded medium.

The "usual" sense for the wave propagation was well defined at the last century by the work coming from nonlinear dynamics of compressible fluids [4,5]. The wave is defined as the disturbance propagating itself into a state constant in time ; the wave as defined above, implies the existence of a wavefront which is the boundary between the disturbed state and the undisturbed state and through which a kinematic discontinuity appears. It is possible with this very elegant approach to solve exactly various nonlinear wave propagation problems associated to arbitrary motions of the source, since the source behaviour is transmitted to the discontinuity which propagates [6]. The Rayleigh wave does not satisfy this usual sense for a wave since the constant state is reached only at an infinite distance from the source, due to the simultaneous propagation of the body waves.

Lamb's approach leads to the solution of linear surface wave propagation problems in the case of a time harmonic lineload. By applying Fourier superposition Lamb presented also explicit expression for the field variables at the surface of the half space for a line load of arbitrary time dependence. This approach was extended by CAGNIARD and DE HOOP by the use of Laplace transforms [7] and by AVRAMESCO [8] for the case of a supersonic excitation.

2. Excitation by an Interdigital Transducer

Two main ways of applying Lamb's theory appears in the modern work about wave propagation in piezoelectric crystals. In the first one an effective permittivity is defined from the relation between the Fourier transform along the propagation direction of charge density and electrical potential at the surface [9]. But this method requires a time consuming numerical computation to find the charge distribution which gives a voltage equal to the known applied voltage on the electrodes ; it is also difficult to distinguish one wave from the others by numerical evaluation of the inverse Fourier transform. In the second approach, the electrostatic potential inside the medium is computed [10] and the mechanical displacement obtained from a forced problem. As satisfactory results are obtained only in the case of low piezoelectric crystals [11,12], the forcing action of the charge density on the electrodes may be used in the case of crystals with high piezoelectric coupling [13,14].

We consider here the excitation by a very wide interdigital transducer deposited on a crystal with low piezoelectric coupling, for example a quartz crystal (Fig. 1).

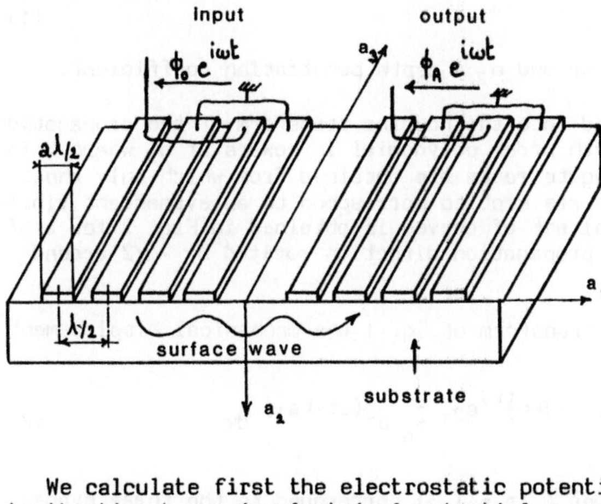

Fig. 1 - Schematic of the SAW delay line

We calculate first the electrostatic potential created inside the crystal by the time harmonic electrical potential ϕ_0 applied to the electrodes [10]. Then the mechanical displacement of the medium comes from a forced problem [11,12]. By using the exponential Fourier transform of all the fields along the propagation direction a_1, the partial derivative system including propagation equations and boundary conditions is transformed into a differential system with the single space coordinate a_2. The Fourier transform $\hat{u}_j$ of the complex mechanical displacement is taken on the form

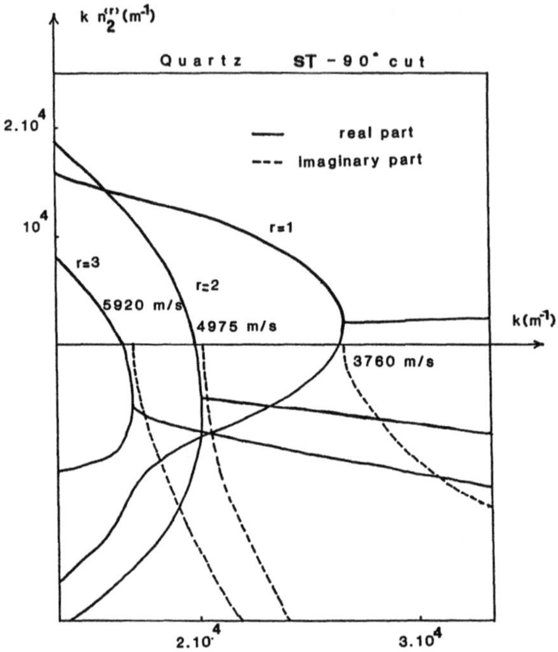

Fig. 2 - Penetration depth coefficients of the three surface modes (r = 1,3) as a function of wave number

$$\hat{u}_j = a_j\, e^{-jkn_2 a_2} \tag{1}$$

where $j^2 = -1$, k is a wavenumber and n_2 a depth penetration coefficient.

The solution (1) is replaced into the Fourier transform of the propagation equations which leads to a sixth order polynomial in powers of n_2 where k is a parameter. Six complex conjugate roots are obtained from which only those with a negative imaginary part are kept to correspond to an evanescent vibration along the depth. A typical set of curves is obtained in Fig. 2 for a ST cut quartz substrate with the propagation direction rotated by $-\pi/2$ around the Oa_2 axis.

Taking the inverse Fourier transform of Eq. 1 the mechanical displacement has the form [12]

$$u_j(a_1,a_2) = \int_{+\infty}^{-\infty} \left(\sum_{r=1}^{3} \frac{B_r}{\Delta}\, a_j^{(r)}\, e^{-jkn_2^{(r)}a_2} \right) \hat{\phi}_o\, e^{j(\omega t - ka_1)}\, dk \tag{2}$$

In eq. 2, the coefficients $n_2^{(r)}$ (r = 1,3) correspond to the three evanescent surface modes ; Δ is the determinant of the boundary condition ; B_r, Δ and $a_j^{(r)}$ are functions of $n_2^{(r)}$ and k. The symbol $\hat{\phi}_o$ denotes the Fourier transform of the electrical potential on the surface.

The integral (2) lies along the real axis in the complex k plane and is calculated from Cauchy's theorem by choosing the integration constant of Fig. 3.

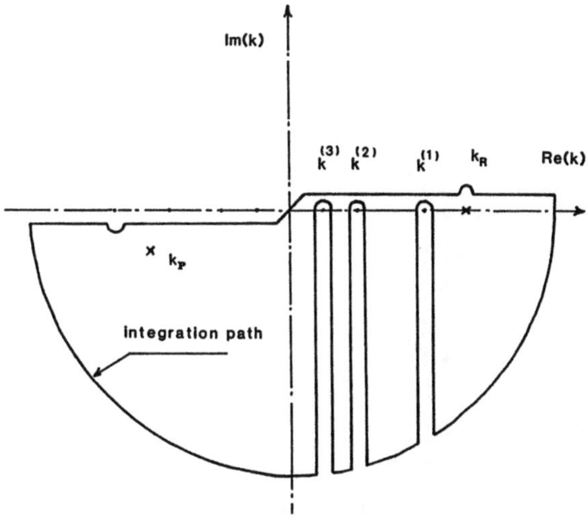

Fig. 3 - Integration path in the complex plane of wave numbers to calculate the mechanical displacement

The pole $k = k_R$ where $\Delta(k) = 0$ is included in the contour and corresponds to the propagation of a Rayleigh wave in Oa_1 direction. At values of the wavenumber k where the imaginary part of the depth penetration coefficient n_2 vanishes, the sixth order polynomial contains a square root and is two valued in the complex plane. Generally three such branch points occur inside the contour and have been excluded by the three cuts shown on Fig. 3. They correspond to three "branch waves" [15] or SSBW's.

The contribution of the pole is obtained by using the residue theorem which yields a closed form for the amplitude of the wave. Radiation admittance and susceptance of the transducer may be calculated. Also shown on Fig. 2 is a new contingent of axis pole $k = k_p$ which corresponds to a pseudo Rayleigh wave. For this wave one of the three partial surface modes has a small growth into the substrate [16].

The contribution due to the branch points is obtained by performing the integration along the paths around the cuts. There are steepest descent paths since the integrand decays most rapidly as one moves away from the branch point along this path. Then one need only consider the contribution near the branch point to obtain the far field representation. The evaluation of the integral uses Whittaker functions [13] and it is shown that the SSBW's amplitude decays as a power law of the form $R^{-1/2}$ gradually increasing to $R^{-3/2}$ at very great distances R from the source [14].

For the case of the quartz ST-90 ° cut of Fig. 2 an efficient SSBW excitation was found [17] at the velocity 4971 m/sec which is 1 ‰ closed to our calculated value. It can be shown that the amplitude of the SSBW is a growing function of the radius of curvature at the branch point of Fig. 2. It is probably why other SSBW modes were not seen by the authors. Nevertheless all SSBW's can be observed in any crystal configuration. An example is seen on Fig. 4 for the case of an ST-cut quartz crystal resonator. We calculate the three SSBWs at the frequencies 115 MHz, 179 MHz and 202 MHz. The first one is not observed since it is masked by the secondary lobes of the Rayleigh wave peak ; the other ones are clearly visible on the spectrum.

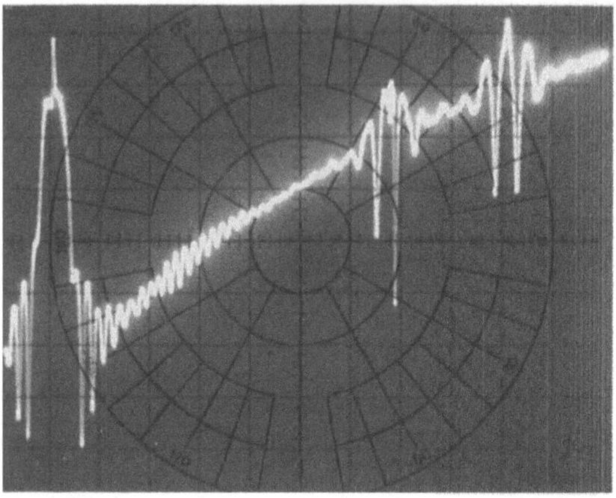

Fig. 4 - Transmission coefficient of a ST-cut quartz resonator (λ = 28.6 μm)

It is worthwhile to mention a simultaneous observation of the character of Rayleigh, pseudo Rayleigh and SSBW's on a 41 ° Y cut lithium niobate [18].

3. Detection

The detection problem is solved by a number of authors by using a theorem of reciprocity [11] but this approach fails for example when the two transducers at input and output of a surface acoustic wave delay line have a different number of fingers. This is due to the fact that, on the input transducer the two acoustic ports are symmetrical, this is not the case on the output transducer. At the detection, signal coming from the first acoustic port is extracted at the electronic port, while mechanical energy reaching out to the second acoustic port is generally absorbed to avoid parasitic reflections on the edge of the substrate. The detection problem was solved by COQUIN [19] for a low piezoelectric medium ; in this approach mechanical fields incident under the comb constitute a source for the electrical potential computed by means of conformal transformation. This approach together with the excitation approach already presented in the previous paragraph was used by the author to calculate dispersion parameters of a Rayleigh delay line on an arbitrary anisotropic substrate [12] ; but it could be extended also to the SSBW delay line. Otherwise, for the case of finite length transducers on highly piezoelectric media a general numerical solution is necessary.

4. Reflexion

A suitable method to reflect a Rayleigh wave is to introduce a network of small discontinuities phase matched with the incident wave. This is known since the first Rayleigh wave resonator was introduced by ASH [20]. In fact the reflector array reflect also the SSBW,s (see Fig. 4). Reflection by shallow grooves or metallic strips was considered by a considerable number of authors. The theoretical problem can be solved in an elegant way by a similar approach to that of the excitation. The groove or strip is considered as a small discontinuity which acts as a surface excitation source for the scattering of the incident wave. Fourier transform techniques and integration in the complex plane leads to a reflexion coefficient at a single discontinuity [21,22]. Then a transmission line technique is necessary to model a resonant cavity with a high reflexion coefficient inside which we use the value of the reflexion coefficient per discontinuity previously calculated [23].

5. Conclusion

Devices using Rayleigh waves have been used as delay lines, filters, resonators, oscillators, sensors and signal processing elements. This is widely due to the growth of planar techniques which reduce cost and size of these systems. The same reason can explain the development of SSBW systems which offers also new advantages as higher frequency for the same transducer and greater insensitivity to external fields. Another promising device [24] uses a corrugated surface to realise the guidance of a shear SSBW, which is temperature compensated by the dispersion effect due to the grating. The reciprocal conversion of Rayleigh, bulk and branch waves at periodic corrugations is a dynamical field of applied physics [25]. As an extension, the fundamental and technical study of surface elastic waves in layered and heterogeneous media will certainly produce new improvements in domains of acoustoelectronics and nondestructive testing.

6. References

1 Lord Rayleigh : "On waves propagated along the plane surface of an elastic solid", Proc. London Math. Soc., 17, 4 (1885).

2 M. Lewis : "Surface Skimming bulk Waves, SSBW", IEEE 1977 Ultrasonics Symp. Proc., p. 744.

3 H. Lamb : "On the Propagation of Tremors over the Surface of an Elastic Solid", Philosophical transactions of the Royal Society, London A203, 1 (1904).

4 J. Hadamard : "Leçons sur la propagation des ondes", Hermann, Paris (1903).

5 R. Courant and K.O. Friedrichs : "Supersonic flow and Shock Waves", Interscience Publishers, Inc. New York (1948).

6 A. Jeffrey, T. Taniuti : "Nonlinear wave Propagation with Applications to Physics and Magnetohydrogynamics", Academic Press, New York (1964).

7 J.D. Achenbach : "Wave Propagation in Elastic Solids", North Holland, Amsterdam (1973).

8 A. Avramesco : "Dynamique des structures élastiques stratifiées à rigidité décroissante avec la profondeur", Thèse, Paris, 1969.

9 R.F. Milson, N.H.C. Reilly and M. Redwood : "Analysis of Generation and Detection of Surface and bulk Acoustic Waves by Interdigital Transducers", IEEE Trans. Sonics Ultrasonics, SU-24, 3 (1977).

10 H. Engan : "Excitation of Elastic Surface Waves by Spatial Harmonics of Interdigital Transducers", IEEE Trans. Electron. Devices, ED-16, 1014 (1969).

11 S.G. Joshi, R.M. White : "Excitation and Detection of Surface Elastic Waves in Piezoelectric Crystals", J. Acoust. Soc. Am., 46, 17 (1969).

12 M. Planat, P. Schiavone : "Efficiency of Surface Interdigital Transducers on Anisotropic Media", J. Appl. Phys., 57, 49 (1984).

13 K. Yashiro and N. Goto : "Analysis of Generation of Acoustic Waves on the Surface of a semi-infinite Piezoelectric Solid", IEEE Trans. Sonics and Ultrasonics, SU-25, 146 (1978).

14 A. Jhunjhunwala, J.F. Vetelino : "Spectrum of Acoustic Waves emanating from an IDT on a piezoelectric half space", 1979 IEEE Ultrasonics Symp. Proc., p. 945.

15 G.F.D. Duff : "Communications on pure and Applied Mathematics", vol. XVII 189 (1964), On Wave Fronts and Bounding Waves.

16 T.C. Lim, G.W. Farnell : "Character of Pseudo Surface Waves on Anisotropic Crystals", J. Acoust. Soc. Am., 45, 845 (1969).

17 K.H. Yen, K.L. Wang, R.S. Kagiwada : "Efficient bulk wave Excitation on ST Quartz", Electronics Letters, 13, 37 (1977).

18 P.D. Bloch, H.G. Doe, E.G.S. Paige, M. Yamaguchi : "Observations on Surface Skimming bulk Waves and other Waves Launched from an IDT on Lithium Niobate", 1981 IEEE Ultrasonics Symp. Proc., p. 268.

19 G.A. Coquin, H.F. Tiersten : "Analysis of the Excitation and Detection of Piezoelectric Surface Waves in Quartz by means of Surface Electrodes", J. Acoust. Soc. Am., 41, 921 (1967).

20 E.A. Ash : "Surface wave Grating Reflectors and Resonators", IEEE 1975 Symp. on Microwave Theory and Techniques, Newport Beach, CA, p. 11.

21 H.S. Tuan, J.P. Parekh : "Theory for SAW Grooved Reflector Arrays", IEEE Trans. Sonics and Ultrasonics, SU-24, 6 (1977).

22 P. Schiavone, M. Planat : "Reflection of Surface Acoustic Waves by Grooves and Strips on Anisotropic Media", Proc. of Ultrason. Int. Conf., London (1985).

23 J.S. Schoenwald, W.R. Schreve, R.C. Rosenfeld : "Surface Acoustic wave Resonator Development", Proc. 29th Ann. Freq. Cont. Symp., 150 (1975).

24 B.A. Auld, D.F. Thomson : "Temperature Compensation of Surface Transverse Waves for Stable Oscillator Applications using Shallow Gratings", IEEE 1884 Ultrason. Symp. Proc., p. 213.

25 A.D. Lapin : "Reciprocal Conversion of Surface and bulk Acoustic Waves at Periodic Corrugations and Inhomogeities of the Boundary of a Solid" (review), Sov. Phys. Acoust., 29, 123 (1983).

Rayleigh Waves and Reflective Periodic Structures

E.G.S. Paige

Department of Engineering Science, University of Oxford, Parks Road, Oxford, OX1 3PJ, U.K.

1. Introduction

While the interdigital transducer is the most important periodic structure for Rayleigh waves, it labours under limitations. To perform its function it must consist of continuous metal strips connected together in a certain configuration. Periodic reflective array structures suffer no such limitations. A rich variety of structures can be conceived and many have been investigated. Examples of periodic arrays are shown in Fig. 1 which may consist of dielectric or metal films or of grooves.

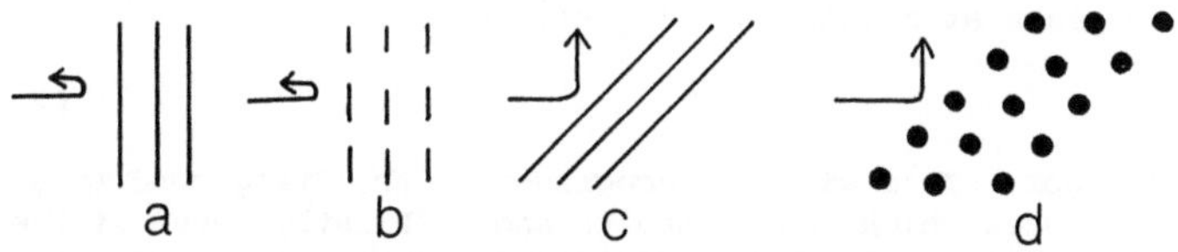

Figure 1. Examples of strip and dot periodic arrays

It is convenient to subdivide the individual reflectors or "elements" into dots or strips. The dot is any surface structure with dimensions of at most a few acoustic wavelengths in any direction. The strip has one dimension many wavelengths long. Strip reflecting arrays in the form of parallel grooves ion beam milled into the surface were first introduced into surface wave devices by WILLIAMSON et al [1]. Rows of dots were invented by SOLIE [2] as a simple means of varying the reflectivity of a row through the density of dots.

Wave propagation in periodic structures is a fascinating subject. The topic makes its appearance in many branches of physics and technology. It is made particularly attractive for Rayleigh waves because of the readiness, using lithography, with which an immense variety of arrays can be created and the accessibility of the reflected and diffracted wave for observation. However, it is their importance within devices which has been the main source of motivation behind the detailed studies of recent years.

To gain a feel for why they are important, consider the situation shown in Fig. 2. An impulse is launched from T_1 to be reflected by the periodic array through an angle χ and received by T_2. Neglecting diffraction, as a first approximation it is evident we may write the impulse response as

$$r_A(t) = \sum_{n=1}^{N} \delta(t-t_D-nT)\otimes r_R(t) \qquad (1)$$

where $r_R(t)$ is the impulse response of a reflecting row (or line), $T = 2p\sin(\chi/2)/v$, v is the surface wave velocity, p is the period of the

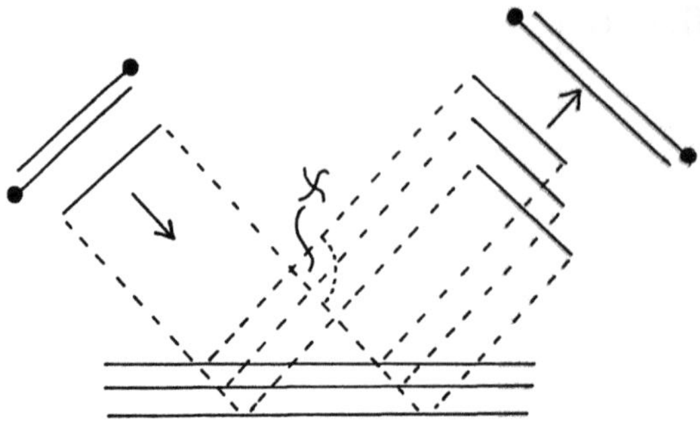

Figure 2.

Impulse response of a periodic array; T_1 and T_2 are transducers

grating, N is the number of reflecting rows, t_D is a time delay and $\delta(t)$ is the Dirac delta function. Fourier transforming using the convolution theorem gives

$$|r_A(k)| = N|r_R(k)|\frac{\sin[Nkp\sin(\chi/2)]}{N\sin[kp\sin(\chi/2)]} \qquad (2)$$

where $k = \omega/v$ has replaced ω as the variable and k is the wave vector $(2\pi/\lambda)$. [Note that here and elsewhere the Fourier transform is denoted simply by the change of variable i.e. the Fourier transform of $h(t)$ is $h(\omega)$]. Equation (2) peaks at a value of $N|r_R(k)|$ whenever

$$kp\sin(\chi/2) = m\pi \qquad (3)$$

where m is an integer. This will be recognised as the Bragg condition. Here are the key features which these results show. Firstly, even if the reflection of an individual row is weak, and this is usually the case, at a given frequency it can be made large by employing many reflecting rows. Secondly, because of the N dependence of the argument of the sinusoid in the numerator, strong frequency selectivity can be achieved. Thirdly, returning to the impulse response, we note great flexibility in the design of the response is possible through varying p between rows and the reflectivity of rows. This implies flexibility in designing frequency response.

The significance of these key features is shown by considering the frequency response $h(\omega)$ of the two port, array device. It can be written

$$h(\omega) = h_1(\omega)h_2(\omega)r_A(\omega) \qquad (4)$$

where h_1 and h_2 are the responses of the transducers. Because the array can be designed to have high frequency selectivity without major loss penalties and, moreover, have its frequency response tailored, then h_1 and h_2 can be broad-band responses. Thus the dual role of performing transduction and determining frequency response is lifted from the transducer. Consequently, the design of the transducer can be simplified and optimised for spurious free generation and detection of waves.

The array, then, can dictate the frequency response in the reflective array device. Apart from the consequences for transducer design, there are other advantages as follows. The array is defect tolerant because breaks in and connection between elements are of only minor local significance. Minimum feature sizes may be larger - thus the width of the strip in the 90° reflection may be $\sqrt{2}$ wider than the $\lambda/4$ of the transducer (see Fig. 1(c)). Spurious signals due to electric circuit generation are absent in the array. In 90° reflection arrays effects of coupling to other modes is reduced because they peak at a different angle. The device length in the direction of Rayleigh wave launch is reduced, typically by a factor of two, by folding the path. In the case of the resonator the array acts as a distributed feed back mirror and the path length is in effect folded thousands of times.

Examples of devices incorporating periodic or near-periodic arrays are shown in Fig. 3. The resonator [3] employs many elements to give near unity reflectivity and to determine the band within which cavity modes are selected. The dual track in-line reflection of Fig. 3(b) has been used as a Reflecting Array Compressor (RAC), the periodicity of the array being slowly varied causing different frequency components to follow different path lengths [4]. In the stop-band filter, reflection from the array produces a narrow stop-band within the broad pass-band of the transducer. In the filter bank, separate arrays of elements of differing periods, frequency select particular pass-bands [5]. Figure 3(e) illustrates a RAC with two 90° reflecting arrays [1] and Fig. 3(f) a band pass filter again using 90° reflecting arrays [6]. Other uses of reflecting arrays include code generators and matched filters and in unidirectional transducers such as the SPUDT [7] and related structures.

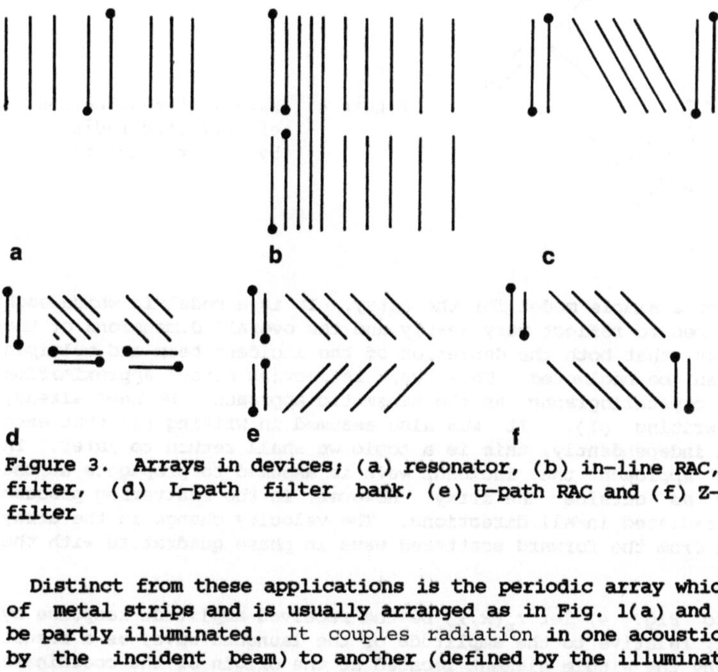

Figure 3. Arrays in devices; (a) resonator, (b) in-line RAC, (c) stop band filter, (d) L-path filter bank, (e) U-path RAC and (f) Z-path band pass filter

Distinct from these applications is the periodic array which must consist of metal strips and is usually arranged as in Fig. 1(a) and is designed to be partly illuminated. It couples radiation in one acoustic path (defined by the incident beam) to another (defined by the illuminated part of the array). In normal use this multistrip coupler [8] couples forward propagating waves; the stop-band when the array is strongly reflecting is avoided. The coupler (MSC) has found numerous uses, including suppression of spurious signal due to bulk waves, compressing the width of acoustic beams, multiplexing acoustic signals and making broad-band unidirectional transducers and mirror structures [9]. While the MSC and components derived from it are of some importance, the emphasis of this paper will be on reflective periodic arrays.

The paper is divided into six sections. In the next a model we shall adopt for the array is introduced. Section 3 considers the behaviour of different types of elements. Sections 4 and 5 are devoted to periodic and near-periodic arrays. Concluding comments are presented in Section 6.

2. Modelling the Array

What properties of the array are of special interest? We shall assume the substrate is piezoelectrically active and waves are launched and detected by interdigital transducers. Figure 4 shows a general arrangement and the coordinate system we shall use, with surface normal to the z-direction. The typical, large aperture transducer is in effect a wave vector selector, picking out that component of the reflected wave with $\underline{k}$ normal to the fingers of the transducer. With reference to Fig. 4, we wish to know the amplitude of the signal which excites the transducer and in some applications - resonators and unidirectional transducers - its phase. Thus we require the amplitude reflectivity of the array $r_A(k,\chi)$. Also required is the velocity in the array so elements can be placed in their correct position. We require, then, a model which leads to adequate predictions of r_A and the velocity change in the array.

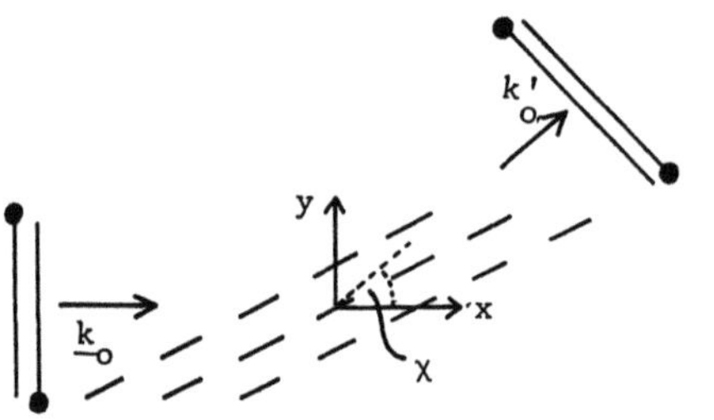

Figure 4. Wave vector selection of reflected radiation by the transducers

We shall adopt a simple model for the array. It is a model in which each element is assumed to reflect very weakly and the overall dimensions of the array are such that both the depletion of the incident beam and multiple reflections can be neglected. This "weak reflecting array" approximation is known in crystallography as the kinematic approach. We have already used it in writing (1). It was also assumed in writing (1) that each element acted independently; this is a topic we shall return to later. In the kinematic approach the incident wave is assumed to propagate at the same velocity as outside the array. However, in the scattering process waves can be radiated in all directions. The velocity change in the array is determined from the forward scattered wave in phase quadrature with the incident wave.

Referring to Fig. 4, let $r_E(k,\chi)$ be the received amplitude response by the transducer, relative to the amplitude of the launched wave, as a result of scattering by the single element located at the origin of the coordinate system. $r_E(k,\chi)$ may be regarded as that component of the spectrum of plane waves generated by the scatterer which travels normal to the output transducer and is normallized by the input amplitude. Identifying the wave vector of the reflected component by a prime we have as the total reflectivity of an array

$$r_A(k,\chi) = r_E(k,\chi)\exp[-j(kx_o+k'r_o)]\Sigma_m\exp[-j(kx_m-k'x_m\cos\chi-k'y_m\sin\chi)]. \quad (5)$$

The summation is over all elements, the mth of which is centred on x_m, y_m. The reflectivity of the array peaks when

$$k'[x_m \cos\chi + y_m \sin\chi] - kx_m = \pi n(m) \tag{6}$$

where n is an integer the value of which varies with subsets of the values of m. A little investigation shows that if all elements satisfy this equation then they lie on parallel lines and (6) becomes again a statement of the Bragg condition among the reflecting elements. The equation for the parallel lines gives their inclination to the incident wave including anisotropy.

Equation (5) serves to show the important factoring of the reflectivity into an element factor and an array factor which stems from the weak reflecting array approximation. We adopt this approximation for the present, returning in Section 4 to a discussion of more accurate models. It is a valuable starting point for Rayleigh waves because usually individual scatterers are very weak scatterers and many arrays used in practice conform to the weak reflecting conditions.

3. Individual Elements in the Array

The most important elements used in periodic reflecting arrays are the groove, the dielectric overlay and the metal overlay. The groove and overlay are shown schematically in cross-section in Figs. 5(a) and (b).

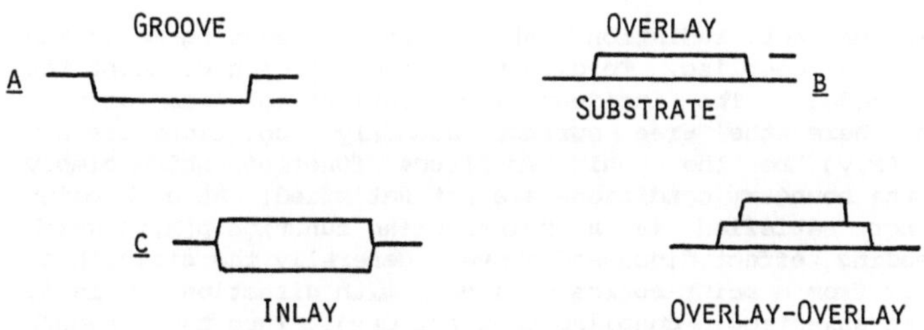

Figure 5. Types of Rayleigh wave reflectors

Combined structures which have received some attention are also shown in Figs. 5(c) and (d). The interaction of these elements with the surface wave have been subdivided into topographic, mass loading and electrical loading [10]. The topographic effect is associated with a local departure of the surface from planar. In the case of the groove, it is the only source of scattering. Mass loading covers effects due to the change in themechanical boundary conditions. Dielectric overlays combine both topographic and mass loading scattering mechanisms. Electrical loading covers effects due to the change in the electrical boundary conditions produced by conducting overlays. Of course, the latter is only relevant if the substrate is piezoelectric. In general the conducting overlay combines all three scattering mechanisms. As an illustration of the differences between the behaviour of these mechanisms, Table 1 lists their behaviour as a function of overlay thickness or step height h. Here we shall take the opportunity of giving a unified approach to modelling of the general behaviour of each. But first, some general remarks.

Scattering results from the incident wave failing to satisfy the local boundary condition. However, because field variables must be continuous

Table 1. Thickness (depth) dependence of reflected amplitude and velocity change

Scattering Mechanism	Amplitude	Velocity Change
Topographic	h	h^2
Mass	h	h
Electric	Constant	Constant

everywhere, the adjustment is spread. For this reason we shall introduce a "point spreading function". In fact it has already been contained in $r_R(t)$ of (1). Furthermore, the surface wave alone is not able to satisfy the boundary conditions. Other propagating and evanescent modes are generated - bulk, leaky surface waves and electric waves. The propagating modes can lead to a loss of power and spurious signals. However, the evanescent waves, which store energy, have three important effects; (1) they produce a phase shift in reflection, (2), in a related way, they change the velocity and (3) they cause "mutual coupling" between elements. To be clear about mutual coupling, we make a careful distinction between interaction among elements via propagating waves which we shall refer to as multiple scattering (and is outside the model we have adopted) and the interaction via evanescent waves which is what we call "mutual coupling". A good example of the latter is the generation of potential on one element by space charge on another due to the evanescent electric wave when electrical loading occurs.

To simplify, we neglect radiation into other propagating modes and initially neglect mutual coupling. To give a unified approach we adopt the following general model. The incident wave provides the local driving force which acts where the free surface boundary conditions are not satisfied. Let m(x,y) be the unit amplitude function which simply identifies where the boundary conditions are not satisfied. At each point where they are not satisfied is a point spread function p(x,y) which carries the spreading effect discussed above. Generally the strength of the radiated wave from a point source will vary with direction. This is incorporated in an anisotropic coupling constant $C(\chi)$. Thus the strength of the source distribution radiating in direction χ, $s(x,y,\chi)$, is given by

$$s(x,y,\chi) = C(\chi)m(x,y)\exp(-jkx)\otimes p(x,y). \tag{7}$$

We shall assume it is normalised by the amplitude of the incident wave, always taken as propagating in the x-direction.

One-Dimensional Problem. In the case of strips normal to the incident wave the problem can be regarded as one-dimensional in that surface variation can be assumed to be restricted to one dimension. If x_1 is some point on the input side of the element and x_2 on the output then the amplitude reflection, referred to as x_1, is

$$\begin{aligned} r(x_1) &= \int_{-\infty}^{+\infty} dx \exp[-jk_o(x-x_1)]s(x,180) \\ &= \exp(jk_o x_1)s(k_o,180) \end{aligned} \tag{8}$$

Similarly the scattered contribution to the forward wave is

$$t(x_2) = \exp(-jk_o x_2)s(k_o,0). \tag{9}$$

Note that in order to avoid confusion with the Fourier transform of s(x) to any value of k, we have identified the specific value of k of the incident wave which is applicable to be $k_o = \omega_o/v$.

Turning to individual scattering mechanisms, we consider the form of s(x) for the groove. For the groove of width d, $m = -\delta(-d/2) + \delta(+d/2)$ where the + and - indicate the direction of the step. Substituting in (7) and (8) gives

$$r(x_1) = -j2\exp(jk_ox_1)C(180)\sin(k_od)p(k_o) \tag{10}$$

$$t(x_2) = 0. \tag{11}$$

On physical grounds we expect the reflectivity to depend on groove depth h and for this to be scaled by the wave length parameter λ i.e Cαh/λ. This is obviously the first order expansion in h/λ. A second order term may exist $(h/\lambda)^2$ for which the direction of the step is removed with the result

$$r_2(x_1) = 2\exp(jk_ox_1)C_2(180)\cos(k_od)p_2(k_o) \tag{12}$$

$$t_2(x_2) = 2\exp[-jk_o(x_2-x_1)]C_2(0)p_2(-k_o). \tag{13}$$

Since there are no loss mechanisms, t_2 must be in phase quadrature with the incident wave, producing an added phase shift which may be interpreted as an effective change in velocity over the length $x_2 - x_1$ given by

$$\left[\frac{\Delta v}{v}\right]_g \approx \frac{-2c_2(0)p_2(-k_o)/j}{(x_2 - x_1)k_o}. \tag{14}$$

More explicitly, writing $C(180) = bh/\lambda_o$ and $C_2(\chi) = jc_2(\chi)(h/\lambda_o)^2$ gives

$$r_g(k) = j2[-bh/\lambda_o\sin(k_od)p(k_o) + c_2(180)(h/\lambda_o)^2\cos(k_od)p_2(k_o)] \tag{15}$$

$$\left[\frac{\Delta v}{v}\right]_g \approx \frac{-c_2(0)(h/\lambda_o)^2p_2(-k_o)}{(x_2 - x_1)k_o} \tag{16}$$

where the reflection has been referenced to the centre of the groove. Note these results are consistent with Table 1. Of course a theory is required to predict magnitude for a, p, c_2 and p_2 [11,12]. The work of WRIGHT and HAUS [12] is particularly worth noting, predicting the second order effect. The value of $bp(k_o)$ for YZ lithium niobate is 0.33, in almost exact agreement with experiment. For ST quartz theory gives a value of 0.44 compared with experiment of 0.27. The theory shows a variation with groove shape for $c_2(0)p_2(-k_o)$. For YZ niobate values in the range 32±8 are predicted and 42 is observed. For ST quartz theory and experiment lie in the range 25-40.

For the mass loading contribution of a strip, ignoring topographical effects, we have $m(x) = \Pi(x/d)$ where $\Pi(x)$ is the gate function. Substituting in (7) and (8) gives a change in velocity of $-C_m(0)p_m(-k_o)/jk_o$ in a film. But it is well known that to first order the change is proportional to h/λ_o and this has appeal on physical grounds. It follows $jC_m(0)p_m(-k_o) = a_{m+}hk_o^2$. Assuming a similar form for $C_m(180)p(k_o)$ gives

$$r_m(k_o) = -ja_{m-}hk_o\sin(k_od). \tag{17}$$

Note this is identical in form to the first order reflection for a groove. The result agrees with the behaviour listed in Table 1. Perturbation calculation yields values of the constants [13].

Finally we turn to the electrical loading neglecting topographic and mass loading effects. Three points should be noted. Firstly, above a thickness of about 100A, a metal film shorts out surface fields irrespective of film thickness. Then the electrical effect is independent of h for typical metal films. Secondly, effects are dramatically non-local. This is because of the long range of the coulomb force and the boundary conditions imposed on total charge. Thirdly, externally imposed boundary conditions (open or short circuit) have a significant effect.

As in the preceding paper we take oscillating space charge as the source of generation. This leads us to rewrite (7) as

$$s_e(x,y,\chi) = C(\chi)\sigma(x,y,k_o) \tag{18}$$

where σ is the space charge set up in the presence of the incident wave. m(x,y) has disappeared because the charge is obviously confined within the metal element. Proceeding as before, we find in the case of a metal strip a fractional change in velocity across its width of $-C'(0)\sigma(-k_o)/k_o d$, where $C = jC'$. Now it can be shown that in a very long film ($d >> \lambda$) that $\sigma(-k_o) = \epsilon k_o$ when ϵ is a permittivity, the value of which is the mean of that for free space and an appropriate value for the substrate. This gives a value to $C'(0) = [\Delta v/v]_{e,\infty}\,\epsilon^{-1}$. The subscript ∞ indicates the effectively infinite value of d. Because the charge will act to launch waves with equal strength in opposing directions then C(0) = C(180) hence

$$r_e(k_o) = j\left[\frac{\Delta v}{v}\right]_{e,\infty}\frac{\sigma(k_o)}{\epsilon} \tag{19}$$

$$\left[\frac{\Delta v}{v}\right]_{e,d} = -\left[\frac{\Delta v}{v}\right]_{e,\infty}\frac{\sigma(-k_o)}{\epsilon k_o d}. \tag{20}$$

In the weak coupling case (with which we are dealing), a major simplification arises because σ may be determined on purely electrostatic grounds. The space charge is that which in the presence of the wave renders the surface of the metal an equipotential, thereby fulfilling the electrical boundary condition. But the potential due to a line charge is known to be $-\ln|x|/\epsilon\pi$, thus we have

$$\frac{1}{2}V_m(1 \pm 1) - V_o e^{-jk_o x} = \frac{-\ln|x|}{\epsilon\pi}\otimes\sigma_\pm(x) \tag{21}$$

where the + sign is for open circuit and the - sign is for short circuit (effectively earthed) conditions.

Figure 6 shows theoretical predictions [14] based on the above equation with experimental results for r for a strip in an array of parallel strips reflecting a Z propagating wave through 180° on Y-cut lithium niobate [15]. Because the piezoelectric coupling is strong for YZ niobate and because the aluminium is thin and well matched mass-wise to the substrate, only electrical loading is important. As the strips were within $\lambda_o/2$ of each other electrostatic mutual coupling effects are included. The agreement shown has used $[\Delta v/v]_{e,\infty}$ found experimentally. This is fortuitous agreement because a more detailed assessment of piezoelectric coupling shows this only to be correct when no other mode than the Rayleigh wave is coupled. It leads to results a factor of 2 in error for YX-lithium niobate when a leaky surface wave is also strongly coupled [16].

The one-dimensional problem has been frequently discussed in terms of a lumped element transmission line equivalent circuit consisting of

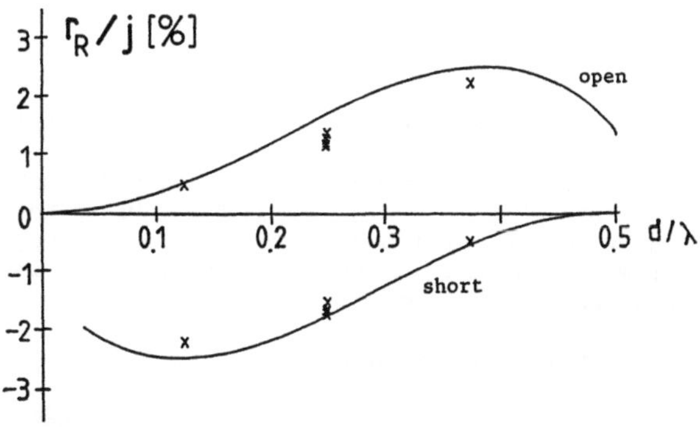

Figure 6. Theory (—) and experiment (xx) for 180° reflectivity per metal strip on YZ lithium niobate as a function of strip width d

alternating sections of cable of impedance Z and Z + ΔZ, shorted by an admittance jB at each junction. If the parameters are regarded as fixed it has limited success. Recently, by replacing B by Bsin(Mπ) and ΔZ by $\Delta Z_o + \Delta Z_1 M$ where M = d/p, WRIGHT [17] has achieved remarkable agreement between the model and experiment for a wide range of overlays, film thicknesses and several substrate materials. Note that for topographic scattering, M at 0 and 1 represents steps which are very close. The spreading functions involving evanescent fields will then overlap and we may expect a simple model to break down. This gives a physical interpretation to the sin(Mπ) factor.

<u>The Two-Dimensional Problem</u>. For strips not normal to the incident wave and for dots, a two-dimensional formulation of the scattering problem is required. We can apply (7) directly to the problem. Making due allowance for the phase shift from sources to the output transducer we have

$$r(\chi) = \int_{-\infty}^{+\infty}\int_{-\infty}^{+\infty} dxdy \exp[-j(k_o' x\cos\chi + k_o' y\sin\chi)]s(x,y,\chi)$$

$$= s(-k'_{ox}, -k'_{oy}) \qquad (22)$$

where $k'_{ox} = k'_o\cos\chi$ and $k'_{oy} = k'_o\sin\chi$, it being understood that $s(x,y,\chi)$ is generated by the incident wave propagating in the x direction.

In the case of topographic effects, the above relation converts into line integrals round the edge of the element. Consideration of hk dependence follows as in the one-dimensional case. Theoretical [18] and experimental results are available for the reflectivity of a groove through 90°. Good agreement is found. For Y-cut lithium niobate the reflectivity from the Z to X-direction is up by 50% of the normal incidence value.

Relaxing our specific concern with a single element, but considering reflection from a single row of elements (elements in a line which reflect through χ in phase) separated from the next rows by p_x (separation between rows measured along x direction), then the reflectivity per row for mass loading may be obtained using (7) in (22) and similar arguments to the one-dimensional case

$$r_{mR}(\chi) = ja(\chi)hk_o^2 m(-k'_{ox} + k_o, -k'_{oy})F \qquad (23)$$

where F is the fraction of the length of row occupied by elements, and

$$\left[\frac{\Delta v}{v}\right]_m \approx \frac{-a(0)hk_o m(0,0).F}{p_x}$$

$$\approx -a(0)hk_o.\ (\text{fraction of surface overlaid}) \tag{24}$$

A theoretical treatment of the problem has been given for a circular dot assuming rigid boundary conditions [19]. SOLIE et al [20], using the functional form for $\Delta v/v$ as given above, find they can successfully apply it to the design of arrays. However, at the present time the constants in the problem have to be determined experimentally.

Electrical loading gives a reflectivity per element of

$$r_e(\chi) = C(\chi)\sigma_E(-k'_{ox}, -k'_{oy}) \tag{25}$$

from (18) and (22). Neglecting anisotropy of permittivity, $\sigma(x,y)$ may be determined from

$$\frac{1}{2}V_m(1\pm1) - V_o e^{-jk_o x} = \frac{1}{2\pi\epsilon(x^2+y^2)^{1/2}} \otimes \sigma_\pm(x,y) \tag{26}$$

where, as before, + and - refer to open and closed circuits respectively. From the work of MILSOM et al [21] the coupling constant may be written as $-j[sd\epsilon_s(\chi)/ds]^{-1}$ evaluated at $s = s_o(\chi) = k_o(\chi)/\omega_o$. Here ϵ_s is the Rayleigh wave component of an effective permittivity. The reflectivity given by (25) is a ratio of potentials but $r.r^*$ is not the power ratio when $\chi \neq 0, 180^o$ because of the difference in electromechanical coupling constant. When this is introduced the reflectivity of a row of elements arranged to reflect through 90°, for example, is [14]

$$r_{eR}(90) = \frac{[G(0)G(90)k_o(0)/k_o(90)]^{1/2}\sigma_E(0,-k_o(90))M}{W} \tag{27}$$

where M is the number of elements in the row illuminated by an input beam of width W. Similarly, the velocity change per row is found to be

$$\left[\frac{\Delta v}{v}\right]_e = \frac{-G(0)\sigma(-k_o(0),0)M}{k_o(0)Wp_x}. \tag{28}$$

Figure 7 shows a comparison of this theory and experiment [22] for aluminium strips on Y-cut lithium niobate arranged to reflect a wave from the Z to X-direction. The period in the Z-direction was λ_o. Theoretical

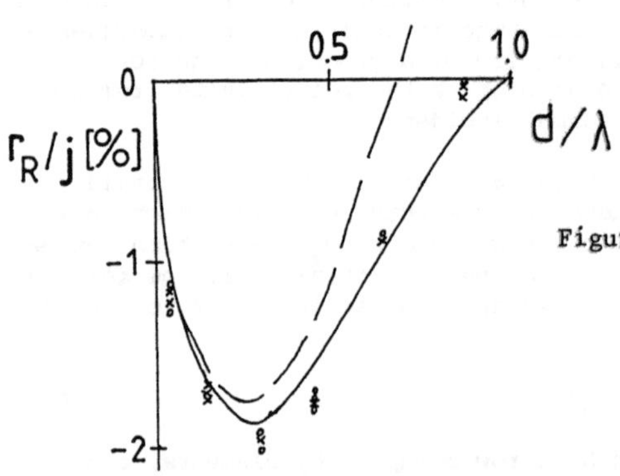

Figure 7. Reflection from Z to X-direction per metal strip on Y-cut lithium niobate; experiment (xx), theory, no mutual coupling (---), theory with mutual coupling (——)

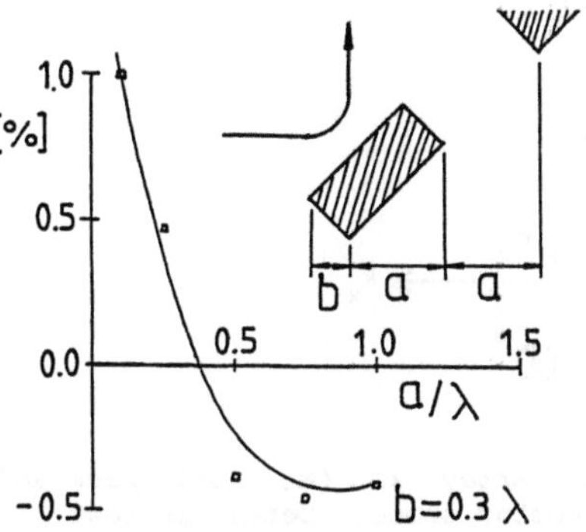

Figure 8. Reflectivity from Z to X-direction per row of metal dots on Y-cut lithium niobate; (__) theory, (oo) experiment

curves are given with and without mutual coupling showing the need to include the coupling in this case for good agreement. Open and short circuit results are the same because the strip inclined to the incident wave front is self-shorting. Experimental results and theory are compared in Fig. 8 when open circuit rectangular dots replace the strip. Parameters are defined on the figure. Note that when the dot is approximately square the 90° reflectivity goes to zero. This is a feature of symmetrical reflectors which can be readily explained by the symmetrical disposition of positive and negative space charge [14]. Good agreement is also obtained between theory and experiment for the change in velocity [14].

It is of value to know the strength of scattering in all directions from an element in an array. Figure 9 shows the predicted scattered potential due to electrical loading by a rectangular metal dot. The very strong angular dependence arises from the asymmetry of the scatterer and the anisotropy of $C(\chi)$.

4. Periodic Arrays

Scattering in periodic structures can be interpreted by transforming the array into lattice vectors, K, of value $2\pi/p$, in wave vector space. The lattice vector can be found by taking the two-dimensional Fourier transform of the periodic array factor (points in real space) assuming the array to extend to infinity. Figure 10 shows lattice vectors for a strip and dot array, the strip array is distinguished by circles around dots. Since this is a construction in wave vector space the incident vector can be included

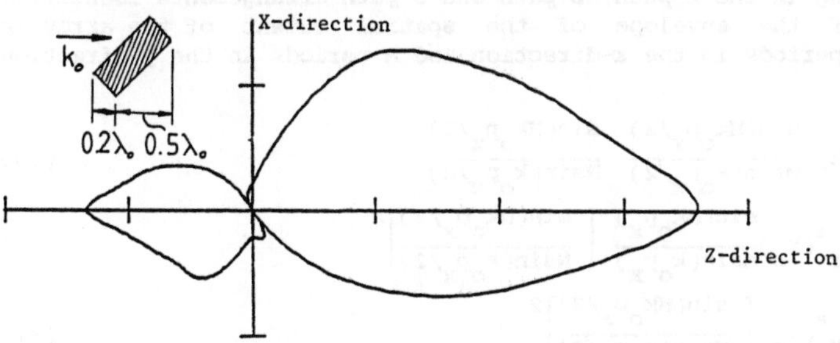

Figure 9. Polar diagram of magnitude of amplitude scattered by a metal dot on Y-cut lithium niobate

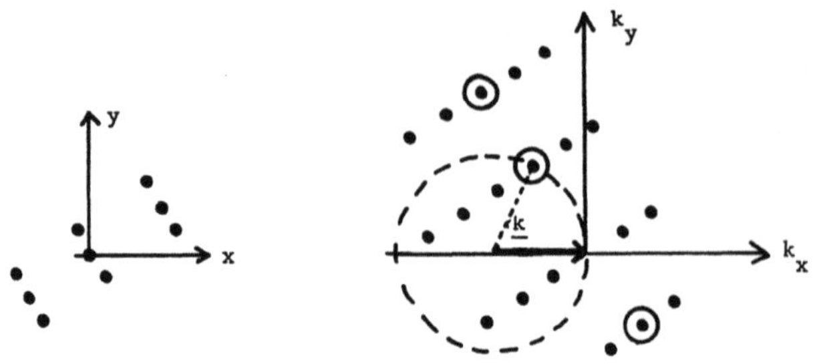

Figure 10. Two-dimensional periodic array in (a) real space and (b) corresponding lattice vectors in wave vector space. Detail in text

together with a construction of the direction of all possible scattered Rayleigh wave vectors under the assumptions that the process is linear and multiple scattering is negligible. If the material is isotropic the locus of scattered vectors lies on a circle (the equivalent of the Ewald sphere in three dimensions), more generally it lies on the slowness surface for Rayleigh waves appropriate to the particular crystal cut and wave frequency. The isotropic case is illustrated in Fig. 10. Where the slowness surface and a lattice vector coincide a Bragg condition for peak reflectivity is satisfied. This will be seen to correspond to "k vector closure" in which the incident and reflected waves combine with lattice vector to form a complete triangle, i.e. $\underline{k} + \underline{K} = \underline{k}'$. It is immediately obvious from Fig. 10 that the dot array requires care in design to avoid unwanted reflections. Close packing in real space assists. The relative coupling strength for a given type of element is revealed by superimposing plots such as that shown in Fig. 9 or Fig. 10. SOLIE [2] has partially overcome the problem of unwanted reflections by randomising the position of dots along the reflecting row.

Off the Bragg condition, the behaviour is determined by the finite dimensions of the array in the weak scattering limit. They can be thought of as introducing a spread in the magnitude and direction of the K vectors. A direct way of obtaining the frequency response in the vicinity of the Bragg condition is to first determine the impulse response as we have done in the introduction. It can often be arrived at by inspection. For a single array, we would only amend (1) by replacing $r_R(t)$, the row response, by the element response $r_E(t)$ times a factor F which is the fraction of the row occupied by elements. The impulse argument may be extended to finite arrays scattering in the L-path, Z-path and U-path arrangements identified in Fig. 3. If the envelope of the spatial extent of the array is rectangular, N periods in the x-direction and M periods in the y-direction then [23]

$$r_L(k_o) = NFr_E(k_o)\frac{\sin(Mk_op_x/2)}{M\sin(k_op_x/2)}\;\frac{\sin(Nk_op_x/2)}{N\sin(k_op_x/2)} \tag{29}$$

$$r_U(k_o) = NMF^2r_E{}^2(k_o)\frac{\sin(Mk_op_x)}{M\sin(k_op_x)}\left[\frac{\sin(Nk_op_x/2)}{N\sin(k_op_x/2)}\right]^2 \tag{30}$$

$$r_Z(k_o) = NMF^2r_E{}^2(k_o)\left[\frac{\sin(Nk_op_x/2)}{N\sin(k_op_x/2)}\right]^2. \tag{31}$$

Phase factors and anisotropy have been ignored.

These expressions apply only in the weak coupling approximation. Several stages can be traced in which the limitations of weak coupling have been thrown off. Firstly, multiple reflections were included in the one-dimensional case [24]. This leads to a simple expression for peak reflectivity when the Bragg condition is satisfied of $j\tanh(NFr_E/j)$ where r_E is the reflectivity of a strip at the k_o satisfying the Bragg condition. Subsequently, OTTO [23] introduced a computational method giving results for L, U and Z-path arrangements including multiple reflections. All these results confirm that a <u>necessary</u> condition for the weak scattering approximation for the arrays to be valid

$NFr_E \ll 1$ for a single array structure

$NMF^2r_E^2 \ll 1$ for a double array structure.

A more elegant approach to non-weak coupling described is to use a coupled mode method to describe the input wave and the principal reflected wave. In a full reformulation of the problem, multiple scattering, depletion of the incident beam and mutual coupling effects would be included. A closed form equation is available for 180° reflection [25],

$$r_A(180) = [(\gamma/C^x) + j(\delta/C^x)\coth(\gamma Np)]^{-1} \tag{32}$$

where $\gamma = \sqrt{C^2-\delta^2}$ and δ is a measure of departure from the Bragg condition $\delta = (k_o - 0.5K)$. The coupling constant C is a characteristic of the array and is proportional to Fr_E. Figure 11 shows the change from the weak coupling behaviour with its characteristic sinc-function dependance to a limiting spectral width determined by the coupling as NFr_E is increased. A similar treatment for 90° reflection is available [26] which points out the surprisingly simple relation for the variation of incident and reflected wave amplitude throughout the array at the Bragg condition. For example, the reflected beam is given by $r_A(x,y) = \sqrt{y/x}\, J_1[C\sqrt{xy}]$ where J_1 is the first order Bessel function. Here x and y are measured in wavelengths where x is the direction of propagation of the incident wave.

A more radical approach to the behaviour of the array is obtained by assuming the array is infinite in size and obtaining the dispersion relation and characteristic impedance in this, effectively new, medium. The behaviour of the finite array is then found by inbedding finite dimensions of the "new medium" into the material without the array and applying appropriate boundary conditions. This procedure has been successfully applied to the one-dimensional model of normal incidence waves

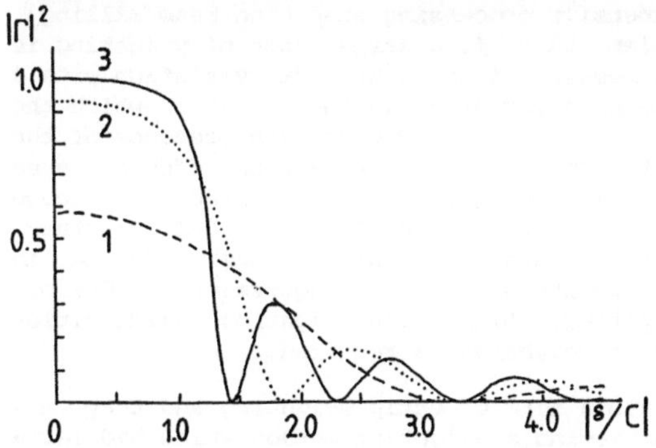

Figure 11. The 180° reflected intensity from a periodic array. Parameter is (CNp)

on metallic strips [27,28]. Mutual coupling is included and interestingly the results are indistinguishable from those of Fig. 6.

Coupling to Other Modes. The Rayleigh wave may excite both propagating and evanescent bulk and leaky surface waves (LSW) in the periodic array as well as electric waves. In satisfying boundary conditions evanescent waves will be excited; their effects have been discussed in Section 3. Coupling to propagating produces energy loss and spurious signals. Fortunately, it is less likely. By their nature, propagating modes are restricted by having to satisfy well-defined phase matching conditions over the surface occupied by the array. Wave vector closure involves the surface wave, the new mode and the lattice vector. Some release from this restriction occurs for bulk waves with $k_z \neq 0$ because the wave vector closure is confined to components parallel to the surface. Because the velocities of other propagating modes are higher than that of the Rayleigh wave, significant energy transfer is only possible when the frequency is above that for Bragg reflection of the Rayleigh wave. Though not a major problem in general, it does influence the design of arrays intended to operate over large fractional bandwidths e.g. RACs. Coupling to LSW has to be considered on those cuts and directions where they propagate and are coupled to, e.g. YX lithium niobate.

5. Near-Periodic Arrays

A near-periodic array is taken to be an array in which the departure from periodic is slow compared to the period of the array or, if more abrupt, the periodic nature of the array is still largely preserved.

Like the interdigital transducer, the full potential for determining frequency response with an array would be severely limited if reflective arrays were constrained to be truly periodic. Referring to (4) we see great flexibility of design is required in the reflective array if the requirements of $h(\omega)$ are to be met with little help from $h_1(\omega)$ and $h_2(\omega)$ because of their supposed broad bandwidth. Lifting the constraint of periodicity enables amplitude and phase encoding to be introduced. We consider these in turn.

Amplitude Weighting. Grooves may be weighted by varying their depth, overlay strips by varying their height, both by varying their length and width. Dots achieve amplitude weighting through density or size. Depth weighting of shallow grooves has been used in the production of dispersive delay lines (RACs). While a good method, it suffers from the necessity of introducing an additional, expensive processing step (ion beam milling). Width weighting has the problem that if a large range of weighting is required, control is difficult because of the sinusoidal variation with d (15 and 17). Length weighting means a variation in the extent to which the aperture of the incident beam is intercepted. Because the presence of the array causes changes in velocity, this leads to refraction effects - even waveguiding. So called "chevron" weighting [30] has been used to overcome this difficulty in 90° reflective arrays. In the "in-line" RAC structure, where two otherwise identical tracks are used, shift of one reflector in one direction and an equal but opposite shift of the equivalent reflector in the other track produces weighting. Like width weighting, difficulties are encountered if a large range of weighting is required.

In effect dots themselves are a form of strip weighting and they were introduced [2] as a means of achieving a weighting method which had large range, occurred uniformly across the aperture of the incident beam and, unlike depth weighting, could be implemented with a single photolithographic and fabrication step. SOLIE and co-workers [2,20] have achieved weighting by varying the number of dots along a reflecting row. The alternative, of using the marked dependence of reflectivity on dot size for metal dots has also been successfully employed [14].

Phase Encoding. Because of its importance in radar pulse compression and in other signal processing functions, the linear FM encoded matched filter in the form of a RAC has received much attention. Two basic designs have been used - the U-path arrangement [1,10] of Fig. 3(e) and the in-line arrangement [4,31] of Fig. 3(b). In both the position of the reflecting elements are varied continuously. In the in-line arrangement it is easy to see great design flexibility is provided by considering its simple impulse response (in the weak reflecting approximation). The physical position of reflectors will control the FM characteristic whose strength is varied by weighting. The U-path arrangement has a more complicated response, but it preserves most of the design flexibility. It offers the advantage that minor adjustments in the phase response can be achieved by a metal overlay of designed width between arrays - a so-called phase correcting plate.

Considerable theoretical effort has gone into the analysis of RACs extending beyond the weak reflection limit [12,24,26]. Noteworthy is the closed form result of WRIGHT and HAUS [12] for the U-path arrangement.

6. Concluding Comments

Outstanding Rayleigh wave device performance has been achieved with reflective arrays. Chief among these has been the large TB product RAC on lithium niobate [32] and the high Q resonator on quartz [33]. Significantly, both employ grooves. As we have seen, alternative reflecting elements are available and they are challenging the position of grooves.

The desirable features one looks for in a reflective array are by now obvious:- (a) ease of design, (b) controllable reflectivity, (c) minimum or at least predictable velocity change, (d) spurious signal free, (e) negligible or avoidable coupling to other propagating modes (f) reproducible and (g) simple and of low cost to implement. No one element combines these features. Pros and cons are set out in Table 2. Cost factors apart, the strength of the groove is evident. However, its

Table 2. Advantages and disadvantages of reflective elements

Mechanism	Element	Advantages	Disadvantages
Topographic	Groove	$\Delta v/v$ second order Large range of weighting	Multistep Fabrication Cost Groove Shape Dependence
Mass loading & Topographic	Strip & dot	Simple Design Rules	Thickness Dependent Combined Mechanisms
Electrical Loading	Strip & dot	Predictable design Dot Weighting Simple Compatible with transducer fabrication	LSW if off Z-direction on Lithium Niobate Strip weighting difficult to control

position is eroded if the cost factor is a major consideration. Application can also be a deciding factor e.g. if the highest temperature stability is required the electrical loading is not a contender as a quartz substrate must be used with its low electromechanical coupling.

The type of array arrangement employed is more directly coupled to the application. But other factors which influence choice are freedom from spurious signals, flexibility in design and availability of phase correction and, again, temperature stability.

A convenient subdivision of reflective array applications is into (I) resonators, (II) unidirectional transduction and (III) two port filters. The key requirement for resonators is low loss, thus arrays are encountered with high reflectivity but, for the normal incidence case, design is straight forward (32). The combination of array and transducer to give unidirectional behaviour began with arrays external to the transducer, but recently they have been incorporated in them [7]. It remains to be seen if low loss, wide bandwidth designs can be achieved. Two port filters incorporating arrays are already established. However, they are designed and operated near exclusively in the weak reflecting approximation. A reason is that synthesis procedures exist only in this regime. Another must be that low loss and maximum design flexibility are mutually exclusive. Multiple reflections see to that. However, it is an important challenge to explore the limits of how far loss can be reduced while retaining adequate design flexibility and it is an equally important challenge to develop synthesis procedures to match.

Finally, it is worth noting some interesting possibilities which have only recently or have yet to be considered. For example, electrical loading effects of semiconducting elements has yet to be investigated (optically excited array?). Electrical effects of a piezoelectric overlay on a piezoelectric substrate have not been studied. New reflecting elements such as composites of a hole-in-dot have only been subjected to a preliminary assessment [34]. Few examples exist where the potentialities of using more than two arrays have been exploited. Superposition of arrays is virgin territory. The potentially rich field of mixed element arrays have only recently been reported. For example, TANSKI et al [33] have used a combination of grooves and pits to weight the reflections of a resonator. More recently the positive and negative reflectivity of open and short circuit metal strips (Fig. 6) has been utilised in a reflective array [35]. And what of time varying arrays?

Acknowledgement It is a pleasure to acknowledge the assistance of David Selviah in preparing diagrams for this paper.

References

1. R. C. Williamson and H. I. Smith: IEEE Trans MTT-21, 195 (1973)
2. L. Solie: Appl. Phys. Letts., 28, 420 (1976)
3. E. A. Ash: IEEE Symposium on Microwave Theory and Techniques, Newport Beach, Calif., 1970
4. R. E. Chapman, R. K. Chapman, D. P. Morgan and E. G. S. Paige: Proc. 1978 IEEE Ultrasonics Symp., 728 (1978)
5. J. Melngalis, J. M. Smith and J. H. Cafarella: Proc. 1972 IEEE Ultrasonics Symp., 221 (1972)
6. R. D. Weglein and O. W. Otto: Proc. 1974 IEEE Ultrasonics Symp., 205 (1974)
7. C. S. Hartmann, P. V. Wright, R. J. Kansy and E. M. Garber: Proc. 1982 IEEE Ultrasonics Symp., 40 (1982)

8. F. G. Marshall and E. G. S. Paige: Electronics Letts., 7, 460, 1971
9. F. G. Marshall, C. O. Newton and E. G. S. Paige: IEEE Trans. SU-20, 134 (1973)
10. R. C. Williamson: Surface Wave Filters; Ed. H. Matthews (Wiley, 1977)
11. O. W. Otto and H. M. Gerard: Proc. 1977 Ultrasonics Symp., 596 (1977)
12. P. V. Wright and H. A. Haus: Proc. 34th Ann. Freq. Control Symp., 1980
13. B. A. Auld: Acoustic Fields and Waves in Solids (Wiley, 1973)
14. F. Huang and E. G. S. Paige: Proc. 1982 IEEE Ultrasonics Symp., 72 (1982) and work to be published
15. G. L. Matthaei, F. Barman and E. B. Savage: Electronics Letts. 12, (1976)
16. F. Huang: Electronics Letts., 20, 637 (1984)
17. P. V. Wright: Proc. 1984 IEEE Ultrasonics Symp., 54 (1984)
18. O. W. Otto and H. M. Gerrard: Proc. 1977 IEEE Ultrasonics Symp. 596 (1977)
19. E. Akcakaya and G. W. Farnell: Wave Electronics 4 (1979)
20. L. Solie and H. van den Vaart: Appl. Phys. Letts., 31, 1 (1977)
21. R. F. Milsom, M. Redwood and N. H. C. Reilly: IEEE Trans SU-24, 147 (1977)
22. T. Kitano and K. Nishikawa: Electronics Letts., 13, 136 (1977)
23. O. W. Otto: IEEE Trans. Sonics and Ultrasonics, SU-22, 251 (1975)
24. E. K. Sittig and G. A. Coquin: IEEE Trans Sonics and Ultrasonics, SU-15, 111 (1968)
25. H. A. Haus: Waves and Fields in Optoelectronics (Prentice-Hall, 1984)
26. P. D. Bloch, E. G. S. Paige and L. Solymar: Proc. 1978 IEEE Ultrasonics Symp., 639 (1978)
27. P. D. Morgan: Previous paper
28. S. Datta and B. J. Huntsinger: J. Appl. Phys. (USA) 51, 4817 (1980)
29. H. A. Haus: IEEE Trans SU-24, 259 (1977)
30. F. G. Marshall, E. G. S. Paige and A. S. Young: Proc. 1974 IEEE Ultrasonics Symp. 202 (1974)
31. R. C. Woods: J. App. Phys. (USA) 54, 6240 (1983)
32. H. M. Gerard, O. W. Otto and R. D. Weglein: Proc. 1974 IEEE Ultrasonics Symp. 197 (1974)
33. W. J. Tanski, M. Block and A. Vulkan: Proc. 1980 Ultrasonics Symp. 148 (1980)
34. E. G. S. Paige and Sun B.S.: Proc. 1984 IEEE Ultrasonics Symp., 72 (1984)
35. K. Yamanouchi and H. Furuyashiki: Electronics Letts. 20, 989 (1984)

The Launching of Rayleigh Waves from Surface Point Sources

C.B. Scruby and B.C. Moss

Materials Physics and Metallurgy Division, AERE,
Harwell, Didcot, Oxon, OX11 0RA, U.K.

1. Introduction

In the 100 years since their discovery, Rayleigh waves have been extensively studied experimentally and theoretically[1,2,3] and have found a wide range of applications[4,5]. A number of techniques is available to excite Rayleigh waves including edge-mounted transducers, wedge probes and interdigital transducers. More recently laser techniques have been applied to the generation of Rayleigh waves, one major advantage being that the method is non-contacting. The pulsed laser will launch Rayleigh waves (in addition to bulk waves) with good efficiency in both the lower power thermoelastic and higher power ablation regimes[6]. The laser beam can be readily focused to a point to act as an omnidirectional source of broadband Rayleigh waves, or focused to a line if a directional beam is required[7].

In the ablation regime the pulsed laser generates a point force [8] normal to the surface whose time dependence may be a short pulse or step-function. This is ideal for studying wave propagation, being very similar to the sources used in theoretical studies of Rayleigh waves[3]. In the lower power regime a different but equally interesting source is generated in which the principal stresses (thermoelastic) are now parallel to the surface[8].

In recent years optical techniques have also been applied to the measurement of ultrasound, including Rayleigh waves[9]. The laser interferometer is an ideal probe for exploring the wave field generated by various ultrasonic sources, because it is both non-contact and a broadband point detector. The combination of laser source and laser detector can be used to confirm and hopefully extend wave propagation theory by experimental means.

In this study the aim was to show how surface waves develop from the displacements generated by two types of laser ultrasonic source, using the laser interferometer to measure the wavefield as a function of distance very close to the source.

2. Experimental Procedure

Surface wave propagation was investigated using a polished steel block of thickness 46 mm (Fig. 1). The lateral dimensions were large enough for any reflections from the edges to be ignored. The surface waves were generated by a Q-switched Nd-YAG laser (1.06 μm) delivering nominally 100 mJ per pulse of duration ~ 30 ns. The energy incident at the surface of the specimen was somewhat less due to absorption in the various optical components. It could also be varied by the insertion of neutral

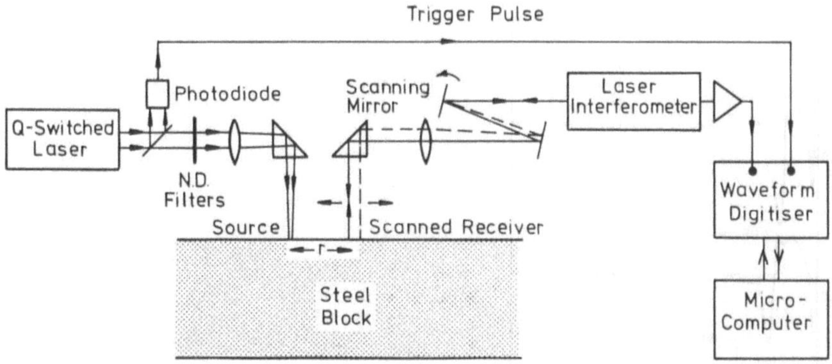

Fig. 1 Apparatus for studying Rayleigh waves generated by pulsed laser

density fitters in the beam. Without N.D. filters, the energy at the surface was 80 mJ. A 140 mm converging lens was used to focus the beam to form an approximately circular source of diameter 0.6 mm as measured by a burn mark on photographic film. The beam was incident at ~ 10° to the normal so that the interferometer beam (at normal incidence) could be used to probe the source region on the specimen. The laser was triggered by computer.

The displacement of the specimen surface was measured with a phase-locked laser interferometer[10]. This was a Michelson interferometer in which a Pockels cell was used to modulate the beam from the 5 mW helium-neon laser, together with phase sensitive detection so that it was stable with respect to low frequency vibration. A computer-controlled parallel beam scanning system was attached to the front of the interferometer, which enabled the wavefield to be measured as a function of distance from the source without realignment between positions. Because a polished specimen was used, the light intensity reflected back into the interferometer was high, so that a relatively low noise level could be obtained. The smallest detectable signal (without averaging) was limited by quantum noise to ~ 0.1 nm over a bandwidth of 20 MHz. An infra-red absorption filter was inserted in the interferometer, in front of the photodiode detectors.

A Tektronix 7612D digitiser was used to record each waveform. At a sampling frequency of 100 MHz each 512 point waveform corresponds to 5.12 μs. The digitiser was triggered 0.16 μs after the start of each record by the laser pulse using a beam splitter and photodiode detector as shown in Fig.1. The data was transferred to a BBC microcomputer via an IEEE 488 bus and stored on disc. The microcomputer enabled data records to be signal averaged to reduce background noise. The number of records for averaging was varied according to the gain of the digitiser that was used, and was chosen to make the noise ≲ 1% of the full-scale signal.

3. Results

In the first experiment the full incident laser energy (80 mJ) was used. A strong plasma was generated at the surface of the specimen. The surface displacement was first measured as close as possible to the source. If the interferometer probe was brought closer than ~ 1.3 mm,

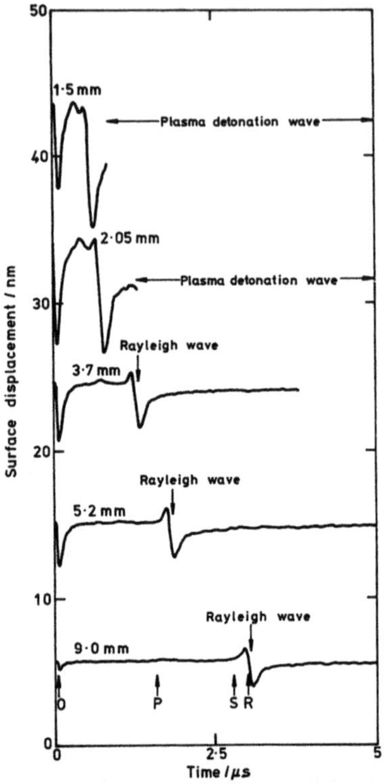

Fig. 2 Displacement as function of distance from ablation source, and showing development of Rayleigh wave. P,S,R are calculated arrival times of compression, shear and Rayleigh waves for r = 9 mm. O is breakthrough of optical pulse

detonation or "blast" waves from the plasma disturbed the interferometer beam, obscuring the surface waves. The surface displacement was measured for values of distance r between 1.3 and 13 mm and some of the waveforms are shown in Fig. 2. The plasma detonation wave is clearly visible for small r. It is not visible at larger r because it travels at a much lower velocity than the surface waves on the specimen.

The first, negative, signal on the waveforms (Fig. 2) is an optical pulse, O, caused by plasma light that is scattered into the photodiodes in the interferometer, and whose amplitude is therefore somewhat variable. There is then a relatively weak positive signal, whose velocity was calculated to be 6.0 mm/μs, and is therefore identified as the surface skimming compression (P) wave. The next signal is principally characterised by an asymmetrical negative pulse which has a small positive precursor at small r, and which changes into a bipolar pulse at larger r. Its velocity was calculated to be 3.0 mm/μs and is therefore identified as the Rayleigh surface wave (R).

The amplitudes of the compression and Rayleigh wave arrivals were then measured. In the latter case it was measured as the difference between the positive and negative peaks. The data are plotted as a function of source- receiver distance in Figs. 3(a) and 4(a). We note that the

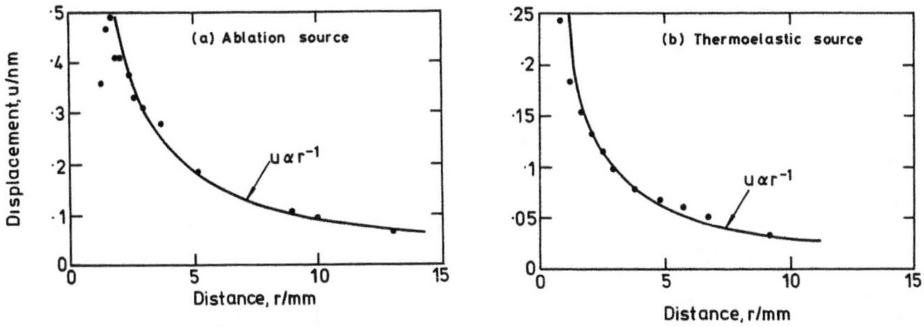

Fig. 3 Amplitude of direct surface compression wave as function of distance. Experimental data (•) fit $u \propto r^{-1}$ except for small r

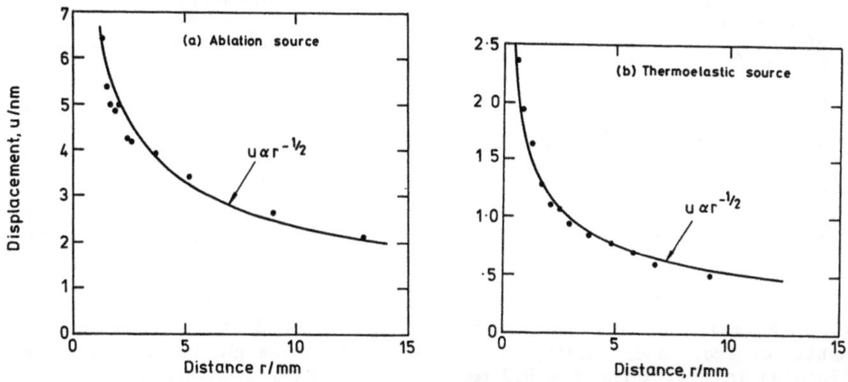

Fig. 4 Amplitude of Rayleigh wave as function of distance, shown to fit $u \propto r^{-\frac{1}{2}}$

compression amplitudes are within experimental error, inversely proportional to distance, r, except at very small r where the compression amplitude appears to fall off (Fig. 3(a)). In the case of the Rayleigh wave the amplitude is inversely proportional to the square root of the distance (Fig. 4(a)).

In the second experiment the incident laser energy was reduced by neutral density filters to 16 mJ which was below the threshold for plasma formation. The source of ultrasound was thus purely thermoelastic. The surface displacements were again investigated as a function of distance (Fig. 5). However, in this case it was possible to make measurements much closer to the source than when there was plasma. For r = 0, the interferometer probe detected the rise of the surface due to thermal expansion in the source. As r was slowly increased, a shoulder appeared on the waveform (r = 0.8–0.9 mm), which later separated as a positive pulse whose arrival time corresponded to a velocity of 6.0 mm/μs and was therefore identified as the direct compression wave. The dip in the waveform immediately following this P wave became eventually a separate bipolar arrival corresponding to a velocity of 3.0 mm/μs, and was therefore identified as the Rayleigh surface wave (R).

It is noted by comparing Figs. 2 and 5 that, while the two types of source generate similar compression waveforms, their Rayleigh waves are

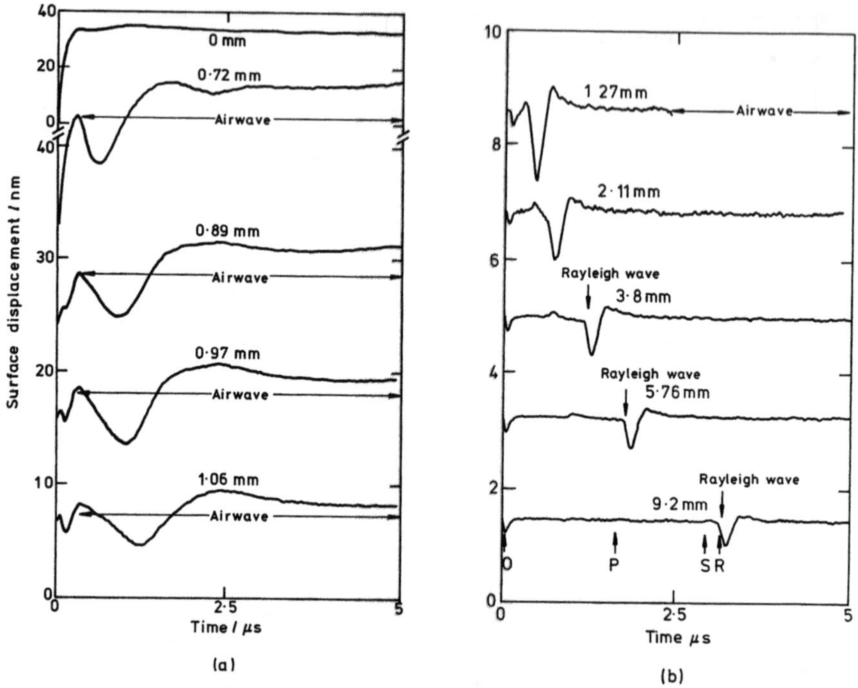

Fig. 5 Normal surface displacement as function of distance from thermoelastic source, and showing development of Rayleigh wave. O,P,S,R defined as in Fig. 2 for r = 9.2 mm

quite different, being of opposite polarity. Fig. 6 further illustrates that when the laser energy incident on a fixed area is increased so that the source transforms from a thermoelastic to ablative source, the polarity of the Rayleigh wave progressively changes. At an intermediate energy (42 mJ) the Rayleigh wave approximates to a unipolar pulse.

The amplitudes of the compression and Rayleigh waves in the thermoelastic regions were measured as a function of distance, and the results plotted in Figs. 3(b) and 4(b). Although the thermoelastic wave amplitudes are all smaller, they show the same dependence on distance as the compression and Rayleigh waves generated by the higher power plasma source.

4. Discussion

The data show a number of features that deserve discussion. First, while the amplitude of the Rayleigh wave is proportional to $r^{-1/2}$ as expected, the amplitude of the direct compression wave is apparently proportional to r^{-1}. Calculations of the compression wavefield generated by point forces as presented, for instance, by MILLER & PURSEY[11] show the far field amplitude to be zero parallel to the surface, implying that in this direction it attenuates more rapidly than r^{-1}. Indeed direct calculation of the surface response to a delta-function source following the work of PEKERIS[3] indicates that the surface compression wave attenuates with distance as r^{-2}. The reason the experimental data show

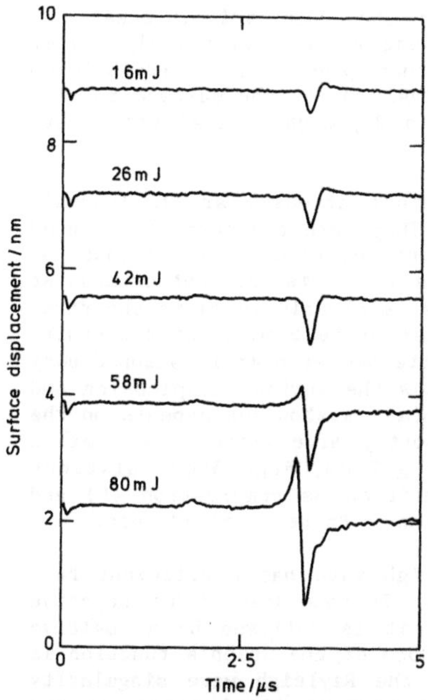

Fig. 6 Displacement measured as function of incident optical energy for r = 10 mm. At low energy source is thermoelastic, at higher energy ablative.

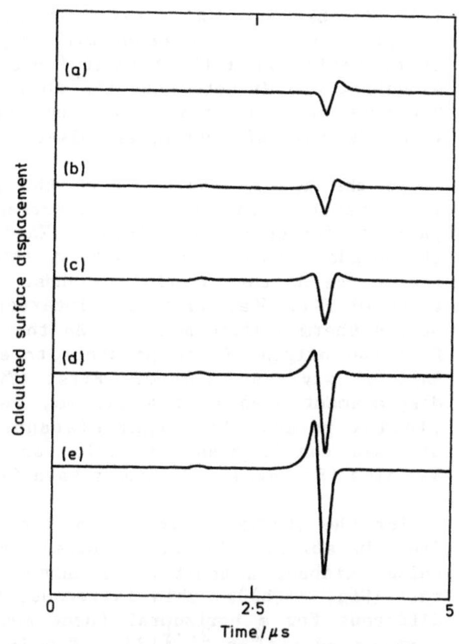

Fig. 7 Displacement calculated as described in text for r = 10 mm, and to be compared with Fig. 6

an r^{-1} dependence is believed to be due to the form of the surface P-wave arrival, which is a short pulse, of small but finite width which increases linearly with r. Thus for small r, this pulse width is much less than that of the source function, resembling a far field delta function arrival. From calculations it is expected that the P-wave amplitude is proportional to r^{-2} for r > 50 mm when the waveform is convolved with a broadening function of width ≳ 100 ns from the finite diameter of the source.

In the ablation regime, the ultrasonic source is dominated by a time-varying force normal to the surface of the specimen[8]. Its time dependence is likely to take the form of an asymmetrical pulse somewhat similar in shape to the optical pulse in Fig. 2. The normal displacement of the specimen surface very close to the source would be expected to be approximately proportional to the applied force. This is confirmed by the first displacement record at 1.5 mm in Fig. 2, although the plasma detonation wave partially obscures the data.

As the source to receiver distance is increased in Fig. 2, the Rayleigh, shear and compression wave arrivals separate. It can be seen that the negative swing of the Rayleigh arrival develops out of the perpendicular surface displacement at the source. Fig. 2 also shows that the shape of the Rayleigh arrival changes with distance. At small r it is almost entirely negative, while at large r it becomes a bipolar pulse.

This is because the separation between the compression, shear and Rayleigh arrivals increases with r, leading to a progressively larger contribution from the term $(t_R^2 - t^2)^{-\frac{1}{2}}$ that gives rise to the Rayleigh singularity. We note that the shear arrival causes the surface to rise but this is soon counteracted by the main Rayleigh arrival which here takes the form of a negative pulse.

In the thermoelastic source the principal stresses are horizontal, i.e. parallel to the specimen surface. They take the form of balanced pairs of forces (dipoles)[6,8]. There is no net normal force because of the boundary conditions. However, there is a net displacement upwards at the source as the material expands. This is clearly shown in the first trace of Fig. 5(a) where the interferometer probe beam is at the centre of the thermoelastic source. As the interferometer beam is scanned away from the origin, it is possible to see how the surface compression and Rayleigh waves are formed. First (Fig. 5(a)) a shoulder appears on the displacement step at $r \simeq 0.9$ mm, and shortly afterwards ($r \simeq 1$ mm) a bipolar pulse. At larger distances ($r \gtrsim 2$ mm, Fig. 5(b)) distinct arrivals can be observed, which are identified as compression (P) and Rayleigh (R) waves. The shear wave is too weak to be resolved here.

For the thermoelastic source the Rayleigh wave has a different form from the ablation (vertical force) source. It commences with a negative pulse, without a positive precursor, but it is followed by a positive tail (Fig. 5(b)). This is because the form of the Green's function is different for a horizontal force source, the Rayleigh wave singularity varying as $t(t^2-t_R^2)^{-\frac{1}{2}}$[3]. The initial surface displacement at the ablation source is negative (downward) and at the thermoelastic source positive, and it is not surprising therefore that the Rayleigh waves have opposite polarity (Fig. 6). Fig. 7 confirms this change of Rayleigh wave polarity from calculation. Following PEKERIS[3] Fig. 7(e) shows the displacement from a vertical force with appropriate time dependence to simulate ablation. For Fig. 7(a) the source was taken to be a horizontal force dipole to simulate thermoelasticity. Figs. 7(b)-(d) are calculated for combinations of horizontal and vertical forces.

To conclude, we have shown how surface compression and Rayleigh waves are launched from two types of surface source, comprising normal and parallel stresses. The form of the Rayleigh singularity is different for the two cases, although both attenuate with distances as $r^{-\frac{1}{2}}$. In both cases the compression wave attenuates as r^{-1} for the range of r chosen, although calculations show that it attenuates as r^{-2} at larger distances.

5. Acknowledgement

This work was funded by the Metrology and Standards Requirements Board of the Department of Trade & Industry through its support of the National NDT Centre.

References

1. Lord Rayleigh: Proc. London Math Soc., 17, 4, (1885).

2. H. Lamb: Phil. Trans. Roy. Soc. (London) A203, 1, (1904).

3. C. L. Pekeris: Proc. Nat. Acad. of Sciences (U.S.A.), 41, 469 (1955).

4. J. D. Maines & E. G. S. Paige: Proc. IEE, 120, 10R, 1078 (1973).

5. I. A. Victorov: "Rayleigh and Lamb Waves", New York, Plenum Press, (1967).

6. C. B. Scruby, R. J. Dewhurst, D. A. Hutchins & S. B. Palmer: Res. Tech. in N.D.T., Vol. V, 281 (1982).

7. A. M. Aindow, R. J. Dewhurst, D. A. Hutchins & S. B. Palmer: "Acoustics '80", Conf. Proc. 277 (1980).

8. R. J. Dewhurst, D. A. Hutchins, S. B. Palmer & C. B. Scruby: J. Appl. Phys. 53 (6), 4064 (1982).

9. R. Kline, R. Green & C. Palmer: J. Acoust. Soc. Am, 64 (6), 1633 (1978).

10. B. C. Moss: "The Harwell Laser Interferometer", AERE Report - R10417 (1982).

11. G. F. Miller & H. Pursey: Proc. Roy. Soc. Lond., A223, 521 (1954).

Quantum Aspects of Rayleigh Waves: Surface- and Interface-Waves at Very High Frequencies

K. Dransfeld

Fakultät für Physik, Universität Konstanz,
D-7750 Konstanz, Fed. Rep. of Germany

This article is concerned with acoustic waves at frequencies above 10 GHz propagating along crystal surfaces or traversing coherent multilayer structures. It is divided into the following sections :

1. Detection of Surface Acoustic Waves (SAW) at THz-Frequencies

2. Possible Photo-Excitation of SAW on Semiconductor Surfaces

3. Generation of Acoustic Waves above 10 GHz by Coherent Multilayers

4. Surface-Guided Acoustic Bulk Waves above 10 GHz

In the high-frequency regime discussed here the quantum nature of acoustic waves becomes more important than at lower frequencies and dominates their interaction with particle beams and with electromagnetic radiation. For acoustic surface waves at lower frequencies including their interesting technical applications the reader is referred to numerous other contributions to this volume.

1 Surface Acoustic Waves at Terahertz-Frequencies

In the past, surface acoustic waves on piezoelectric crystal have mainly been excited by interdigital electrodes, but also by other means. The interdigital technique allows the generation and detection of surface acoustic waves with a considerable efficiency but only up to frequencies of a few GHz. For still higher frequencies the acoustic wavelength becomes very small and therefore the production of interdigital structures having lateral dimensions of less than a micrometer would be required which becomes increasingly difficult. In addition, the thickness of the metallic fingers should be smaller than the acoustic wavelength but cannot be reduced too much below the electrical skin depth without seriously limiting the electrical conductance of the fingers.

Surface acoustic waves can also be detected by the scattering of light. The Brillouin scattering of light is a quantum process in which the total energy and the momentum of the interacting particles, i.e. the phonon and the photon, are conserved. By this extremely sensitive optical technique it has become possible to observe the weak thermally excited surface waves, even below room temperature. But according to the quantum mechanical conservation law, acoustic surface waves can only be "seen" by the scattering of light if the acoustic wavelength is at least as large as half the optical wavelenth of the scattered light. Consequently surface acoustic waves at frequencies well above 30 GHz cannot be detected by the scattering of light from a free surface. Only in the exceptional case that the surface is not free but carries a submicron superstructure which imposes a lateral variation of the optical or elastic properties, surface acoustic wave at frequencies well above 30 GHz might still be detected by the scattering of light.

It would be of great interest for the study of the dynamics of free surfaces and surfaces covered by adsorbates to experiment with surface acoustic waves at frequencies above 100 GHz.

Since electromagnetic fields (photons) cannot be used for the excitation or detection of such high frequency waves (neither by using interdigital electrodes nor by Brillouin scattering) beams of heavier particles such as electron beams or atomic beams are more suitable for probing surface acoustic waves at THz-frequencies on account of their shorter de Broglie wavelength.

Let us first consider the elastic scattering of low energy (about 50 eV) electrons from crystalline surfaces. Davison and Germer were the first to demonstrate the wave nature of the electron from the diffraction pattern they obtained with their prototype of a modern LEED system. Good descriptions of the modern state of low energy electron diffraction have been given by PENDRY /1/, and by HEINZ and MÜLLER /2/.

The scattering of electrons from surfaces does not only give structural information about the surface: If the surface atoms are not at rest but in motion due to an acoustic surface wave the electronic scattering process becomes inelastic: The incident electron either loses or gains an energy corresponding to the creation or destruction of a surface phonon. At the same time also the momentum of the scattered electron is changed due to the scattering process. In the first experiment of this type IBACH /3/ detected the optical phonons of ZnO- and dipolar Si-surfaces. During the scattering process the electrons interact mainly with the strong alternating electric fields of the optical phonons which extend out of the crystal surface. The sensitivity of this new type of electron spectroscopy for optical phonons is particularly large for surface phonons of long wavelength and surpasses the sensitivity of infrared detection by several orders of magnitude. The interaction with non-polar acoustic surface waves is, however, much weaker. Similar dipolar scattering experiments on GaAs-surfaces were later on performed by LÜTH /4/, and by MATZ and LÜTH /5/. Again mainly optical phonons whose energy is well known /6/ and surface plasmons as well /5/ were seen.

In 1980 IBACH and BRUCHMANN observed for the first time acoustic Rayleigh waves on Ni(111) surfaces by means of the same electron energy loss spectroscopy /7/ . More recently LEHWALD et al./8/ succeeded to measure for each wavevector within the whole Brillouin zone the energy of the corresponding surface wave. A good review of this new type of surface phonon spectroscopy has been given by IBACH and MILLS /9/. See also /10/ for further references.

At about the same time TOENNIES and his group, experimenting in Göttingen with atomic beams, were able to observe acoustic surface phonons on LiF(001) by the inelastic scattering of He-atoms /11/. Since He-atoms are neutral, the strong electric fields of the optical phonons do not interfere at all with the sattering process. TOENNIES /12/ and his collegues could measure the dispersion relation for Rayleigh waves on a polar LiF-crystal surface in the whole Brillouin zone. For a summary also of the theoretical aspects see BENEDEK /13/.

There is still another way to detect THz - surface waves by interaction with atomic particles : The photo-acoustic detector. Let us , for example, consider a monochromatic Rayleigh wave (surface velocity V) propagating along a crystalline surface and the crystal is surrounded by air (M is the mass of a single air molecule). Both the crystal and the surrounding air have exactly the same temperature. Nevertheless one can show that the molecules after collision with the vibrating crystalline surface gain an average energy MV^2 above the thermal energy of the gas /14/. Whereas in the usual photoacoustic spectroscopy /15/ a solid surface is periodically heated by optical radiation and subsequently heats the surrounding air, here the air adjacent to the crystalline surface is mechanically heated by interaction with the THz-surface wave although the crystal remains cool. Therefore if on a crystal embedded into a photoacoustic cell a surface wave is periodically excited - for example by electromagnetic radiation - a photoacoustic signal will appear even though the crystal temperature remains constant. The photoacoustic signal does give information only about the surface velocity V of the vibrating crystal but - unfortunately - not about the frequency of the vibration or surface wave.

2. Possible Photoexcitation of SAW on Semiconductor Surfaces

So far we have mainly discussed how THz-surface waves could be detected, mainly by particle radiation. Another interesting question is whether surface acoustic waves are excited on surfaces of semiconductors if these are illuminated by intense bandgap radiation. The non-thermal driving mechanism would be the non-radiative electron-hole recombination.

It is a well-established fact that the non-radiative recombination of electrons and holes leads to an emission of acoustic bulk waves in light emitting diodes /16/ and in semiconductors in general /17 - 21/. Although the mechanism of phonon emission is not well understood /22,23/, the phonon emission seems to be responsible for the considerable degradation of light emitting diodes. The presence of dislocations, of deep levels /24/ and of a high piezoelectric constant /22/ probably enhances the acoustic emission.

As far as the mechanism of acoustic emission is concerned, it has been pointed out by BISWAS and AMBEGOAKAR /25/ ,that electronically excited semiconductors show a well-defined softening of their elastic constants. Therefore in the presence of local charge carrier fluctuations and the corresponding time-dependent elastic constants, parametric phonon generation can be expected.

Although it is an established fact that phonons are excited in optically pumped semiconductors or when charge carriers are injected it is less clear what the frequency and the polarisation of the created dominant phonons might be. There is, however, an interesting theoretical prediction by ORBACH /26/,(see also /27/) that due to the phonon dispersion the phonons of the lowest acoustic branch are expected to have at T=0 the longest lifetime. In very good agreement with this prediction the population of the bulk phonons excited by optical pumping has been found to be highest for transverse bulk phonons near the zone edge.(MCWHAN et al./28/) What has not been mentioned before is the interesting fact that for the same reason surface phonons - having the lowest phonon branch - should exhibit an even longer lifetime at T=0 than the transverse bulk phonons, at least for a perfect surface. Therefore after illumination one would - according to Orbach's argument - expect a strong photogeneration of SAW.

Can THz-surface acoustic waves be excited on semicondutor surfaces by illumination ? In my view there are two reasons why this seems to be likely: Firstly, according to ORBACH's argument /26/ the lifetime of surface acoustic waves should be even longer than for transverse bulk waves. Secondly, for a direct gap semiconductor, such as GaAs, the optical absorption length is only about 100 nm, i.e. the density of the recombining electron-hole pairs is highest close to the surface. Experimentally at the present time very little is known about the possible photo-excitation of THz- acoustic surface waves on semiconductor surfaces.

A competing photoinduced process which is known to occur in bulk semiconductors is the recombination-enhanced migration of defects and atoms /28,30 - 32/. Such a photoinduced atomic diffusion (photocorrosion of surfaces) indeed also occurs on surfaces: According to preliminary and not yet published electron scattering experiments by Prof.Lüth at Aachen a perfect GaAs-surface cleaved under ultrahigh vacuum conditions becomes corrugated on an atomic scale after moderate illumination.

In summary, it seems rather interesting to study by means of electron scattering the still rather unknown surface dynamics of an illuminated semiconductor surface.

3 Generation of Sound above 10 GHz by Coherent Multilayers

The very interesting electronic and optical properties of multilayers built up of alternating layers of two semiconductors have been described in several recent reviews /33 - 37/. The two semiconductors can be either chemically different, (for example GaAs/GaAlAs) forming a "heterostructure", or they can be differently doped parts of the same semiconducting material. A sequence of alternate n-type, intrinsic, p-type, intrinsic, etc. layers is often called a "nipi-structure".

We want to point out here that these man-made artificial structures - apart from their well known electronic and optical properties - are well suited for the efficient generation of ultrasonic waves at very high frequencies. Let us for simplicity discuss here only one structure : A nipi-structure made up in a coherent way by about 100 alternate layers of n- and p-type silicon, doped by 10^{19} donors and acceptors respectively. By the technique of molecular beam epitaxy or liquid phase epitaxy the thickness of the individual layer can be programmed, usually between 50 and 5000 Angström. In general, the multilayer structure is carried by a thin planar substrate of the same material, for example silicon. Within the nipi-structure the n-type layers are positively charged (the positive charges being the immobile ionized donors) and the p-type layers carry a negative charge (due to the acceptors).

If an electric field of a suitable high frequency is applied perpendicular to the layers, longitudinal acoustic waves can be excited in the structure propagating through the layers, being partially reflected inside the periodic structure and certainly from the top and bottom surfaces of the whole sample.

If V is the longitudinal acoustic velocity for propagation perpendicular to the layers and if each layer has a thickness d, maximum generation of acoustic waves occurs for an electric field at the resonance frequency

$$\omega = (V/2d),$$

neglecting possible upper harmonics. The strain amplitude σ and the particle velocity v in the wave grow to a value such that the acoustic energy loss per second and per unit volume is equal to the energy gain from the electric field E in the same time and the same volume.

$$\rho v^2 \omega / Q = ENev \qquad (1)$$

with Q being the mechanical quality factor of the wave (about 1000 at low temperatures), ρ the density, N the doping level (10^{19}/cc), e the elementary charge. From (1) we derive the particle velocity

$$v = ENeQ/\rho\omega = \sigma V$$

or a strain amplitude

$$\sigma = ENeQ/V\rho\omega = x/2d$$

with x being the vibrational amplitude. For the electrical polarisation P we find

$$P = Nex = \chi\varepsilon_0$$

resulting in a dielectric susceptibility

$$\chi = (\varepsilon - 1)/\varepsilon_0 = Q(Ne)^2/\rho\omega^2\varepsilon_0.$$

Using the following numerical values :

$N = 10^{25}$ pro cbm, $Q = 1000$ (at low temperature),
$\rho = 1000$ kg/cbm, $V = 1000$ m/s, $E = 10$ Volt/mm and
$\omega = 10^{11}$ Hz

the strain amplitude of the acoustic wave becomes

$$\sigma = 10^{-4}$$

corresponding to an acoustic intensity of

$$\rho V^3\sigma^2 = 1\ \mathrm{W/cm^2}$$

and for the loss angle δ one finds

$$\tan\delta = (QNe)^2/\varepsilon_0\rho\omega^2 ,$$

also a rather large value. In summary, it looks promising to use multilayer structures for the generation of ultrasonic waves above 10 GHz. An equally interesting application of nipi-structures or coherent heterostructures is the realisation of resonant rf filters for frequencies above 10 GHz (up into the THz-range), for which the periodicity of the structure is equal to the acoustic wavelength.

The acoustic waves we have discussed can also be considered as optical phonons of the superlattice at zero wavevector with the strongly enhanced dielectric constant and loss angle mentioned above. Thus multilayers represent effective dielectric resonators for submillimeter radiation.

4 Surface Guided Acoustic Waves

Lithium niobate has become one of the most important materials for optoelectronic applications. One can - by heating the crystal in benzoic acid - exchange in a surface layer a sizeable fraction of all Li-ions by protons. It is well known that this proton exchange process leads to an increase of the refractive index by a few percent.

What is less known, and what we like to point out here, is our observation /38/ that the effect of proton exchange on the elastic properties is even stronger : In the proton exchanged surface layer - extending up to a few micrometers - the acoustic velocity decreases by about 20%, corresponding to a softening of the corresponding elastic constants by as much as 40%. Thus by proton exchange it is posssible to build acoustic waveguides adjacent to the surface. For further details of the preparation and for several other applications see /38/.

Acknowledgements

I am grateful to the Deutsche Forschungsgemeinschaft for support and to Prof.C.Quate for an encouraging discussion.

References

/1/ J.B.Pendry : "Low Energy Electron Diffraction", (Academic, London (1974))

/2/ K.Heinz and K.Müller : "LEED-Intensities, Experimental Progress and New Possibilities of Surface Structure Determination" in Springer Tracts, Vol.91, Springer, Berlin, Heidelberg (1982)

/3/ H.Ibach : "Low Energy Electron Spectroscopy - A Tool for the Study of Surface Vibrations" J.Vac.Science, 9, 713, (1972)

/4/ H.Lüth : "High Resolution Electron Loss Spectroscopy on Semiconductor Surfaces" in Festkörperprobleme, 21, 117, (1981)

/5/ R.Matz and H.Lüth : "Conduction Band Surface Plasmons in the Electron-Energy-Loss Spectrum of GaAs(100)" in Phys.Rev.Lett., 46, 500, (1981)

/6/ see for example : S.M.Sze : "Physics of Semiconductor Devices" (Wiley-Interscience, New York (1970))p.51

/7/ H.Ibach and D.Bruchmann : "Observation of Surface Phonons on Ni(111) by Electron Energy Loss Spectroscopy" Phys.Rev.Lett., 44, 36, (1980)

/8/ S.Lehwald et al.: in Proc.2nd Int.Conf.Vibr.at Surfaces (Plenum, New York(1982)),p.137

/9/ H.Ibach and D.Mills : "Electron Energy Loss Spectroscopy and Surface Vibrations" ,(Academic, New York(1982))

/10/ Proc.3rd.Int.Conf.Vibr.at Surfaces, (Elsevier,Amsterdam 1983)

/11/ G.Brusdeylins, R.Bruce Doak and J.P.Toennies : "Observation of Surface Phonons in Inelastic Scattering of He-Atoms from Li(001) Crystal Surfaces" Phys.Rev.Lett., 44, 1417,(1980)

/12/ G.Brusdeylins, R.Bruce Doak and J.P.Toennies : Measurement of the Dispersion Relation for Rayleigh Surface Phonons of LiF(001) by Inelastic Scattering of He-Atoms" Phys.Rev.Lett., 46, 437, (1980)

/13/ G.Benedek, Journ.El.Spectrosc.and Rel.Phenomena, 30, 71,(1983)

/14/ F.Tsuruoka and K.Dransfeld : "The Vibration and Rotation of Piezoelectric Particles in High Frequency Electric Fields Observed by a Thermoacoustic Method" in Appl.Phys.A, 36, 125, (1985)

/15/ C.K.N.Patel et al.: "Optoacoustic Spectroscopy" in Rev.Mod.Phys.53, 517, (1981)

/16/ T.Ikoma, M.Ogura and Y.Adachi : "Acoustic Emission Study of Defects in GaP Light Emitting Diodes" in Appl.Phys.Lett.,19, 414 (1978)

/17/ J.Shah : "Hot Electrons and Phonons under High Intensity Photoexcitation of Semiconductors" in Solid State Electronics, 21, 43, (1978)

/18/ K.Wasa, K.Tsubouchi and N.Mikoshiba :"Investigation of Non-Radiative States in GaAs and InP by Photoacoustic Spectroskopy" Jap.J.Appl.Phys.,19, L 653 (1980)

/19/ K.Wasa, K.Tsubouchi and N.Mikoshiba : "Photoacoustic Measurement of Non-Radative States and Defects in Cds and Si with ZnO Transducer" in Jap.J.Appl.Phys., 19, L 475 (1980)

/20/ N.Mikoshiba, K.Wasa and K.Tsubouchi : "Photoacoustic Measurement of Non-Radiative States and Defects in Semiconductors with ZnO Transducer" IEEE Ultrasonics Syposium Proceedings, (1980), p.658

/21/ I.Suemune, M.Yamanishi and N.Mikoshiba : "Investigation of InGaAsP/InP Lasers by the Measurement of Current-Injection-Induced Acoustic (CIA) Signals" Jap.J.Appl. Phys., 20, L 631 (1981)

/22/ H.J.Queisser : "Recombination at Deep Traps" Solid State Electronics, 21, 1495, (1978)

/23/ C.H.Henry and D.V.Lang :"Non-Radiative Capture and Recombination by Multiphonon Emission in GaAs and GaP" Phys.Rev.B , 15, 989 (1977)

/24/ H.Sumi : "Multiphonon Non-Radiative Recombination due to Successive Electron and Hole Capture by Deep Level Defects in Semiconductors" Phys.Rev.Lett., 47, 133,(1981)

/25/ R.Biswas and Vinay Ambegoakar : "Phonon Spectrum of a Model of Electronically Excited Silicon" Phys.Rev. B, 26, 1980, (1982)

/26/ R.Orbach , Phys.Rev.Lett., 16, 151, (1965)

/27/ M.Lax, P.Hu and V.Narayanamurti, Phys.Rev.B, 23, 3975, (1981)

/28/ D.B.McWhan, P.Hu, M.A.Chin and V.Narayanamurti : "Observation of Optically Excited Near-Zone-Edge Phonons in GaAs by Diffuse X-Ray Scattering" Phys.Rev.B, 26, 4774 (1982)

/29/ D.V.Lang and L.C.Kimerling : "Observation of Athermal Defect Annealing in GaP" Appl.Phys.Lett., 28, 248 (1976)

/30/ J.D.Weeks, J.C.Tully and L.C.Kimerling : Theory of Recombination-Enhanced Defect Reactions in Semiconductors" Phys.Rev.B, 12, 3286 (1975)

/31/ G.A.Baraff, M.Schlüter and G.Allan : "Theory of Recombination-Enhanced Migration of Interstitial Aluminum in Silicon" Phys.Rev.Lett.,50, 739 (1983)

/32/ G.A.Baraff and M.Schlüter :"Enhanced Migration of Interstitials in Silicon" Phys.Rev.B, 30, 3460 (1984)

/33/ G.Bauer, F.Kuchar,H.Heinrich (eds.): "Two-Dimensional Systems, Heterostructures and Superlattices" Springer Solid State, Vol.53,(Springer, Berlin, Heidelberg 1984)

/34/ G.Döhler : "Solid State Superlattices" Scient.Amer. 249, 118 (1983)

/35/ R.Dingle : "Quantum States in Heterostructures" Festkörperprobleme, 15, 21 (1975)

/36/ see : "The 4th Int.Conf. on Two-Dimensional Systems, Electronic Properties" Surf.Science, 43, (1982)

/37/ see also : "Physics Today, Special Issue on Materials Science, Semiconducting Heterostructures", Oct. (1984)

/38/ V.Hinkov, M.Barth and K.Dransfeld : Acoustic Properties of Proton Exchanged Lithium-Niobate Investigated by Brillouin Scattering" Appl.Phys.A (1985) (in print)

Surface-Acoustic Wave Generation by Thermoelasticity

W. Arnold, B. Betz, and B. Hoffmann

Fraunhofer-Institute for Non-Destructive Testing,
Bldg. 37, University, D-6600 Saarbrücken, Fed. Rep. of Germany

We have experimentally verified the conditions for maximal generation of surface-acoustic via the thermoelastic effect using a pulsed laser.

1. Introduction

Non-destructive testing of materials and components by surface-acoustic waves (SAW) is an important and well-known technique. In certain cases it is desirable to use a non-contact technique for generation and detection of SAW. For generation such a non-contact technique is the pulsed thermoelastic source using laser pulses. This has been demonstrated previously /1/, however, no systematic studies have been performed on the conditions required for maximal efficiency or lowest "insertion loss" for SAW. In this paper we report on this question.

2. Experimental Results

We generated SAW on a polycrystalline aluminium sample by a pulsed ruby laser whose pulse-width was 50 nsec (gaussian shape, full-width at 1/e points). The laser beam was focussed either to a spot by a circular lens or to a line by a cylindrical lens. We used the laser only in the TEM_{00} -mode to assure spatial gaussian intensity profiles. The optical intensity was kept below 1 MW/cm^2, so that the data were only taken in the thermoelastic regime. The signals were detected by a wedge transducer whose center frequency was 10 MHz (6-dB-bandwidth: 6 MHz). The transducer was located ∿ 1 cm from the laser spot/line. In the case of the line source, we first

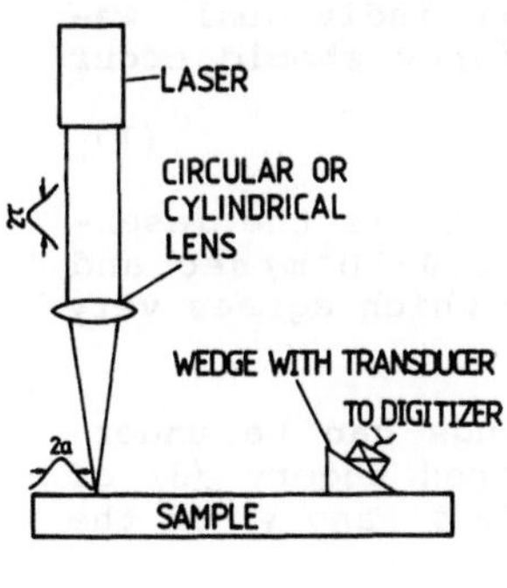

Figure 1: Experimental set-up

expanded the laser beam by another cylindrical lens whose axis was orientated perpendicular to the first one in order to obtain negligible laser-intensity variation over the width of the wedge transducer. A schematic representation of our experimental set-up is shown in Fig.1.

Figure 2 shows the intensity of the SAW in relative, arbitrary units as a function of the full-width 2a of the laser spot or line. As can be seen very clearly, there is a pronounced maximum at a certain diameter or width $2a_{max}$ of the laser spot-source (solid line) or line-source (dashed line), respectively.

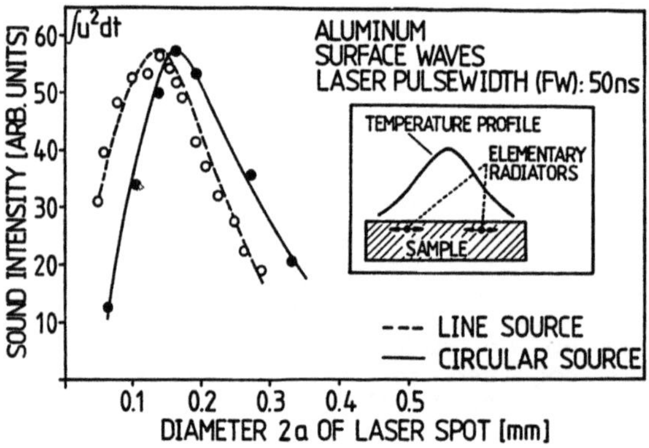

Figure 2: Intensity of the SAW as a function of the laser-pulse diameter 2a

3. Discussion

In a qualitative manner this behavior can be understood quite easily. One should keep in mind that the gradient of the temperature is responsible for sound generation /2/. When a is large the gradient of the lateral temperature profile is small, and hence the efficiency of SAW generation. In the limit of a lateral infinite laser pulse profile only longitudinal waves are produced because there is then only an axial gradient of the temperature. When a is small, however , the efficiency is low because a point (line-) source is obtained whose extent is small compared to the wavelengths generated. Because from each expanding surface element an elementary SAW is launched, maximal efficiency is achieved if there is mainly constructive interference between the individual wavelets. From these arguments maximal efficiency should occur if

$$2a = 2\tau v_R \qquad (1)$$

holds. Here, v_R is the Rayleigh-velocity and 2τ is the laser-pulse width. Inserting into Eq.(1) $v_R = 2.9 \cdot 10^3$ m/sec and $2\tau = 50$ nsec, we obtain for $2a_{max} = 0.145$ mm which agrees very well with our experimental results.

In a more quantitative manner our findings can be understood in the framework of a recently developed theory /3/ of this problem. In this theory, the potentials Φ and Ψ of the

SAW were derived from which its displacements can be calculated. Besides the relevant thermal parameters, the potentials contained the Fourier-transform of the lateral laser-profile. The authors predicted that maximal efficiency should occur if $k = \sqrt{2}/a$ for 100% amplitude-modulation of the laser beam at a given frequency with corresponding wave-vector k of the SAW. In our case, however, we generated a spectrum of SAW due to the finite laser pulse-width. This can be incorporated into the theory in a simple way by multiplying the potentials with the Fourier transform of the temporal laser pulse profile. It is then straightforward to calculate the maximal efficiency as a function of the parameters a and τ because the SAW-intensity is proportional to the prefactors squared of the potentials /4/. As a final result Eq. (1) is obtained/5/.

References

1. A.M. Aindow, R.J. Dewhurst, D.A. Hutchins, and S.B. Palmer: J. Acoust. Soc. Am. 69, 449 (1981)
2. R.M. White: J. Appl. Phys. 34, 3559 (1963)
3. V.V. Krylov and V.I. Pavlov: Sov.Phys.Acoust. 28, 493 (1982)
4. S.R. Seshadri: J. Appl. Phys. 54, 1699 (1983)
5. W. Arnold, B.Betz, and B.Hoffmann, to be published in Applied Physics Letters (L6256)

Rayleigh Waves on Gallium Arsenide

R.T. Webster and P.H. Carr

Electromagnetic Sciences Division, Rome Air Development Center, Hanscom AFB, MA 01731, USA

1. Introduction

Since gallium arsenide (GaAs) is both piezoelectric and semiconducting, it offers the opportunity of integrating passive Rayleigh wave components with active circuits. Judged by its Rayleigh wave properties alone, GaAs is not an attractive substrate material. Its piezoelectric coupling is slightly less than quartz and its attenuation slightly higher. In spite of these shortcomings, much effort has been expended to develop surface acoustic wave (SAW) components on GaAs for integration with electronics. This paper will review surface acoustic wave technology on gallium arsenide considering SAW propagation characteristics, transduction techniques, and unique device structures afforded by the piezoelectric semiconductor properties.

2. Material Constants and SAW Propagation

Gallium arsenide is a III-V compound semiconductor in the cubic class $\bar{4}3M$. Table I is a collection of the independent elastic, piezoelectric, and dielectric constants with the mass density and thermal expansion coefficient. The temperature coefficients of the elastic constants are included and were obtained from a least squares fit to data in COTTAM and SAUNDERS [1].

The constants in Table I were used to calculate the SAW properties of a selection of gallium arsenide plates [2]. The plate normals were chosen to form a map of the SAW properties as a function of crystal orientation. Table II lists the SAW properties of a small sample of orientations to illustrate the trends observed in the full set of data.

For SAW applications, high electromechanical coupling and low temperature coefficient of frequency (TCF) are often desirable. As the data in Table II indicate, the coupling on GaAs ranges from a high of 6.4 x 10^{-4} to zero while the computed TCF varies from 46.2 parts per million (ppm) to 34.8 ppm. Unfortunately, the higher coupling orientations also have high TCF and the low TCF orientations have low or zero coupling. Experimental values for several orientations also appear in Table II. The measured SAW velocities agree with the computed velocities within one percent but measured temperature coefficients are consistently higher than the computed values. The discrepancy may be due to the improvement in the quality of gallium arsenide crystals since the COTTAM and SAUNDERS [1] measurements were published in 1973. The remeasuring of these constants on recently grown and presumably higher quality GaAs may be warranted.

Although no temperature compensated orientation exists for SAW on bare GaAs, temperature compensated delay lines have been realized by overlaying the GaAs with thin films [3,4]. The technique uses a two layer structure of silicon dioxide (SiO_2) sputter deposited on the GaAs and a gold layer

Table 1. Gallium arsenide material constants

ij	$C_{ij}(25)$ ($\times 10^{11}$ N/m^2)	$TC^{(1)}_{ij}$ ($\times 10^{-4}$/°C)	$TC^{(2)}_{ij}$ ($\times 10^{-7}$/°C^2)	$TC^{(3)}_{ij}$ ($\times 10^{-9}$/°C^3)
11	1.18404	-1.0197	0.13622	0.69266
12	0.53733	-1.29207	-1.4983	1.4848
44	0.59107	-0.98098	0.64093	0.86992

Relative Permittivity = 13.1
Piezoelectric Constants $e_{14} = e_{25} = e_{36} = -0.16$ (C/M^2)
Thermal Expansion Coefficient = 6.86×10^{-6}
Mass Density = 5.32×10^3 (Kg/M^3)

$$C_{ij}(T) = C_{ij}\,[1 + TC^{(1)}_{ij}(T-25) + TC^{(2)}_{ij}(T-25)^2 + TC^{(3)}_{ij}(T-25)^3]$$

$C_{ij}(T)$ = Elastic Stiffness Constant at Temperature T

$TC^{(n)}_{ij}$ = nth Order Temperature Coefficient of C_{ij} Evaluated at 25C.

Table 2. Surface Acoustic wave propagation characteristics on selected cuts on gallium arsenide

Euler Angles	Temperature Coefficient of Frequency (PPM) Theory	Exp	Electro-Mechnical Coupling $k^2(\times 10^4)$	SAW Velocity(m/sec) Theory	Exp
45, 0.00, 0	-46.2	-52.0	6.4	2856	2841
45, 54.74, 30	-44.1	-50.7	2.0	2596	2598
45, 54.74, 0	-34.8	--	→ 0	2421	--
45, 35.26, 90	-44.2	-50.8	2.88	2611	2621
45, 90.00, 90	-45.8	-50.2	1.80	2812	2804
45, 90.00, 0	-35.5	--	→ 0	2392	--
15, 30.00, 60	-38.9	--	3.06	2609	--

deposited on the SiO_2. The SiO_2 has a positive TCF which offsets the negative TCF on the GaAs. The gold layer is required to load the SiO_2 surface, preventing the wave from leaking into the GaAs substrate.

The great activity in gallium arsenide electronics has made high quality crystals readily available. The electronics industry has standardized the crystaline orientation to the 100 cut crystal wafers. Fortunately the standard wafer includes the (100) cut [110] propagating orientation which has the highest available electromechanical coupling to the SAW mode on gallium arsenide.

A second advantage of the (100) cut wafers is offered by the {110} cleavage planes. Rectangular substrates are easily cleaved from the (100) cut wafers along these perpendicular {110} planes. The cleavage planes provide a near perfect reference for SAW transducer alignment.

The propagation mode on the (100) cut, [110] propagating substrate is an isolated SAW mode on a pseudo-surface wave branch [5]. The SAW velocity of 2860 M/Sec is considerably higher than the 2468 M/Sec velocity of the shear horizontally polarized bulk wave in the same propagation direction. This shear horizontal wave satisfies the free surface boundary condition but is not piezoelectrically coupled. The surface wave on the other hand exhibits the highest electromechanical coupling of any SAW mode on gallium arsenide with $k^2 = 6.4 \times 10^{-4}$.

The normalized mechanical displacements and electric potential of the SAW mode are shown in Fig. 1. The mode is fairly tightly bound to the surface, being nearly fully confined to 1.5 wavelength penetration depth. The mechanical displacement is confined to the sagital plane defined by the propagation direction; there is no transverse horizontal displacement.

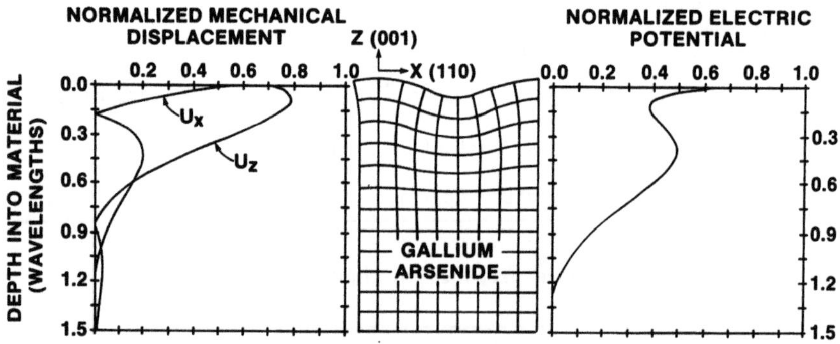

Fig. 1. Normalized mechanical displacements and electrical potential of a SAW on (100) cut, [110] propagating gallium arsenide

Several other interesting modes of propagation are discussed in FELDMAN, et.al. [6]. However, considering the virtually exclusive use of the (100) wafers by the electronics industry and the absence of any temperature compensated cut, it is unlikely that other GaAs SAW orientations will see practical application.

3. Transducers

On semi-insulating GaAs conventional interdigital transducers are possible but the low coupling leads to an unfavorable bandwidth insertion loss tradeoff. The maximum bandwidth - minimum insertion loss break point occurs at 2.8% fractional bandwidth, compared to 4.5% for ST-Quartz and 23.6% for $LiNbO_3$ [7]. Trading insertion loss for bandwidth is straightforward and compatible with fabrication of electronic devices. Bandwidths of 30 MHz at 307 MHz are reported for simple series tuned transducers at the cost of 21 dB of conversion loss per transducer [8]. Bandwidth enhancement has been attempted primarily by the addition of a stronger piezoelectric deposited on or bonded to the GaAs. For example, a zinc oxide layer can be deposited on the GaAs surface to enhance the coupling. Bandwidths of 8.8% at 57 MHz with 10 dB insertion loss have been reported [9]. OATES and BECKER [10] report bandwidths as wide as 91% using an edge bonded transducer. Their technique uses a $LiNbO_3$ bulk shear wave transducer bonded to the edge of the GaAs. The shear waves are efficiently converted to surface waves in the GaAs.

To date, both of these bandwidth enhancement techniques have been of limited use. In principle, the zinc oxide film deposition can be extended to higher frequencies with wider bandwidths, but in practice the thin film deposition is difficult to control and films become lossy. The edge-bonded technique presents fabrication difficulties. A processing sequence which includes both edge-bonded transducers and active electronics remains to be devised.

The semiconducting properties which provide the motivation for generating SAWs on GaAs can also complicate the transduction problem. A high density of free carriers under the transducer electrodes screens the applied electric potential, leading to low transduction efficiency. The transduction efficiency can be recovered by using depletion mode transducers [11]. The transducer electrodes form a Schottky contact to the GaAs while a nearby ohmic contact completes a diode structure. Reverse biasing the diode depletes the region under the transducer of mobile carriers thereby eliminating the electric screening effect.

4. Gallium Arsenide SAW Devices

There are two approaches to integrating SAW devices with electronics on GaAs. In the first approach, the piezoelectric SAW device is separate and distinct from the electronic device even though the two devices are electrically connected on the same chip. In the second approach, the fields generated by the piezoelectric device interact directly with the semiconductor charge carriers. The second approach may also include separate electronic or piezoelectric devices in addition to the interacting device.

An example of the first device would be an integrated SAW oscillator including a SAW delay line or resonator, an amplifier, and coupling network on a single chip. As a step toward a monolithic microwave integrated circuit SAW oscillator, 1.5 GHz SAW delay lines have been fabricated on semi-insulating GaAs [12]. To achieve this frequency, the transducer structure used three electrodes per wavelength at a fundamental frequency of 760 MHz. Each transducer consisted of 173 electrodes with 0.62 micrometer lines and spaces, and an acoustic aperture of 350 micrometers.

The untuned delay line insertion loss versus frequency is shown in Fig. 2. The loss at the 760 MHz is 33 dB while at the 1520 MHz overtone the loss is 55 dB. The higher loss at the overtone is due in part to the

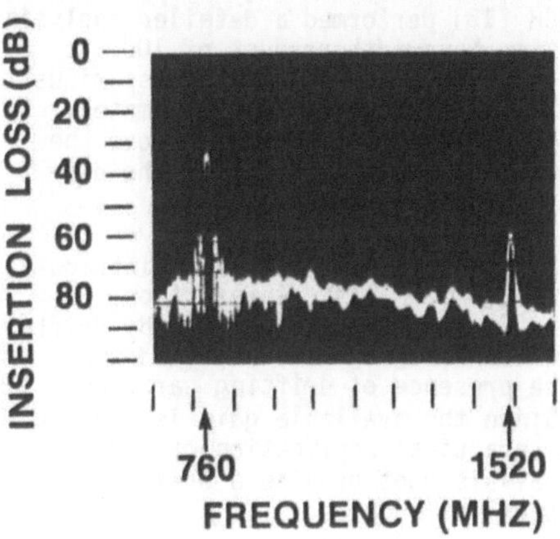

Fig. 2. Gallium arsenide delay line insertion loss

propagation loss increasing as the square of frequency, and in part to the greater effect of parasitics at higher frequencies. These rather high losses would be significantly improved in a monolithic implementation. The monolithic SAW oscillator would not suffer from bond wire nor bond pad parasitics. With further reductions in metalization parasitics, optimization of transducer radiation conductance and appropriate matching, the delay line insertion loss at 1.5 GHz could be reduced to ~ 30 dB. This value of insertion loss would allow fabrication of an oscillator with useful phase noise performance.

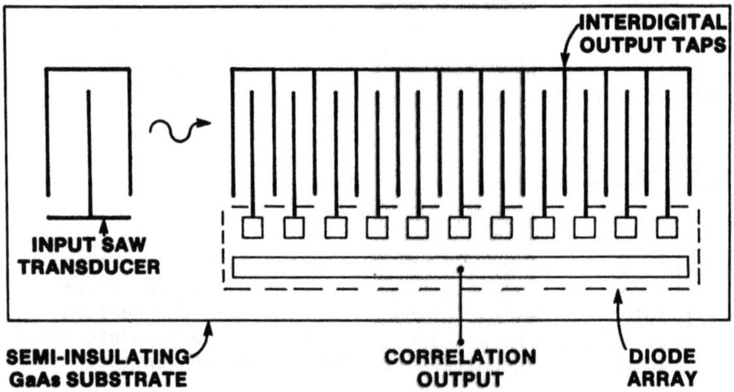

Fig. 3. Gallium arsenide strip coupled memory correlator

Another example of the separated device approach is offered by the GaAs strip coupled memory correlator [13]. This monolithic correlator, shown schematically in Fig. 3, uses conventional interdigital transducers to launch the SAW which then propagates through an interdigital output tap structure. The output taps are connected to an array of Schottky diodes lying outside the acoustic beam. To operate the device, a reference waveform is applied to the device input. As the reference passes under the output array, the diodes are momentarily biased causing a memorization of the reference signal. Subsequently, a signal applied to the input transducer will correlate with the reference; the correlation signal appears at the device output. WAGERS and MELLOCH [13] performed a detailed analysis of this structure and concluded that a time bandwidth product of 1000 is possible. Actual performance was limited by internally generated spurious signals which degrade the dynamic range. The limitations are ultimately traceable to the low electromechanical coupling. Attempts to improve the coupling by zinc oxide overlays have not met with success due to the dispersion and high propagation loss in those films [14].

The direct interaction of SAWs with GaAs charge carriers is an intriguing, though theoretically difficult phenomenon. A basic interaction which offers some theroetical insight is the amplification of SAW by drifting carriers. SWIERKOWSKI, VAN DUZER and TURNER [15] have analyzed the problem of surface waves amplified by the presence of drifting carriers for (100) cut [110] propagating GaAs. Since the available gain is low and transistor amplifiers readily available, practical application of SAW amplifiers is unlikely. However, the analysis does provide a basis for study of more complicated and more practical structures.

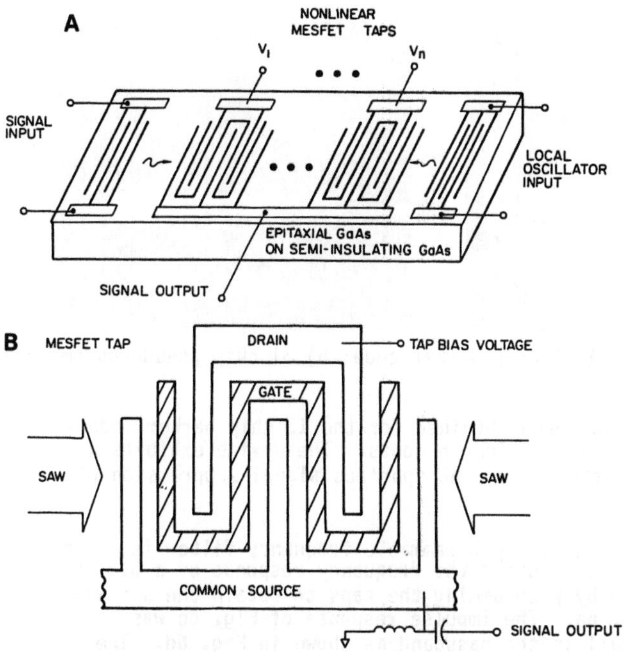

Fig. 4. a) Gallium arsenide programmable filter; b) FET tap structure

One such structure has a SAW propagating through a field effect transistor. The analysis of this configuration remains incomplete, however sophisticated components have been demonstrated. For example, a versatile programmable transversal filter was formed by integrating an array of FET taps in the path of counterpropagating SAWs as shown in Fig. 4a [16]. Depletion mode transducers are used to excite the SAWs with one wave generated by the signal and the other generated by a CW local oscillator. The piezoelectric fields associated with the SAWs interact directly with the fields in the FETs. The FET taps, shown schematically in Fig. 4b, were fabricated on an n-type epitaxial layer with a carrier concentration of $1 \times 10^{16} cm^{-3}$ and a thickness of 1.3 microns. Source and drain contacts are ohmic and the floating gate electrode forms a Schottky barrier. A DC bias voltage applied between the source and drain establishes a quiescent current. The mixing of the counterpropagating SAWs within the FETS produces components of the drain current at the sum and difference of the two SAW frequencies. The nonlinear signal power is proportional to the power carried by the SAWs and can be varied over at least 50 dB by varying the drain bias voltage. Reversing the polarity of the bias voltage shifts the phase of the output signal by 180°. Center-to-center tap spacing corresponds to a sampling frequency of 10 MHz. The taps have a common source electrode for summing the nonlinear signals from the individual taps, but drains and gates are segmented, allowing each tap to be independently programmed. The signal transducer has a center frequency of 127 MHz and a bandwidth of 9.5 MHz while the local oscillator transducer has a center frequency of 110 MHz and a 4 MHz bandwidth. An external bandpass filter selects the output signal at the difference frequency.

When used as a matched filter for biphase signals, the coded waveform is correlated with the code programmed on the taps. Fig. 5 shows near

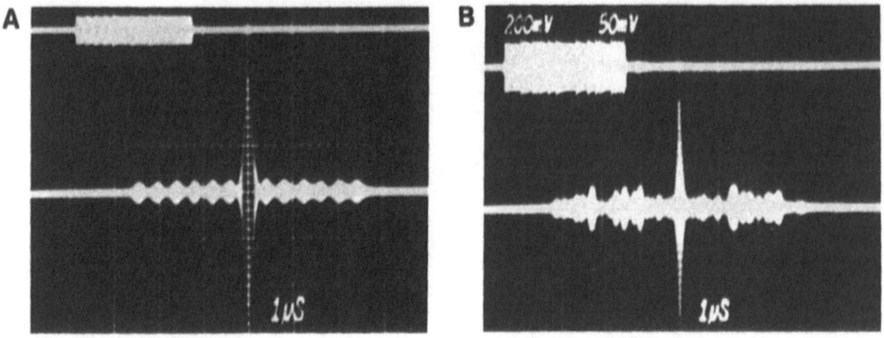

Fig. 5. Correlation of a) 13 chip Barker code; b) 31 chip pseudorandom code

ideal sidelobe response has been obtained for the 13 chip barker code and for 31 chip pseudorandom maximum length codes. The device exhibits a dynamic range of greater than 65 dB and spurious signal suppression of at least 50 dB.

Operation of this device as a programmable frequency filter is demonstrated in Fig. 6. Figure 6b shows the frequency response of a 3.8 MHz bandpass filter realized by programming the taps to obtain the sin x/x impulse response of Fig. 6a. The impulse response of Fig. 6c was calculated to place a null in the passband as shown in Fig. 6d. The 40 dB null would be very effective in suppressing narrowband interference and demonstrates the accuracy that can be achieved with this device.

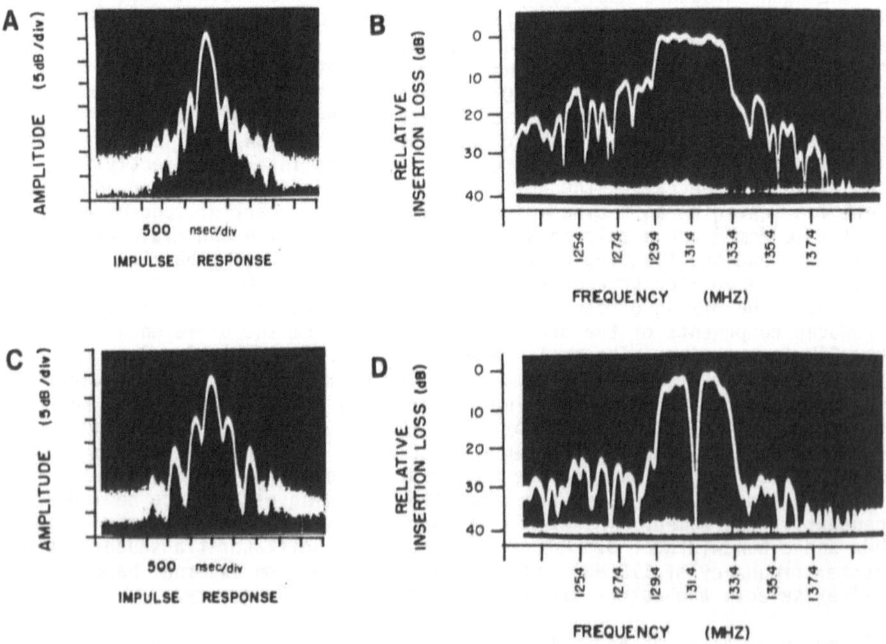

Fig. 6. Programmable filter impulse responses and frequency responses

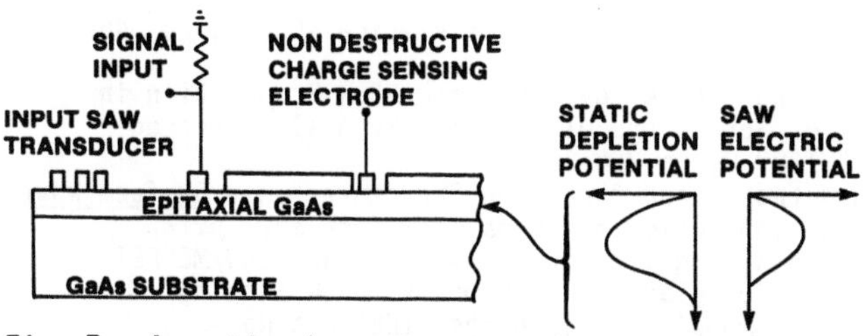

Fig. 7. Acoustic charge transport structure

This technology has been extended to achieve 30 MHz bandwidth at 300 MHz center frequencies [8]. A quadrature arrangement of two programmable delay lines on the same substrate allows programming of each tap signal for amplitude and arbitrary phase. Thus a generalized transversal filter can be formed.

As a final example interacting SAWs and semiconductor charge carriers, consider acoustic charge transport [17]. This new device structure combines features of both SAW devices and charge coupled devices. The acoustic charge transport (ACT) structure is illustrated by Fig. 7. The traveling wave electric potential of the SAW propagating through the depleted semiconductor forms potential wells. A signal injects charges into the well through an input diode which spans the acoustic beam. The traveling potential wells provide a smooth and uniform drift field for the charge packets and the CW sinusoidal SAW creates very little interference or "clock noise" in the signal bandwidth. Charge packets can be nondestructively sensed by capacitive taps deposited on the channel region. A very attractive aspect of this device is that the signal bandwidth is determined by the potential wells sampling the signal at the input diode; the bandwidth can be as high as one-half the SAW frequency. Therefore, the SAW transducer can be very narrowband and it can excite a strong acoustic signal. These features theoretically permit high charge transfer efficiency with wide dynamic range.

5. Conclusion

The piezoelectric semiconductor gallium arsenide is an important Rayleigh wave substrate because it allows monolithic integration SAW devices with electronic devices. The attractiveness of integrated devices inspires research into bandwidth enhancement and temperature compensation techniques in order to overcome inherent limitations in material properties. Even though theoretical understanding of the direct interaction of SAWs with semiconductor devices is immature, high performance prototype devices have been demonstrated.

References

1. R.I. Cottam and G.A. Saunders: J. Phys. C, Sol. State Phys. 6 (13) pp 2105-2118 (1973)
2. R.T. Webster: "Temperature Coefficients of Surface Acoustic Wave Delay on GaAs", in 1981 Ultrasonics Symposium Proceedings, IEEE Cat No 81 CH1689-9, pp 360-363
3. T.W. Grudkowski and M. Gilden: Appl. Phys. Lett. 38 (6), pp 412-413 (1981)
4. J.M. Saurel, J. Lassack, W. Shahab, G. Cambron, and J. Attal: Electronics Letters 17 (10), pp 344-345 (1981)

5. T.C. Lim and B.W. Farnell: J. Acous. Soc. of America 45 (4), pp 845-851 (1968)
6. M. Feldmann, J. Henaff, and M.A. Kirov, "SAW and SSBW Propagation in Gallium Arsenide" in 1981 Sonics and Ultrasonics Symposium Proceedings, IEEE Cat No 81 CH1689-9, pp 264-267
7. P.B. Snow: "Matching Networks and Packaging Structures", in Surface Wave Filters, H. Mathews, ed. John Wiley and Sons, New York 1977 p.227
8. S.W. Merritt, G.K. Montress, and T.W. Grudkowski: "GaAs SAW/MESFET Programmable Asynchronous Correlator with Complex Tap Weighting", in 1983 IEEE Ultrasonics Symposium Proceedings, IEEE Cat No 83 CH1947-1, pp 181-184
9. G.R. Adams, J.D. Jackson, and J.S. Heeks: "Monolithic ZnO-GaAs Acousto Electric Devices" in 1980 IEEE Ultrasonics Symposium Proceedings, IEEE Cat No 80 CH 1602-2, pp 109-112
10. D.E. Oates and R.A. Becker: $LiNbO_3$ Surface Acoustic Wave Edge-Bonded Transducers on ST Quartz and <001> - Cut GaAs", in 1980 IEEE Ultrasonics Symposium Proceedings, IEEE Cat No 80 CH 1602-2, pp 367-370
11. S. Takada, H. Hayakawa, T. Ishiguro, and N. Mikoshiba: Japan J. Appl. Phys 11 (1) pp 24-30 (1972)
12. R.T. Webster: to be published in IEEE Trans Microwave Theory and Techniques (Sep 1985)
13. R.S. Wagers and M.R. Melloch: "GaAs Strip Coupled Memory Correlators", in 1983 IEEE Ultrasonics Symposium Proceedings, IEEE Cat No 83 CH1947-1, pp 377-386
14. R.A. Becker, R.W. Ralston, and P.V. Wright: IEEE Transactions on Sonics and Ultrasonics SU-29 (6), pp 289-298, (1982)
15. S.Swierkowski, T. VanDuzer, and C.W. Turner: IEEE Transactions on Sonics and Ultrasonics SU-20 (3), pp 260-267 (1973)
16. T.W. Grudkowski, G.K. Montress, M. Gilden, and J.F. Black: IEEE Trans on Microwave Theory and Techniques MTT-29 12), pp 13481355 (1981)
17. M.J. Hoskins and B.J. Hunsinger: J Appl Phy 55(2), pp 413-426 (1984)

Part IV

Components and Systems

Rayleigh Wave Linear Filters

*R.S. Wagers**

Central Research Laboratories, Texas Instruments Incorporated, Dallas, TX 75265, USA

1.0 INTRODUCTION

Waves passing through physical structures provide a convenient means for achieving analog approximations to the mathematical operations of convolution and correlation. Because the waves propagate from one position to another, they provide a natural scan through the structure synchronous with the flow of time. It is up to the ingenuity of the designer to provide for interaction between the wave and distributed structures within or on the medium. He must also devise means for carrying out spatial integration of that interaction. Rayleigh waves have proven extremely applicable to this operation. The Rayleigh wave can be excited at one position on the surface of a piezoelectric crystal and allowed to propagate through a succession of electrodes distributed on the surface of the crystal. The electrodes are easily arranged to provide a sampled approximation to the function to be convolved with (or correlated against). For a broad range of parameters, the interactions (products) of the electrodes with the passing Rayleigh wave can be summed at a single bus bar thus approximating integration over a finite range. With carrier frequencies in the UHF range, crystals on the order of a few centimeters in length with metallization features on the order of a few microns can easily accommodate structures capable of performing powerful signal processing functions. It is because of this simplicity of operation, ease of fabrication, reliability in operation, high signal processing performance, low cost, and lack of requisite adjustment that these devices have come into such popular use during the last decade. For high performance convolution and correlation in the VHF to L-band range with many time samples (wide bandwidth), the Rayleigh wave linear filter typically provides results superior to those of any other technical approach.

2.0 RAYLEIGH WAVE FILTERS

2.1 INTRODUCTION

Rayleigh wave linear filters, now commonly referred to as surface acoustic wave filters, are approximations to the transversal filters described by KALLMANN and BLUMLEIN [1], [2]. While Kallmann and Blumlein had no idea of the potential of

---

**Presently: Director, Measurement Instrumentation Technology, Fisher Controls, Austin, Texas 78754.*

Rayleigh wave filters; they did anticipate the practical value of obtaining delays from acoustic media. Observing that the low velocity of sound can yield large delays in short distances, KALLMANN [1] made passing reference to the fact that mechanical analogs should be considered. Today, the most successful mechanical analog is the Rayleigh wave linear filter. These filters provide high performance filtering which is vital to modern signal processors.

Figure 1 illustrates an unapodized interdigital transducer (IDT) and an apodized (weighted) IDT on the surface of a crystal polished to permit high frequency Rayleigh wave propagation. The electrode overlaps of the apodized IDT set the magnitudes of the weighting coefficients of the signals received at the electrodes. The bus bars on either side of the array of electrodes carry out summations of the signals, and each buss bar prescribes a sign for its half of the tap weights. Bipolar summation of the output signal is obtained by connecting one buss bar to one side of the output load resistance and the other buss bar to the opposite side of the resistor.

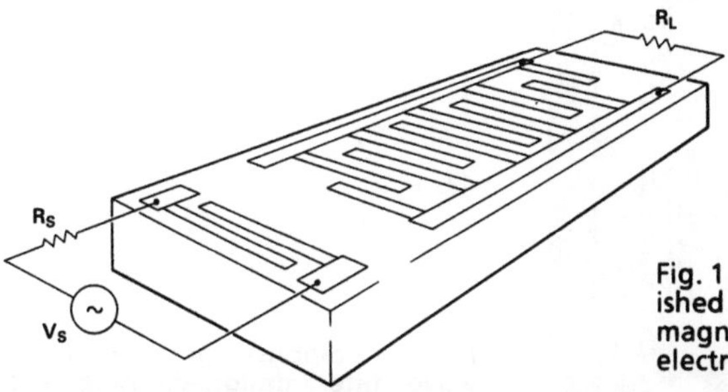

Fig. 1 Transducers on a polished crystal. Tap weight magnitudes are set by the electrode overlaps.

2.2 IMPULSE MODELS

Among the first models of Rayleigh wave linear filters were those described as *impulse* models. TANCRELL and HOLLAND [3] put forth a model in which the signal was assumed to be derived from a sequence of impulses positioned at the edges of the electrodes of Fig. 1. The overall filter response was then expressable as a sum of contributions with frequency-dependent phases corresponding to the propagation delays. Outstanding filter performance was obtainable using this synthesis procedure (Fig. 2) [4]. Note that for the response of Fig. 2 nearly -60 [dB] sidelobes were obtained.

The other popular impulse response model, attributable to HARTMANN et al. [5], represented the effect of the transducer, in the excitation and reception of Rayleigh waves, by placing half cycles of sinusoid between electrodes of the IDT. Thus, in the case of alternating polarity electrodes, the sine wave exciting term would go through one-half cycle from the center of one electrode to the center of the next electrode. If the electrode positions were not uniformly spaced, then the frequency of the sinusoidal excitation was varied such that the pi points of the excitation occurred at the centers of the electrodes. This approach also gave remarkably good experimental results in the synthesis of Rayleigh wave linear filters. Figure 3 shows results obtained from a filter designed by this procedure [6]. With only 10 [dB] of insertion loss and sidelobes 70 [dB] below the peak

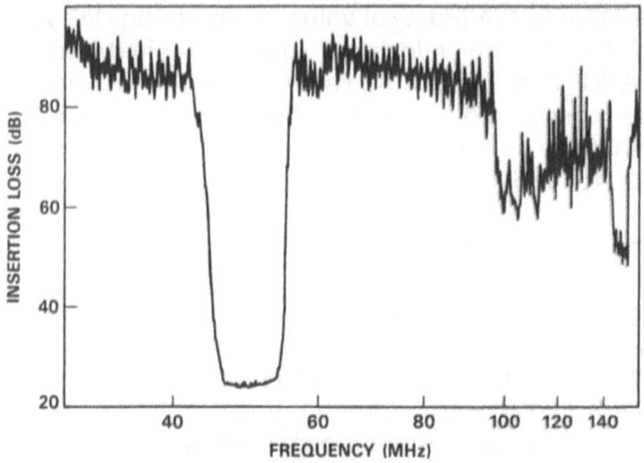

Fig. 2 Insertion loss of a Rayleigh wave filter designed using the delta-function impulse model during the early development of Rayleigh wave filter technology. (Adapted from Tancrell and Engan [1973] © 1973 IEEE).

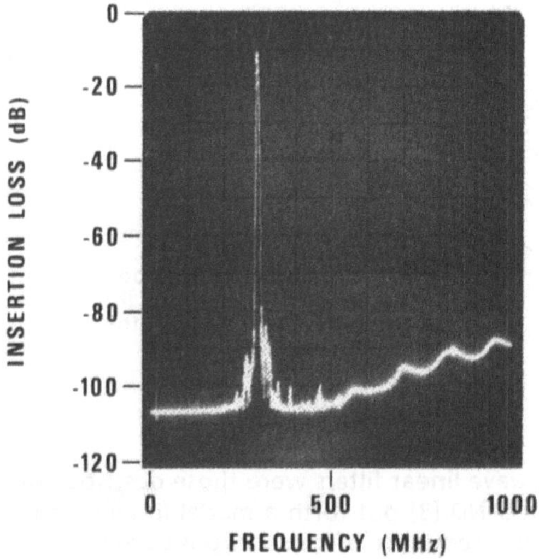

Fig. 3 Response of a Rayleigh wave filter designed using the *sinusoidal impulse* model during the early years of SAW filter technology development (Adapted from Hays and Hartmann [1976] © 1976 IEEE).

response, this filter had truly remarkable properties. To be sure, the filter of Fig. 3 was designed using more than the simple concepts of the impulse response model; the designers applied the art and knowledge that accrues from years of application of any theory. Nevertheless, the fact that a theory as simple as the impulse model can serve as the basis for designing such impressive filters is remarkable.

Enormous mathematical complexity is required to describe the fields associated with a Rayleigh wave propagating in the highly anisotropic and piezoelectric crystals used for modern Rayleigh wave filters. Yet the very simple *impulse* models, described above, have been shown to be relatively accurate in predicting the excitation of Rayleigh waves by IDTs. This has occurred because those simple models have approximated the assumptions implicit in the transversal filter theory. Those two approximations are that the taps not perturb the waves propagating on the delay line, and that the taps approximate spatial impulse sampling. The first of these assumptions has been met in Rayleigh wave filters by making the insertion loss of the filter high. This has been especially true of very high performance Rayleigh wave filters where phase linearity and sidelobe control have been essential. Insertion loss is often sacrificed in order to achieve the stability required. It is not hard to imagine how one makes the coupling weak enough that very minor perturbations occur; spatial impulse response sampling seems a little less obvious, though, when one considers that the taps placed in the acoustic beam are comparable to a wavelength. In fact, what is more important than achieving the equivalent of spatial impulse response sampling is that the local sampling have a spatial resolution comparable to the reciprocal of the information bandwidth of the signal rather than the reciprocal of the instantaneous frequency.

2.3 COMPARISON OF *IMPULSE* MODELS TO NORMAL MODE THEORY

One can show on a rigorous basis by the normal mode theory [7], [8] that the excitation of Rayleigh waves can be described by an integral equation of the form

$$a_+(x) = e^{-j\beta x}\,\frac{j\omega D_+^*(0)}{4P}\int_{-\infty}^{x} e^{j\beta\xi}\phi(0,\xi)\,d\xi \quad , \tag{1}$$

where a_+ is the amplitude of the Rayleigh wave propagating in the x direction; D_+ is the value of the normal component of electric displacement of the Rayleigh wave; P is the magnitude of power flow density in the Rayleigh wave; Φ is the self-consistent electric potential at the surface of the crystal throughout the propagation path of the wave; and x is the space coordinate where the Rayleigh wave amplitude is evaluated. Integration begins at $-\infty$ on a formal basis and is carried out through the entire spatial extent of the propagation path. In reality this integration need originate only at the negative edge of the IDT.

For many Rayleigh wave filters, the first assumption of the transversal filter theory, that of very weak perturbations by the taps, is met through satisfaction of what is called the "weak-coupling approximation." That is, the electric potential associated with the Rayleigh amplitude, a_+, of (1) is very small in comparison to the electric potential imposed by the IDT. In most cases, it is so small that in evaluating the integral of (1), the total self-consistent electric potential, ϕ, can be replaced by the potential due to the IDT alone. If that is done, the integral of (1) can be carried out without recourse to more advanced mathematic formalism such as the solution of Fredholm integral equations.

Both analytic and numerical solutions exist for the electric potential in a number of IDT structures [9] - [12]. Figure 4 shows the exact solution to the electric potential existing at the surface of a crystal with an IDT that has $\lambda/4$ wide electrodes on it. Also shown is the fundamental component of that self-consistent electric potential distribution. Considering (1) and the spatial variation of electric potentials illustrated in Fig. 4, one can appreciate why the *sinusoidal impulse* model [5] has worked as effectively as it has. One knows that the electric potential of an IDT pattern with uniformly placed electrodes, or with slowly varying electrode periodicity, will be represented essentially as shown in Fig. 4. We also know that consistent with the weak-coupling approximation, the

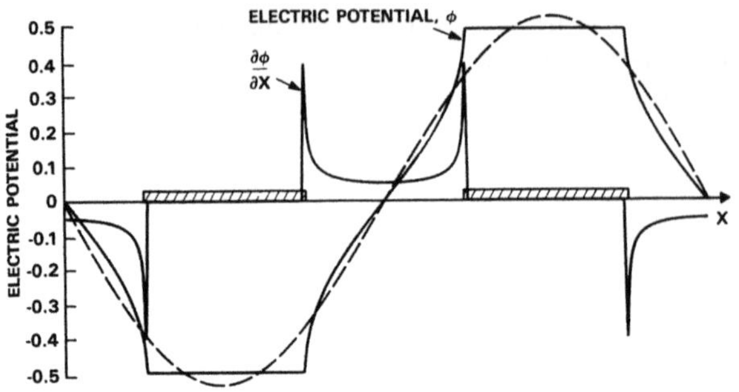

Fig. 4 Electric fields in an interdigital transducer. Solid curves show electric potential, ϕ, and $d\phi/dx$ for one period of the transducer. Dashed curve is fundamental component of electric potential.

amplitude of the Rayleigh wave excited by that interdigital pattern will be rigorously given by (1). To find the Rayleigh wave amplitude, one need perform an integration over the electric potential distribution.

The *sinusoidal impulse* model approximates the rigorous steps described above by calling for the placement of half-sinusoids of signal positioned synchronously with the interdigital pattern. This sinusoidal pattern is then Fourier transformed to obtain the signal frequency response. Notice that (1) is in the form of a Fourier transform of the actual electric potential, and that for $\beta \simeq \beta_0$, the integral of (1) will be approximately the same whether $\phi(0,\xi)$ is represented by the exact field distribution or the fundamental component of that field distribution.

It is clear from (1) and the electric potential illustrations of Fig. 4 why the *sinusoidal impulse* model is effective in the design of some filters, such as that illustrated in Fig. 3. One deficiency of this model can also be seen from (1). Equation (1) shows that the actual amplitude of the Rayleigh wave requires an integration over the actual electric potential distribution that exists throughout the IDT. For frequencies off synchronism, the value of the integral is not well approximated by replacing ϕ by the fundamental component of the distribution. The exact shape of the electric potential distribution is required. Accordingly, the design of filters, such as a TV IF filter which has a relatively large fractional bandwidth, is inaccurate using this model whereas, this model works successfully in the design of narrow fractional bandwidth filters, such as that shown in Fig. 3.

The *delta function impulse response* model can also be understood by considering (1). Using the dispersion relation, $\omega/\beta = v$, and the Fourier transform theory, we can rearrange (1) to obtain

$$a_+(x) = e^{-j\beta x} \frac{vD^*_+(0)}{4P} \int e^{j\beta\xi} \frac{d\phi(0,\xi)}{d\xi} d\xi \ , \qquad (2)$$

where the integration is over the gradient of the electric potential taken along the propagation direction. That gradient is also shown in Fig. 4. Note that the gradient becomes impulsive at the edges of the electrodes. It does not vanish in the regions between the electrodes but goes to a relatively low value. Clearly the integral of (2) is strongly influenced by the value of the potential gradient near the edges of the electrodes, but it is not solely given by that value. One approximation to the potential gradient distribution shown in Fig. 4 is to replace it by a sequence of impulses positioned at the edges of the electrode. That, of course, is the *delta function impulse* model [3] that produced the filter results of Fig. 2. Again, this is an approximation that yields good results near synchronism but becomes less accurate away from synchronism because of the spatial excitation that occurs at positions other than the periodic sampling sites at the edges of the electrodes.

Figure 5 illustrates how both these models depart from the ideal for frequencies away from synchronism. To obtain these results, the integrals of (1) and (2) were evaluated over characteristic lengths of the IDT. In the terminology of antenna theory, the *element factor* was calculated. In the case of the *sinusoidal impulse* model, the integral was carried out from the center of one gap to the center of an adjacent gap with the origin of coordinates taken at the center of the subtended electrode. The integral of the exact electric distribution was calculated as well as the integral of the fundamental component of electric potential. The ratio of the two integrals is what is illustrated in Fig. 5. At synchronism, $\beta = \beta_0$, the two integrals give the same answer and the ratio is equal to unity as Fig. 5 shows. However, for $\beta = 0.5\ \beta_0$ or $\beta = 1.5\ \beta_0$, the *sinusoidal impulse response* gives predictions that are in error by ~ 2%.

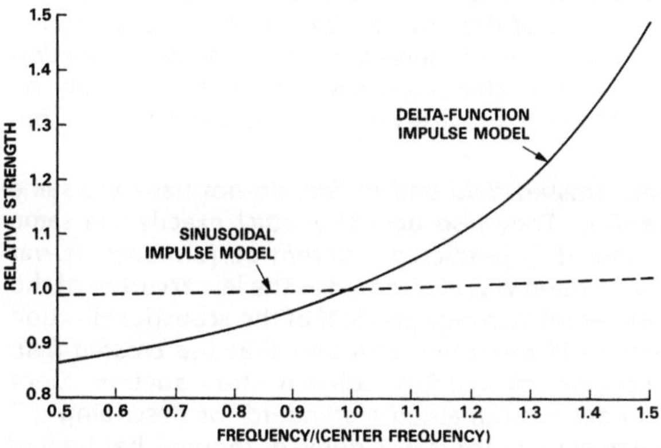

Fig. 5 Frequency variation of element excitation functions (Eqs. 1 and 2) using the impulse models relative to using the exact electric potential.

For the *delta function impulse* model, the integral of (2) was carried out from the center of an electrode to the center of the adjacent electrode with the origin of coordinates at the gap between the two electrodes. The integral of (2) was computed both for the exact potential gradient and for the impulse approximation. In computing the impulse approximation, the impulse amplitudes were normalized such that the integrals computed at synchronism gave the same value. Figure 5 illustrates how the ratio of integrals varies as a function of frequency. At the two extremes of the Fig. 5 plot, $\beta = 0.5\ \beta_0$, and $\beta = 1.5\ \beta_0$, the *delta-function impulse* model [3] gives even more erroneous predictions than does the *sinusoidal impulse response* model. That is because as the frequency is changed, the contribution to the exact integral from regions between the impulses becomes a more significant portion of the total integral. Considering the extreme case where $\beta = 2\beta_0$, the contribution to the integral of

(2) from the impulses at the edges of the electrodes goes to zero; the value of the integral over a single gap is determined solely by values of potential gradient within the gap.

2.4 Bulk Wave Modeling

While the impulse models have been popular and effective for Rayleigh wave filter synthesis, the most popular models of the IDT come from bulk wave analysis. With these models, the IDT is broken up into a cascade of sections. Typically, one might represent an electrode as an individual section and each of the gaps in the transducers as individual sections. Within each section, the fields are regarded as constant, and bulk wave analysis is employed to derive an equivalent circuit for the isolated section. Then the IDT properties are described as a summation of all the individual sections working in concert. In the earliest models [13], the fields were taken as being either parallel to the propagation path (*in-line* model) or normal to the surface of the crystal and to the propagation path (*crossed-field* model). These two models invoke entirely different phenomenological consideration, yet both were relatively successful in the description of Rayleigh wave filter properties. Quite likely, much of the initial success derived from the fact that narrow band filters were examined; these filters have many electrodes in them, and thus, the overall device properties were more a consequence of the *array factor* then the characteristics of an individual element. We have already observed that with the *delta function impulse* model one can obtain good characteristics for the filter simply by considering each element of the filter to provide an impulse of spatial excitation.

The two bulk wave models, *crossed-field* and *in-line*, do not have the same equivalent circuit representation. They also do not predict exactly the same terminal device properties. Since they predicted different performance, it was only a matter of time until one of them was eliminated as the less accurate of the two models. A detailed experimental investigation [14] of the acoustic reflection characteristics of Rayleigh wave IDT operation revealed that the *crossed-field* model was indeed a better approximation to Rayleigh wave transduction. Since that time, the *crossed-field* model has remained a standard for describing IDT operation. The semiempirical way in which it was derived, though, has limited the accuracy of device description. To help remedy this situation, in 1976 the model was improved [11] by eliminating the simple physical approximation of constant fields within any one section; the mathematics was generalized to allow for actual field distribution within the IDT. This greatly added to the accuracy of the model but still would not permit a quantitative description of the IDT properties. As a final step in making the model quantitatively accurate for broadband description of IDT operations, the normal mode theory was used to provide frequency scaling and amplitude normalization for the admittance derived from the more generalized bulk wave *crossed-field* model.

It is truly amazing that physical assumptions as simple as those of the *crossed-field* model could lead to predictions of device performance that are as good as the record shows. One knows that the fields within a section are not constant nor unidirectional and that the lattice motion of the disturbed crystal is much more complex than a simple shearing motion or a simple compressional motion as is usually assumed in the analyses of specific crystals with bulk wave models. Indeed the surface of the crystal executes somewhat elliptical motions that are not even necessarily in the sagittal plane, and the disturbance decays rapidly within the crystal. However, we know that in describing Rayleigh wave excitation all this physical and mathematical complexity can be reduced to a simple integration along the surface of the crystal over the self-consistent electric potential (Eq. 1). Since this complexity can be reduced to very simple forms, it is perhaps not surprising that a simple bulk wave model has been found successful in describing the excitation of Rayleigh waves. In fact, it can be shown [15], [16] that the *crossed-field* bulk wave model leads to excitation mathematics that is identical in form to that predicted by the normal mode theory. Even though the physical processes assumed in deriving the bulk wave models are different from reality and different from those of the normal mode theory, the excitation mathematics implied by that physical model is identical to that which is known to be rigorously true.

3.0 LIMITING FACTORS IN LINEAR RAYLEIGH WAVE FILTERING

3.1 INTRODUCTION

In spite of the success in modeling IDT operation for the excitation of Rayleigh waves, there are limits to the accuracy with which a designer can specify the performance of a filter. Ultimately, all of the models of IDTs that are employed in filter design have accuracy limitations which derive from having made too simple an approximation to the process and/or from neglecting other factors. In general, one finds that control of frequency sidelobes tends to be limited to the -60 [dB] level. Stated in other terms this means that the composite residual error in exciting and detecting the various phasors comprising the filter response is highly resistant to improvement when the error approaches one part in a thousand. This is not an absolute limit, though. For some architectures, shape factors, and insertion losses, the sidelobe levels could be worse than -60 [dB]. Some of the factors that contribute to this error are:

- Diffraction from the finite aperture of the IDT
- Inaccurate modeling of the aperture excitation function
- Insufficient accuracy in modeling the interaction of electrodes with Rayleigh waves
- Neglect of relevant parasitic effects such as ohmic losses and shunt capacitance
- Excitation of other acoustic modes.
- Insufficient control of the crystal axes alignment relative to the crystal surfaces
- Insufficient control of the photomask alignment relative to the crystal

Most of these subjects have been examined in the literature. These inquiries have served to advance the understanding of the subjects and to improve control over the processes. It is another matter, though, to illustrate where fundamental limits lie with regard to the processes that contribute to Rayleigh waves in transversal filters. Limits do exist in the ingenuity of the designers to conceive of physical models for the relevant processes along with quantitative mathematical models that have tractable solutions amenable to their use in synthesis procedures, and then to carry out these modeling activities consistent with the fabrication art and processes of the time.

3.2 PRACTICAL FILTER EXAMPLE

Commonly today high performance Rayleigh wave filters employ multistrip couplers (MSC) between the two IDTs. A typical configuration is illustrated in Fig. 6. The MSC serves to transfer the majority of the Rayleigh wave energy laterally across the crystal from the input transducer to the output transducer while allowing the bulk wave energy launched by the input transducer to continue in a straight path thus missing the output IDT. The MSC also serves to integrate the input acoustic beam which has a nonuniform profile and transmit to the output IDT a beam which has been smoothed in amplitude and phase. This allows the output IDT to be weighted by apodization just as the input IDT had been weighted. In so doing, both IDTs can be pressed to the task of achieving the desired impulse response of the filter.

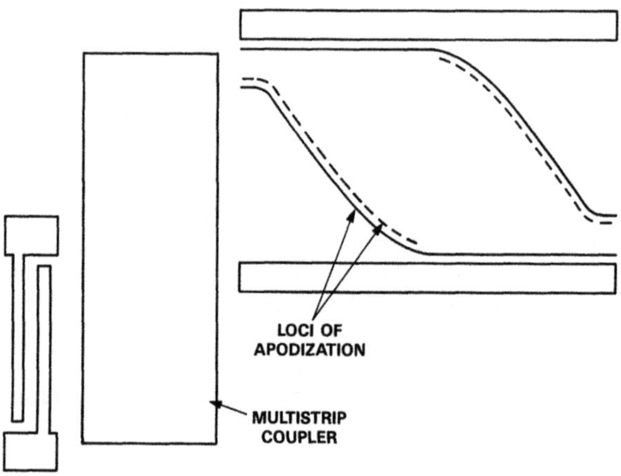

Fig. 6 Layout of Rayleigh wave filter used for diffraction studies and frequency response compensation.

The output signal radiated from the MSC contains all of the errors in modeling and control mentioned above. To the extent that the signal radiated by the MSC does not have a spatially uniform amplitude, or straight-crested phase, or frequency dependence as per the models chosen for the synthesis procedure, then the output characteristics of the filter are degraded. Figure 7 illustrates, for the simplest possible case, how the signal radiated by an MSC departs from the ideal models employed in Rayleigh wave filter design. For the results illustrated in Fig. 7, a filter geometry such as illustrated in Fig. 6 was employed. The radiating IDT was unapodized and had double electrodes. Measurements were made at 94 [MHz] which was the center of the passband of the IDT. The results of Fig. 7 were obtained by a laser probe measurement of the signal both at the input to the MSC and the output of the MSC. The wavelength of the radiated acoustic signal was ~36 microns while the diameter of the laser beam used for measurement of the spatial energy distribution was ~26 microns; thus a small degree of smoothing of the data occurred through convolution of the laser spot with the acoustic profile.

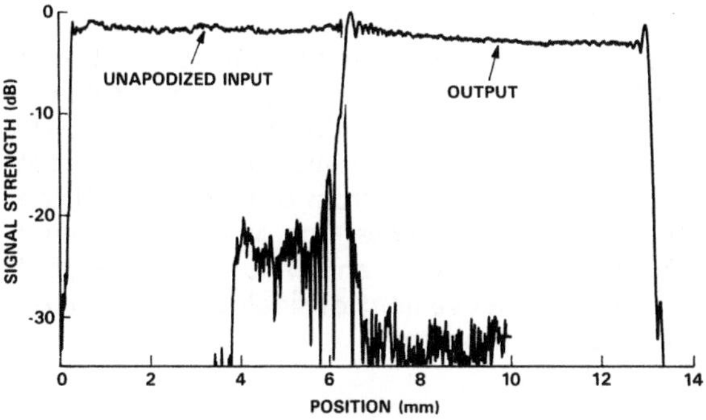

Fig. 7 Acoustic beam profiles at the input and output of an MSC. The input IDT was not apodized.

The left-hand curve of Fig. 7 shows the signal radiated by the input transducer as it enters the MSC. The signal shows peak-to-peak amplitude variation of ~1 [dB]. The phase variation across the input beam (not shown) was less than 10° ptp within the central 90% of the beam. The dynamic range of the laser probe system was ~35 [dB] as can be seen in Fig. 7 by the signal to the right of the input acoustic beam. The output acoustic beam is smoothed by the MSC as expected but is by no means uniform in amplitude. Ringing near the edges of the output acoustic beam is several decibels in amplitude and is not adequately comprehended in any available design procedures. Similarly, the slope in the output acoustic beam, as well as the wiggles evident in the amplitude, are not comprehended in available theory. Yet that profile is the one which excites the output transducer and produces the device frequency response. Uncertainties and errors in that acoustic profile translate directly into lack of control in the frequency sidelobes of a bandpass filter.

The more serious case of acoustic beam profiles that are inadequately comprehended in the current design procedures is illustrated in Fig. 8. These data were obtained from a filter such as illustrated in Fig. 6 where the input transducer had the skewed apodization evident in the figure. The skewing of the apodization was intended to make the exciting acoustic beam more uniform than would occur if the electrode overlaps were arrayed along the center line of the IDT. As the left-hand curve in Fig. 8 shows, the input acoustic profile is not uniform. The amplitude ripple on the input beam is ~3 [dB] ptp (phase ripple was ~20° ptp on the input beam and ~5° ptp on the output beam). Contrasting this is the output acoustic profile, where the spatial variation of signal intensity

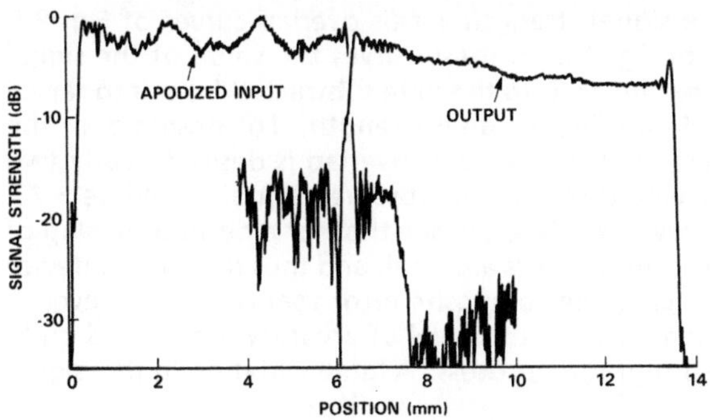

Fig. 8 Acoustic beam profiles at the input and output of an MSC . The input IDT was apodized as illustrated by the right-hand IDT in Fig. 6.

has indeed been smoothed by the MSC. But the slope introduced by the MSC is very much greater than that in the case of the unapodized transducer illustrated in Fig. 7. There is approximately a 5 [dB] roll off across the output acoustic beam. Quite obviously if the output IDT were skewed and apodized like the input IDT then the signal available for exciting the taps in the two small-overlap regions at the ends of the IDT would be substantially different from one another. If the output IDT is apodized symmetrically about the center line of the IDT, then a more nearly ideal relation exists between apodization and impulse response, but still it is sufficiently in error to make the achievement of sidelobe levels lower than -60 [dB] an exceedingly difficult design problem.

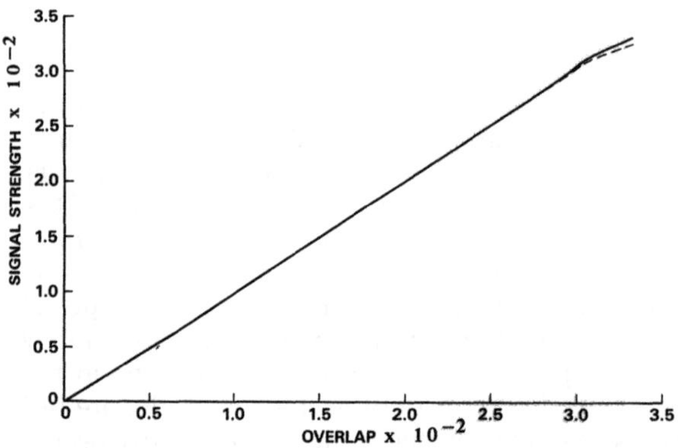

Fig. 9 Integrals of output acoustic beams shown in Figs. 7 and 8. The integration limits were increased symmetrically from the center line of the output beam.

Figure 9 illustrates the signal strength available to an output electrode pair as a function of electrode overlap. For this calculation, the output overlap was symmetrically increased from the center of the output beam. This calculation neglected end effects in the overlapping electrodes and phase ripple in the output acoustic beam. Two results are shown in Fig. 9. One result came from integrating the output signal shown in Fig. 7, and one came from integrating the output beam shown in Fig. 8. Each result was scaled such that a least squares fit to the signal had a slope of 1 in the coordinate system of Fig. 9. Note that even though the two output acoustic beams of Figs. 7 and 8 have severe spatial slopes and not insignificant ripple, the integrals of these acoustic beams from their center lines outward are extremely close to linearly related to the overlap. To the eye no difference can be observed between the unapodized and the apodized results until the edges of the acoustic beam are reached.

The degree of error in the signal strength versus overlap curves of Fig. 9 is illustrated in the two curves of Fig. 10. In these curves the value of the signal strength has had a least squares linear fit to the data subtracted from it to reveal the deviation between actual and linear signal strength. For example, in the case of input with an apodized IDT, if a 75-unit overlap is desired nearly two units of signal strength error would result; the observed signal would be ~73. For larger overlaps, the error reverses. One can see that for a desired overlap of 250, ~252 would be achieved. These errors are small and thus the overall device error introduced is small. Typically the composite error that can be achieved is one part in a thousand. To some extent this level of accuracy is achieved quite fortuitously by the electrode integration process. A large variation in the signal

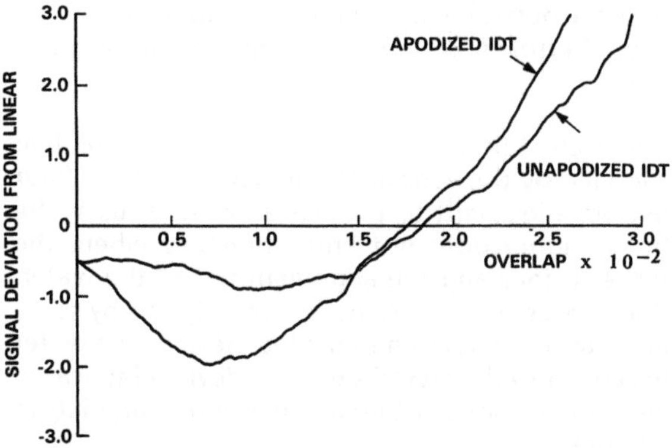

Fig. 10 Deviation of the integrals shown in Fig. 9 from a linear dependence on electrode overlap.

profile (as illustrated in Figs. 7 and 8) still leads to exceedingly linear output tap characteristics.

The sequence of illustrations, Figs. 6-10, exemplify how the collection of effects described above (as limitations to analysis and design) lead to imperfect device performance. But they have not illustrated how each effect has contributed to the overall error. Thus one cannot tell from the results shown here how much crystal or photomask misalignment contributed to the *ears* evident at the edge of the output acoustic beam. Nor can one tell directly what role diffraction played in producing the output results, or to what extent distributing the apodizations to two sides of the acoustic beam (rather than symmetrically about the center line) played a role in improving or hindering the spatial uniformity of the radiated acoustic beam. One knows that for low-performance filters such as those with 40 [dB] sidelobes, these various effects are not significant because of such fortuitous events as those illustrated in Figs. 9 and 10 by the integral of the output acoustic beam. In the more demanding filters exemplified by high shape factor, low in-band ripple, low insertion loss, and 70 [dB] sidelobes, each of the modeling and control factors mentioned above plays a role in setting device performance. These are not fundamental limits; rather, they are practical limits. To assess the role of each of these problems requires the application of perceptive personnel with state-of-the-art resources and a substantial budget. These requirements are difficult to satisfy simultaneously for an extended period of time. Thus, we find that the practical limits have not been substantially improved for some time.

4.0 MISALIGNMENT OF IDT FROM CRYSTAL AXES

One practical limit that has not been addressed in the previous literature is the effect of crystal axis misalignment from the IDT electrodes. A great deal of literature has been written on the effects of diffraction from finite aperture IDTs, and it is well known that if the IDTs are not aligned parallel to a symmetry axis, then the acoustic energy will not travel perpendicular to the phase fronts. It seems that in all previous work on Rayleigh wave filters and their compensation for the effects of diffraction, the crystal axes were aligned as closely as possible to the propagation direction of the acoustic beam during device fabrication, but analyses which assess the impact of error in achieving this alignment have not appeared in the public literature.

Acoustic beam misalignment from the crystal axes can occur due to two effects. First, the crystals, as supplied by the vendor to the Rayleigh wave filter designer, are cut with an imperfect alignment to the crystal axes. Thus at the outset the Rayleigh wave filter designer does not know precisely where the crystal axes lie with respect to the surface and the alignment flat of the wafer. Secondly, during the fabrication process the photomask is misaligned by some amount from the alignment flat. Not only does an individual device on a wafer have imperfect alignment with respect to the crystal axes, but devices fabricated on separate wafers will be different from one another due to nonrepeatability in the photomask alignment procedure.

Figure 11 illustrates how significant this problem of misalignment can be. The results shown there come from a filter with 40 wavelength wide apertures and electrode withdrawal weighting only. The minimum propagation path between inner electrodes of the two IDTs was 35 wavelengths, while the maximum propagation path between outer extremes of the two IDTs was 535 wavelengths. Design procedures for this filter indicated theoretical sidelobes in excess of 60 [dB] below the main response on the low-frequency side of the passband and greater than 70 [dB] down on the high side of the passband. In fact, what was found was that for devices with alignment of the propagation

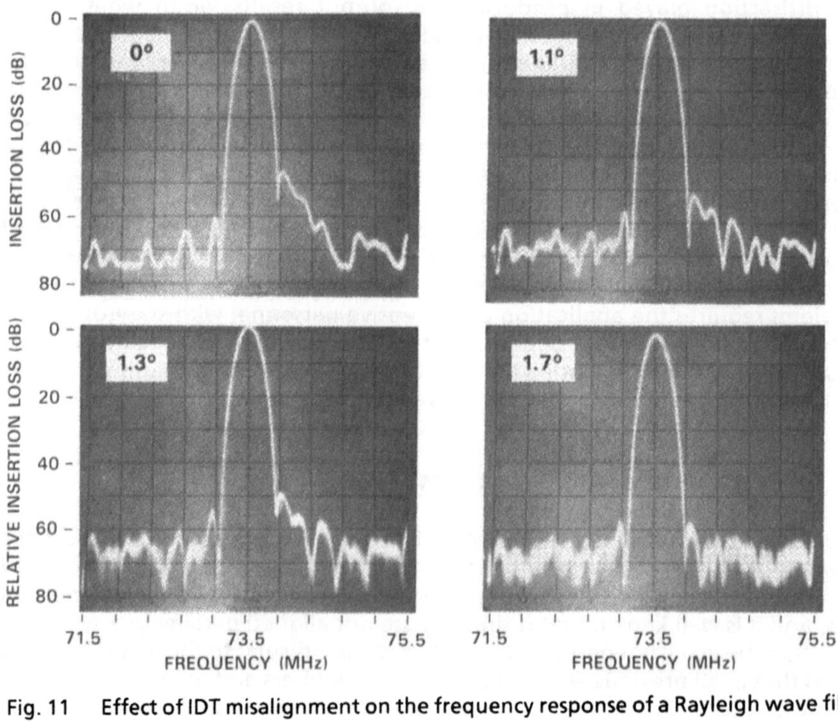

Fig. 11 Effect of IDT misalignment on the frequency response of a Rayleigh wave filter

path and the crystal axes (as near as possible), the high-frequency sidelobes tended to be on the order of -45 [dB] due to diffraction effects [17]. This can be seen in the upper left-hand photograph of Fig. 11. The other three illustrations in Fig. 11 come from filters fabricated on the very same wafer and from the very same photomask pattern. In the fabrication of each filter, the photomask was reregistered a specific number of degrees away from the original alignment that produced the results indicated at 0° in Fig. 11. One can see that the sidelobes on the high-frequency side are altered at all of the misalignments shown in Fig. 11 and that by the time 1.7° of misalignment has occurred, what used to be -45 [dB] sidelobes on the high-frequency side are now -60 [dB] and lower everywhere. Thus, the design of this filter, which was totally unacceptable with proper alignment of the photomask, would become an acceptable design with only slight variation in the manufacturing process.

Intuitively, one can see how photomask misalignment from the crystal axes can lead to altered tap weights; the acoustic beam simply tends to miss the output IDT by a small amount. Quantifying this effect, though, can be exceedingly difficult because, not only is there diffraction to account for, but the variation in tap weight occurs due to changes in the received signal at the ends of the electrodes where electrode end effects play a significant role. Nevertheless, if the electrode end effect is neglected and the diffraction pattern integrated for various degrees of acoustic beam misalignment relative to the crystal axes, then one can obtain guidelines for the magnitude of the effect. In Figs. 12 and 13, this is illustrated for 40 wavelength wide transmitting and receiving IDTs on ST-cut quartz. The abscissa indicates misalignment of the photomask in degrees. In Fig. 12 the ordinate displays change (in percent) in the amplitude of the signal from that at perfect alignment. The effect is illustrated for a succession of propagation distances from 100 wavelengths up to 800 wavelengths. Notice that at 300 wavelengths propagation and 1.7° photomask misalignment, the amplitude error for a 40 wavelength wide tap is on the order of 0.7% of the perfectly aligned amplitude. This means that there is a phasor amplitude error ~43 [dB] down from that of the aligned response. This is consistent with the experimental results illustrated in Fig. 11. For shorter propagation distances and similar misalignments, the errors are smaller but still not insignificant when trying to achieve -60 [dB] sidelobes. With longer

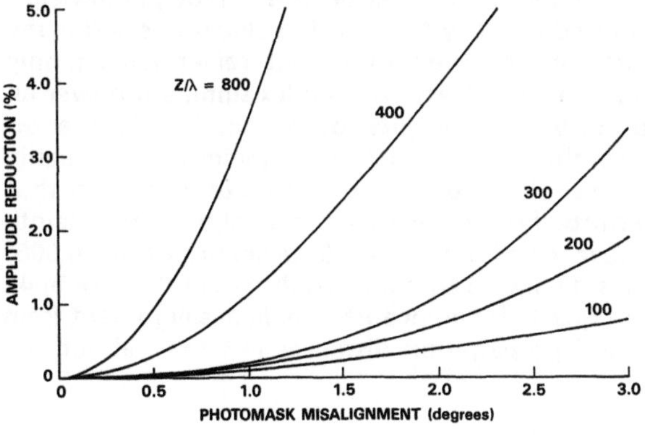

Fig. 12 Theoretical amplitude change in the signal transmitted between a 40 wavelength wide exciter/receiver pair on ST-cut quartz as a function of misalignment from the X-axis of the crystal. Propagation distances of 100, 200, 300, 400 and 800 wavelengths are shown.

propagation distances like 400 wavelengths and 800 wavelengths, even minor misalignments lead to rather substantial errors in the amplitude of the tap signal.

Figure 13 illustrates the phase error concomitant with photomask misalignment from the crystal axis. For propagation distances up to 300 wavelengths, the phase error is small but not insignificant. Even 0.1° phase error in the placement of a phasor produces an error phasor having a magnitude at the -55 [dB] level relative to the desired phasor. As one can see from Fig. 13, for photomask misalignments up to ~2° the tap phase errors associated with propagation distances of less than 300 wavelengths are limited to approximately ± 0.1°. For larger propagation distances such as the 400 wavelength distance illustrated in Fig. 13, the phase error can become quite significant. Note that the phase error associated with the 400 wavelength propagation path reaches 1° at ~1.3° of photomask misalignment. This error in angular placement of the phasor introduces an error phasor in quadrature with, and 35 [dB] below, the desired phasor.

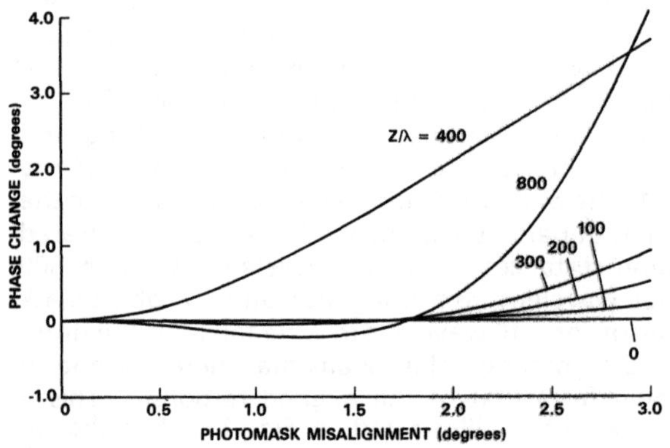

Fig. 13 Theoretical phase change in the signal transmitted between a 40 wavelength wide exciter/ receiver pair on ST-cut quartz as a function of misalignment from the crystalline X-axis. Propagation distances of 100, 200, 300, 400 and 800 wavelengths are shown.

The preceding discussion about phasor errors has been for a single phasor propagated between transmitting and receiving structures. In fact, in Rayleigh wave filters the IDTs are comprised of many transmitting electrodes and many receiving electrodes. The errors from any one transmitter/receiver pair can only indicate the potential magnitude of the effect. A complex summation over all contributions must be carried out to evaluate filter distortion. The fabrication technology at most places is such that 1° of photomask alignment error is easily within their fabrication tolerances. However, if one wanted error phasors that were 80 [dB] below each unperturbed tap value, then the amplitude uncertainty and phase uncertainty would have to be less than 0.0001 in amplitude and 0.006° of phase. One can see from Figs. 12 and 13 that this is only met in the case of 40 wavelength wide IDTs on ST-cut quartz for an acoustic beam misalignment from the x axis of less than 0.25° and propagation distances less than about 300 wavelengths.

5.0 COMPENSATED RAYLEIGH WAVE FILTERS

In spite of the inaccuracy in Rayleigh wave filter modeling, and the limited accuracy in device fabrication, one could extend Rayleigh wave filter performance beyond current levels by employing perturbation techniques [18], [19]. By this approach the Rayleigh wave IDT is physically perturbed and the impact of this perturbation on the filter response is recorded. If a sufficient number of perturbations are performed to create a kind of *basis,* then the difference between the desired device performance and the initially achieved device performance can be expanded in terms of this basis. Subsequent implementation of a new set of perturbations on the filter then yields the desired device performance.

This type of device development has been carried out on filters of (YZ) $LiNbO_3$ [19]. The filters operated at ~94 [MHz] and had a metallization layout similar to that of Fig. 6; for these filters both transducers were apodized with the skewed electrode distribution like that of the upper right-hand IDT in Fig. 6. The design called for IDTs with an impulse length of 0.169 microseconds. The initial electrode overlaps were made substantially larger than those prescribed by TAYLOR [20] weighting (n = 14, 43 [dB] sidelobes). After the Taylor weights were calculated, the taps were set according to the relation T = 2TW/(1 + TW), where T is the final tap weight, TW is the Taylor weight, and the weights are normalized to a maximum of unity. The multistrip coupler, which employed 1/8 wavelength electrodes, was designed by a conventional technique [21]. No special care was taken in the fabrication of this device. To be acceptable for further processing, though, the electrode patterns produced on the $LiNbO_3$ had to be error free.

The unmatched and untuned transmission characteristics of a typical filter are illustrated in Fig. 14. The device has approximately -52 [dB] sidelobes. These levels are approximately consistent with the design models employed; however, in the sidelobe regions there are differences, in both amplitude and phase, between the as-fabricated device results and the analysis predictions. The filter has a moderately wide fractional bandwidth filter of ~11%.

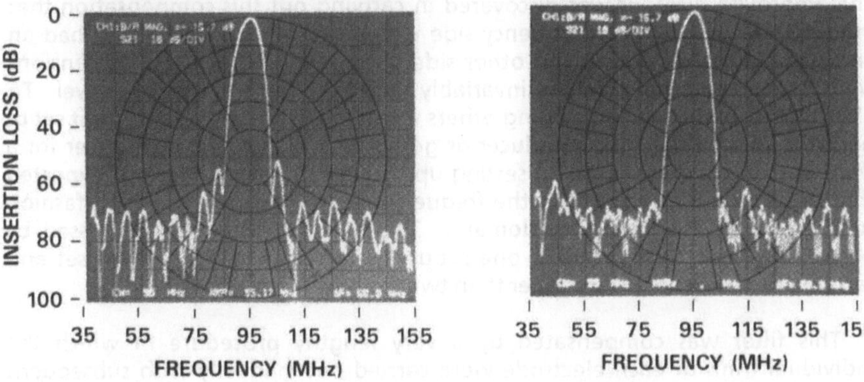

Fig. 14 Rayleigh wave filter frequency responses. Left) Response of filter as initially fabricated. Right) Response after inscribing specified perturbations to the electrode geometry.

The filter then had half of the electrodes in each transducer shortened 5 wavelengths (3% of the 167 wavelength aperture). The locus of trims is shown in Fig. 6 by the dashed lines. After each electrode was shortened, the frequency response of the filter was measured with an automatic network analyzer, and the data were stored in a computer. Each new frequency response of the perturbed IDT was subtracted from the previous frequency response to produce a frequency domain *basis* for expansion of the error function. For this filter, the error function was defined to be the near-in sidelobes of the transmission response. Expansion of the error function was carried out using a weighted least squares optimizer. The optimizing procedure would select contributions from each of the *basis* functions to the total expansion such that the difference between the expansion and the Rayleigh wave error function was driven to a minimum. In the ideal case, this would produce a filter with uniform sidelobes throughout the range of expansion.

Figure 14 shows the filter performance after compensating both input and output IDTs. Notice that, with only one exception, the near-in sidelobes have been uniformly driven down to approximately -70 [dB]. The improvement of ~18 [dB] was accomplished by inducing changes in the tap weights that brought them back near to that prescribed by Taylor weighting. In the case of the filter illustrated in Fig. 14, which had two apodized IDTs, there were no obvious trends evident in the difference between the final apodizations and the apodization prescribed by Taylor weighting. However, for other filters, with an unapodized input IDT and a layout identical to that of Fig. 6, only those taps nearer the margin between input and output beams of the MSC returned to approximately the value predicted by Taylor weighting. Those taps near the outer edge of the acoustic beam were larger by approximately the value needed to overcome the losses shown in Fig. 7.

The one sidelobe that extends above the -70 [dB] level could probably have been compensated too had a different compensation procedure been employed. In the work conducted here, one transducer was compensated to achieve an improved filter and then the second transducer was compensated to achieve the final improved filter. It was discovered in carrying out this compensation that the sidelobe on the high-frequency side which extends up to -64 [dB] had an unfortunate correlation to the other sidelobes and the *basis* set. Expansions which suppressed that sidelobe invariably raised the general sidelobe level. To alter that sidelobe without raising others would have required a different set of perturbations to the first transducer or going back to the first transducer for a third pass of compensation. In setting up this experiment, it was not expected that there would be regions of the frequency domain correlated in this fashion to the rest of the frequency domain. This problem could be addressed by perturbing both transducers in one sequence to create a larger *basis* set and carrying out one expansion rather than two expansions in sequence.

This filter was compensated by a very lengthy procedure in which the individual trims of each electrode were carried out manually with subsequent network analyzer measurements. In the preferred approach to compensation of Rayleigh wave IDTs, the perturbations and measurements would all be conducted automatically. One way to carry out the perturbation to IDTs that could perhaps be automated is to employ nonthermal photochemical deposition of metal [22] on the surface of the crystal. With this approach, the Rayleigh wave filter would be placed in a chamber containing the carrier gas and a laser would be focused on the surface to designate the region where metal should coalesce. Because the laser stimulates the organometallic gas molecules to release metal atoms by a nonthermal means, the surface of the crystal is not warmed and thus, no phase errors are introduced in the device. In this manner, the filter could be altered while being measured. This would not only enable the rapid creation of a *basis* set but would permit a validation of the current *basis* set as compensation progressed through various iterations.

7.0 CONCLUSION

There has not been a high demand for Rayleigh wave filters with rejection ratios greater than 1000:1. That may explain why the general performance level of these filters has not progressed much beyond that point. Whatever the case, the practical limit does turn out to be ~ 60 [dB] for the passband-to-sidelobe ratio and precisely which processes control this level have not been defined. It is likely that many processes contribute to setting this and that they may contribute in comparable measures. While there is certainly room for further refinement of the models used in the design of Rayleigh wave filters, we feel that the most tractable approach to the control of error signals in high performance Rayleigh wave filters is to apply perturbation techniques. We have illustrated [19], that even when performing these procedures manually, improvements of 10-20 [dB] in the sidelobe levels can be obtained. There is no question that this procedure can be automated at this time with mechanical encoders. With the rapidly advancing fabrication art, it seems likely that direct-writing capabilities could be integrated with the compensation procedure in the future.

---

[1] H.E. Kallmann, Proc. IRE **28,** 302 (1940).
[2] A.D. Blumlein, British Patent No. 517516.
[3] R. Tancrell and M. Holland, Proc. IEEE **59,** 393 (1971).
[4] R.H. Tancrell and H. Engan, *Proc. **IEEE** Ultrasonics Symposium*, Monterey, California, 1973 (IEEE, New York, 1973), pp. 419-422.
[5] C.S. Hartmann, D.T. Bell, Jr., and R.C. Rosenfeld, IEEE Trans. Microwave Theory and Techniques **MTT-21**, 162 (1973).
[6] R.M. Hays and C.S. Hartmann, Proc. IEEE **64,** 652 (1976).
[7] R.S. Wagers, IEEE Trans. Sonics & Ultrasonics **SU-25,** 85 (1978).
[8] R.S. Wagers, in *Precision Frequency Control*, Vol. 1, E. Gerber and A. Ballato, eds. (Academic Press, New York, 1985), p. 228.
[9] H. Engan, IEEE Trans. Electron Devices **ED-16,** 1014 (1969).
[10] C.S. Hartmann and B.G. Secrest, *Proc. **IEEE** Ultrasonics Symposium*, Boston, MA, 1972 (IEEE, New York, 1972), p. 413.
[11] W.R. Smith and W.F. Pedler, IEEE Trans. Microwave Theory and Techniques **MTT-23,** 853 (1975).
[12] R.S. Wagers, IEEE Trans. Sonics & Ultrasonics **SU-26,** 105 (1979).
[13] W.R. Smith, H.M. Gerard, J.H. Collins, T.M. Reeder, and H.J. Shaw, IEEE Trans. on Microwave Theory and Techniques **MTT-17,** 856 (1969).

[14]W.R. Smith, Jr., *Proc.* ***IEEE*** *Ultrasonics Symposium*, Monterey, California, 1973 (IEEE, New York, 1973), pp. 410-413.
[15]R.S. Wagers, "Physical Processes in Surface Acoustic Wave Interdigital Transducers - Part I: Normal Mode Analysis," Texas Instruments, Central Research Laboratory Technical Report No. 08/80/29 (29 September 1980).
[16]R.S. Wagers, "Physical Processes in Surface Acoustic Wave Interdigital Transducers - Part II: Bulk Wave Modeling," Texas Instruments, Central Research Laboratory Technical Report No. 08/80/36 (17 December 1980).
[17]R.S. Wagers, IEEE Trans. Sonics and Ultrasonics **SU-23**, 249 (1976).
[18]R.S. Wagers and R.W. Cohn, *Proc.* ***IEEE*** *Ultrasonics Symposium*, Atlanta, Georgia, 1983 (IEEE, New York, 1983), pp. 5-10.
[19]R.W. Cohn and R.S. Wagers, Appl. Phys. Lett. (Submitted for publication).
[20]T.T. Taylor, IRE Trans. Antennas and Propagation **AP-3**, 16 (1955).
[21]F.G. Marshall, C.O. Newton and E.G.S. Paige, IEEE Trans. Microwave Theory and Techniques **MTT-21**, 206 (1973).
[22]V.S. Dolat, D.J. Ehrlich, and J.Y. Tsao, ***IEEE*** *Ultrasonics Symposium*, Dallas, Texas, 1984 (IEEE, New York, 1984), Paper No. I-3.

Rayleigh Wave Resonators and Oscillators

J.J. Gagnepain

Laboratoire de Physique et Métrologie des Oscillateurs du C.N.R.S., associé à l'Université de Franche-Comté-Besançon, 32, avenue de l'Observatoire, F-25000 Besançon, France

1. Introduction

A resonator is a device, which can store energy by means of steady waves. Its quality factor is directly related to the stored energy over dissipated energy ratio. In many applications, and in particular in oscillators, high Q-factors are desirable since they have a large influence on the frequency stability. Steady waves suppose wave reflections. But unfortunately Rayleigh waves cannot be reflected with efficiency at the edge of a plate, since they do not verify such a boundary condition, and are mainly transformed into bulk waves by mode conversion. Therefore a different principle must be used.

It was known that interdigital transducers could reflect a surface wave, but the first publication regarding grating reflectors was presented by E.A. ASH in 1970 [1], with the first resonator. The reflection can come from electrical perturbation of the surface, i.e., the shorting out of the piezoelectric field. But for materials with low electromechanical coupling factors stronger effects are obtained by mechanical perturbations : mass loading strips [1], grooves [2], ion implanted strips [3], metal diffused strip [4] have been used.

Two identical reflectors placed on the surface define a SAW cavity which is coupled to an external circuit by using an interdigital transducer with electrodes positioned at the peaks of the electric potential. The Q-factor of this device will be limited by different loss mechanisms, due to the acoustic attenuation of the substrate, the diffraction of the finite width beam, the imperfectly reflecting grating. In fact the propagation loss of the substrate corresponds to the ultimate limit of the Q-factor.

In an oscillator, the resonator reduces the influence of the phase fluctuations, induced in the oscillating loop by the different components, the electronics and the resonator itself. The phase-to-frequency conversion maintains the 2π phase condition of the loop. This conversion is inversely proportional to the resonator Q-factor. Therefore higher Q-factors lead to lower frequency fluctuations of the output signal. Spectral purity, short term stability, long term aging are the main characteristics of the oscillator. However the resonance frequency of the resonator can be directly affected by thermal and mechanical perturbations. The sensitivities to temperature and temperature graidents, to accelerations, to forces, to pressures, remain today important problems, for instance in airborne applications. Desensitizing SAW oscillators is a large part of the efforts, but as a natural byproduct applications to sensors also are in progress. SAW oscillators and resonators cannot be considered independantly of the equivalent BAW devices. Both have their advantages and inconvenients. Comparisons will be made, and partial conclusions will be proposed.

2. Resonator Configurations

A surface grating reflector is a corrugated structure, which therefore is dispersive [5,6]. The theoretical dispersion curve of an infinite array of grooves is represented in Fig. 1. When the wave frequency is in the range corresponding to the stop band (for instance the π stop band) the structure behaves as a reflector. For an infinite array this band does not have leakage into bulk waves. For a finite one, leakage occurs near the beginning and the end of the array, but it is minimized.

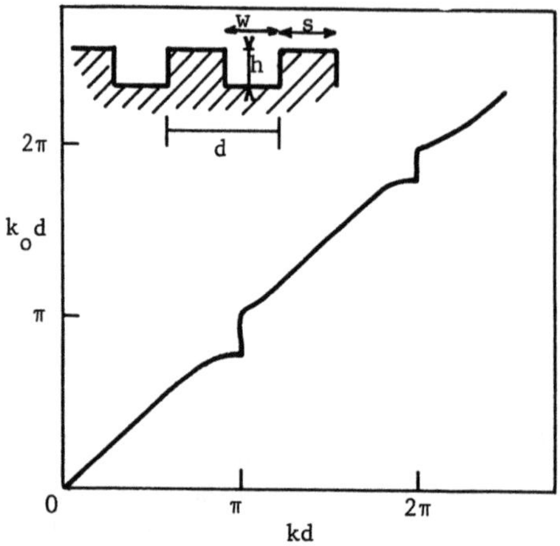

Fig. 1. Theoretical dispersion curve for an infinite array of grooves of period d. k_o is the unperturbed Rayleigh wave wave number, and k is the wave number of the wave on the periodic structure.

The stop band property of the array was shown by using a delay line made of two interdigital transducers with a single grating reflector between. The stop band effect is illustrated in Fig. 2 where it appears as a notch on the regular $(\sin x/x)^2$ type response of the delay line [2]. The reflection coefficient is maximum at a frequency $\hat{f}$, which is close to the synchronous frequency $f_c = V/2d$, where V is the velocity of the unperturbed wave, as shown in Fig. 3, and decreases to zero over a frequency range corresponding to a fractional bandwith of the order of 2/N, where N is the number of elements of the structure. On the same figure is given the phase of the reflection coefficient. This phase value is important, since in a resonator configuration it determines the position of the two reflectors and the transducer [7,8,9]. Different techniques were used for the elements of the reflector. Interdigital electrodes, made of a metallic layer, which are electrically shorted out are easy to achieve, but will have a good reflection coefficient only on substrates with high electromechanical coupling factor. If not, the efficiency can be improved by using the mass loading effect of thicker metal strips. Ion implantation modifies the elastic properties and the density of the substrate. Therefore the wave velocity change and the surface step at the boundary of the implanted zone were used for obtaining periodic gratings made of implanted strips [3]. Metal diffused gratings were also studied [4]. Today grooves, made by ion etching, seem to have the main advantages, in terms of efficiency and of lower aging mechanisms.

Different geometries of Fabry-Perot type cavities are possible for resonators, as shown in Fig. 4. The one-port or two-port geometries of Fig. 4a and

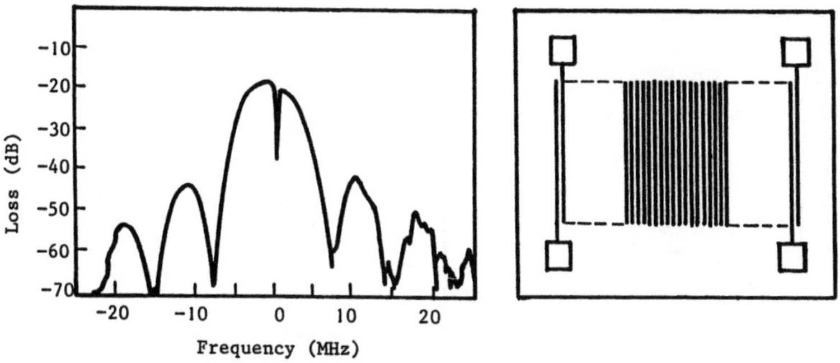

Fig. 2. Experimental response of a delay line with a grating reflector between the transducers (ref. 2)

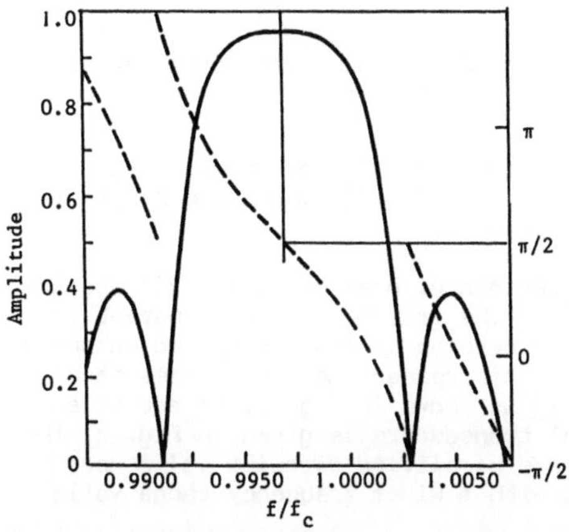

Fig.3

Fig.4

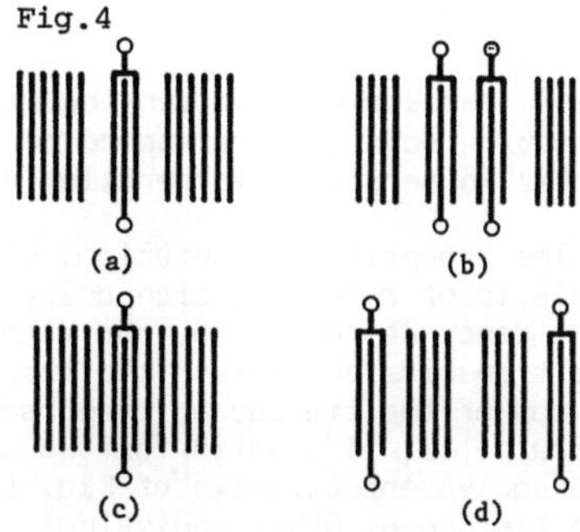

Fig. 3. Amplitude and phase of the reflection coefficient of a groove reflector on Y-Z $LiNbO_3$ (h/λ = 0.015, w/s = 1.0), from ref. 6

Fig. 4. Resonator geometries. (a),(b) : conventional one-port and two-port resonators. (c),(d) : unconventional

4b are most commonly used. Fig. 4c corresponds to a single grating ; the transducer induces an additional phase shift (due to mass loading and electrical shorting) and thus an additional resonance in the stop band. Fig. 4d shows a configuration which, in particular, enables testing of the characteristics of the cavity, when used as a delay line.

3. Resonator Characteristics

A typical admittance curve of a one-port resonator is shown in Fig. 5. This admittance is similar to the bulk wave resonator's one and presents resonance and antiresonance. The resonance occurs when the phase interference is constructive ; this is when the length of the cavity (the distance between the effective centers of the reflectors) is an integral number of half wavelengths. The one-port resonator can be represented by an equivalent circuit (Fig. 6) composed of a series resonant circuit L_1C_1, with a capacitance C_0 in parallel corresponding to the interelectrode capacitance of the transducer.

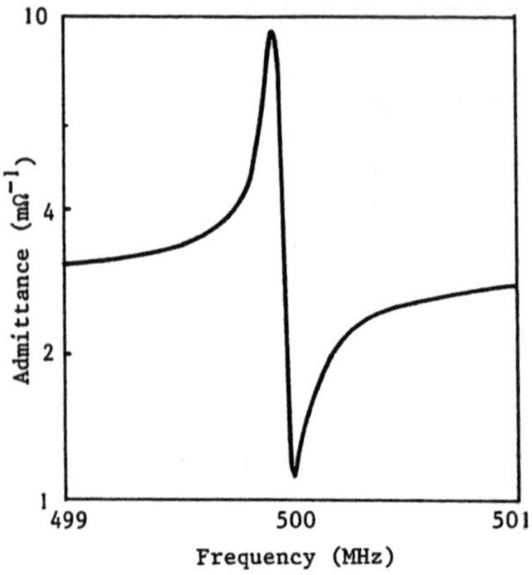

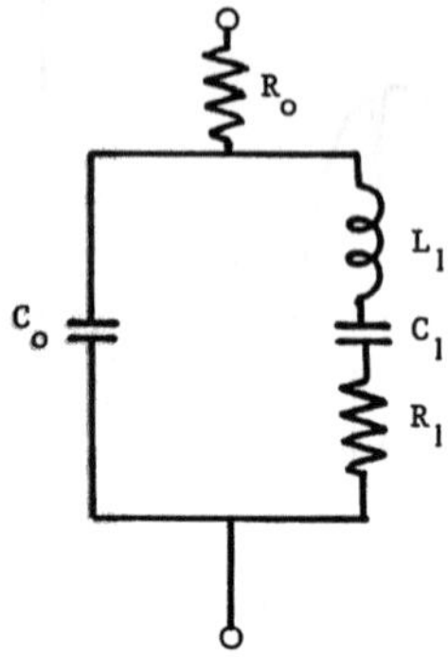

Fig. 5. Admittance curve of a one-port resonator

Fig. 6. One-port resonator equivalent circuit

The R_1 resistance involves the different losses, and R_o represents the acoustic radiation resistance of the transducer [7, 10], which can be placed either in series or in parallel with the circuit.

The two-port configuration, when off resonance, has the $(\sin x/x)^2$ characteristic of a transmission delay line since the gratings are transparent in that case. At resonance, the standing wave sets up in the cavity and enhances the transmission between the transducers. This appears as a sharp resonance on top of the transducer response [10, 11] as shown in Fig. 7. An equivalent circuit [12, 13], valid for two identical transducers is given in Fig. 8. The two equivalent circuits of Fig. 6 and 8 are simplified circuits valid near the resonance. Other equivalent circuits with a wider frequency range valid have been also proposed [10, 14].

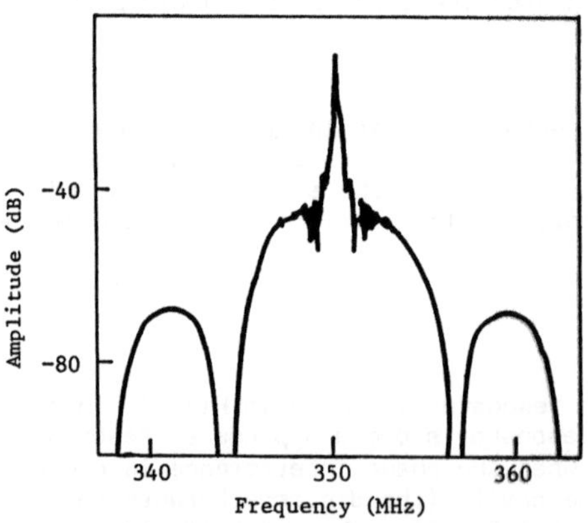

Fig. 7

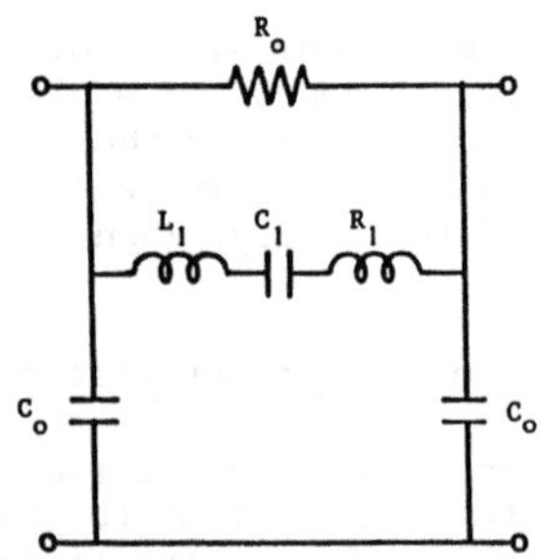

Fig. 8

Fig. 7. Frequency response of a two-port resonator

Fig. 8. Equivalent circuit of a two-port resonator with identical transducers, near resonance

The Q-factor of the resonator, which is the main parameter, is limited by different loss mechanisms. These losses can be inherent to the resonator structure or related to the coupling with the outside. This leads to the definition of unloaded and loaded Q-factors. Since the different losses are added, one can write [15]

$$\frac{1}{Q} = \frac{1}{Q_m} + \frac{1}{Q_d} + \frac{1}{Q_r} + \frac{1}{Q_b} \tag{1}$$

for the unloaded Q-factor. With our definition the loaded Q will correspond to the value when loading the resonator with an external electrical circuit.

- Q_m corresponds to the acoustic attenuation of the material and can be considered as the ultimate limitation of Q-factor. If the device is not under vacuum, air loading must be added to $1/Q_m$ [16]. Air loading losses are frequency independent, whereas the viscous acoustic attenuation is proportional to f^2. For the last one the product Q.f depends only on the crystallographic orientation of the material and the wave propagation direction. This is illustrated in Fig. 9. But in fact the devices are sealed in enclosures under vacuum and air loading is eliminated.

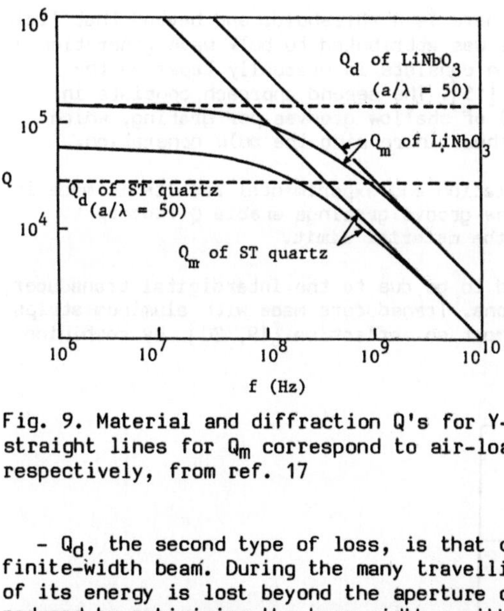

Fig. 9. Material and diffraction Q's for Y-Z $LiNbO_3$ and ST quartz. Curved and straight lines for Q_m correspond to air-loaded and evacuated conditions, respectively, from ref. 17

- Q_d, the second type of loss, is that due to the diffraction of the finite-width beam. During the many travellings, the beam spreads and a part of its energy is lost beyond the aperture of the reflectors. This can be reduced by optimizing the beam width a with respect to the wavelength λ. It has been shown that Q_d is proportional to $(a/\lambda)^2$. Values of Q_d are also given in Fig. 9. Diffraction can also be generated when the wave is propagating on the grating. This will depend on the type of grating, i.e. on the relative speed of the wave when on the grating, with respect to the unperturbed Rayleigh wave velocity, and higher transverse modes can be excited [17, 18].

Two other important sources of losses due to the gratings arise from the leakage of energy, when they are imperfectly reflecting (Q_r), and from the mode conversion into bulk waves (Q_b). The reflecting coefficient can be improved by using deeper grooves in a short grating, or by using a larger number of shallow grooves. As shown in Fig. 10 the Q-factor is improved when

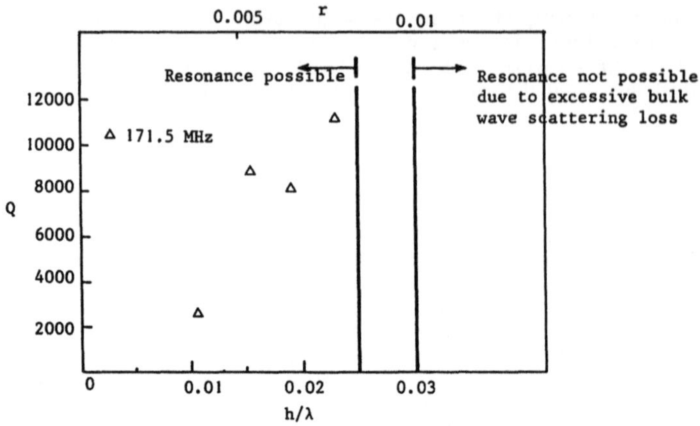

Fig. 10. Measured Q's at 170 MHz and bulk-wave limit on groove depth for resonators on Y-Z $LiNbO_3$, from ref. 15

increasing the groove depth. But there is a threshold, and beyond that limit the resonance can even vanish. This was attributed to bulk wave generation at the front of the grating. A solution consists in gradually tapering the groove depth up to its final value [15]. The second approach consists in using a much larger number (≈ 1000) of shallow grooves per grating, which could minimize the leakage loss without increasing the bulk generation.

Comparison of theoretical calculation and experimental results is made in Fig. 11. This indicates that shallow groove gratings enable Q-factors to be achieved with values closer to the material limit.

Another source of loss was found to be due to the interdigital transducer, which can induce spurious reflections. Transducers made with aluminum strips recessed in quartz grooves give almost no reflection [19, 20]. By combining

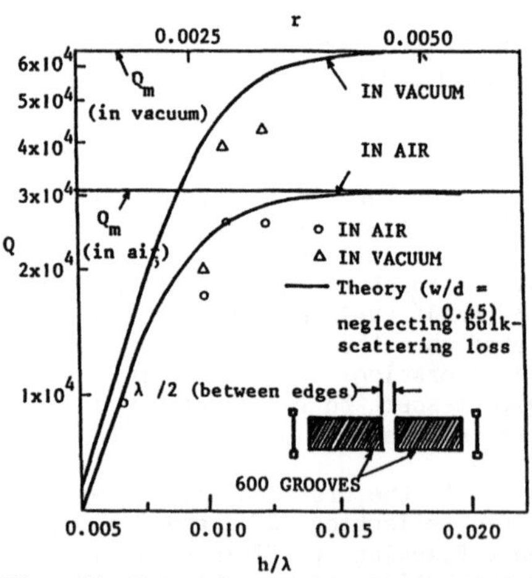

Fig. 11. Comparison of measurement and calculation of Q-factors on ST quartz, from ref. 5

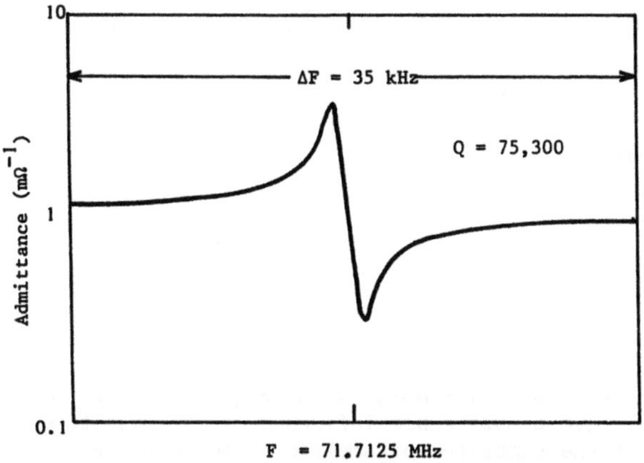

Fig. 12. Admittance characteristic about resonance frequency of one-port resonator on quartz. The series resonant Q is 75 300, from ref. 21

shallow groove gratings and recessed transducers, Q-factor as high as 75 000 at 70 MHz were achieved [21] as shown in Fig. 12.

Surface wave and bulk wave resonators were complementary on the frequency range. However, recently the use of ion etching or chemical etching for processing thin membrane enabled to achieve bulk wave resonators up to 1 GHz. Therefore the frequency ranges of the two kinds of devices now overlap, and comparison of their characteristics can be made, as shown in Table I.

Table I. Comparison of the Q-factors of BAW and SAW resonators

frequency		5 MHz	100 MHz	1 GHz
Q	BAW	3×10^6	10^5	$1\text{-}3\times10^4$
	SAW		8×10^4	$0.3\text{-}1\times10^4$

4. <u>Oscillator</u>

The oscillator can be simply represented as composed of an oscillating loop, which contains a resonator and a sustaining amplifier, and of an output or buffer amplifier, as shown in Fig. 13. Different noise sources are localized in the electronics and in the resonator.

The noises are classified into internal noise, i.e. noise sources in the oscillating loop, and external noise, i.e. noise sources in the output amplifier.

Also distinction is made between additive noise, which is generated by white thermal noise sources and directly added to the signal, and parametric noise which modifies the values of the parameters defining the phase or the frequency.

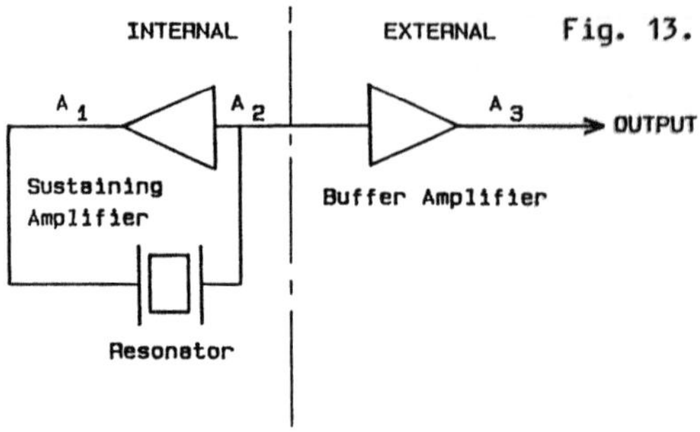

Fig. 13. Schematic diagram of an oscillator

Therefore only the components of additive noise with frequency close to the oscillating frequency will have a noticeable influence on the amplitude and the phase of the signal. At the opposite, parametric noise is a low frequency noise, which induces phase and frequency modulation of the signal.

a) Internal additive noise

In response to any phase perturbation in the oscillating loop there will be a frequency change in order to maintain the 2π phase condition of the loop. Such a readjustment will be mainly made by the element which presents the maximum phase-frequency sensitivity, i.e., the resonator. This phase-frequency characteristic can be written as

$$\phi \simeq -2Q \left(\frac{f - f_o}{f_o}\right) \quad \text{for} \quad f \simeq f_o \tag{2}$$

where f is the driving frequency, f_o the resonator's resonance frequency and Q its loaded quality factor.

Wide band thermal noise sources are located on each side of the resonator, as indicated on Fig. 1 by points A_1 and A_2. The behaviours of these two sources will present some differences because the noise of the first source (point A_1), called input noise, will entirely be injected in the loop, whereas the noise of the second source (point A_2), called output noise, partly is injected in the loop and partly goes directly to the buffer amplifier.

The contribution to the phase power spectral density of the output signal is given by [22, 23]

$$S_\phi(F) \propto \frac{kT}{PQ^2} \left(\frac{f_o^2}{F^2} + 4Q^2\right) \tag{3}$$

where F is the Fourier frequency, T the temperature, k the Boltzmann's constant and P the power in the loop.

In time domain the corresponding Allan variance of the fractional frequency fluctuations $y = \Delta f/f_o$ is

$$\sigma_y^2(\tau) \propto \frac{kT}{PQ^2} \left(\frac{1}{2\tau} + \frac{3Q^2F_h}{\pi^2 f_o^2 \tau^2}\right) \tag{4}$$

τ is the measurement time interval and F_h is the bandwidth of the output amplifier and/or of the measurement system.

The phase spectrum (relation 3) is composed of two parts, the one in F^{-2} and the other one in F^{0}. They intersect at $F_c = f_o/2Q$, which corresponds to the half bandwidth of the loaded resonator. The corresponding stability $\Delta f/f = \sigma_y(\tau)$ presents a $\tau^{-1/2}$ and a τ^{-1} dependence versus time. The $\tau^{-1/2}$ dependence law is characteristic of internal additive noise.

b) External additive noise

This noise is associated with the output circuits, outside of the oscillating loop. If the total noise power is small in comparison with the total signal power, half the noise power appears as phase modulation and half as amplitude modulation.

The broadband thermal noise voltage (with spectral density S_o) will be filtered by the limited bandwidth of the buffer amplifier, and can be represented at the output (point A_3 of Fig. 13) by the relation

$$S_N = \frac{S_o}{1 + 4\left(\frac{F - f}{F_h}\right)^2} \tag{5}$$

where F_h is the bandwidth of the circuit.

The corresponding spectrum of the output signal is

$$S_\phi(f) = \frac{2}{\pi F_h}\,\frac{P_N}{P_S}\,\frac{1}{1 + 4\,(F^2/F_h^2)} \tag{6}$$

P_S is the total signal power, and P_N the total noise power.

In time domain

$$\sigma_y(\tau) = \frac{\sqrt{2}}{2\pi f_o \tau}\left(\frac{3\,P_N}{\pi\,P_S}\right)^{1/2} \qquad \text{for} \qquad \tau > \frac{1}{\pi F_h} \tag{7}$$

and

$$\sigma_y(\tau) = \frac{1}{2\pi f_o}\left(\frac{\pi\,F_h\,P_N}{\tau\,P_S}\right)^{1/2} \qquad \text{for} \quad \tau < \frac{1}{\pi F_h}\quad . \tag{8}$$

If not using a very narrow band filter, only relations (7) will generally apply for usual values of τ.

c) Parametric noise

This noise is present as well in electronics, where it induces phase modulation, as in the resonator, where it is responsible for direct modulation of its resonance frequency.

Phase noise

As for thermal additive noise, these parametric phase fluctuations are transformed into frequency fluctuations in order to maintain the phase condition of the loop. Distinction can be made again between input and output phase noises. These low frequency fluctuations, essentially 1/F fluctuations, give therefore $1/F^3$ and $1/F$ phase spectra at the oscillator output. The two curves intersect at $F_c = f_o/2Q$. In time domain $\sigma_y(\tau)$ will respectively show a flicker floor and a $1/\tau$ variation law.

Frequency noise

The sensitive element generally being the resonator, frequency parametric noise results from a modulation of the resonance frequency of this last one by internal or external perturbation.

If $S_{y_0}(F)$ is the power spectral density of the fractional frequency fluctuations of the resonator, the phase spectrum of the output signal is given by

$$S_\phi(f) = \frac{f_0^2 \, S_{y_0}(F)}{F^2} \quad . \tag{9}$$

Two kinds of frequency noise generally are encountered : flicker noise proportional to $1/F$ and random walk noise proportional to $1/F^2$. The corresponding phase spectra respectively are proportional to $1/F^3$ and $1/F^4$.

In time domain the Allan variance $\sigma_y(\tau)$ therefore will present again a flicker floor and a $\tau^{1/2}$ variation law.

d) Frequency drift

Large efforts have been spend for many years for modeling frequency drift, in fact without any definitive success because too many uncontrolled parameters are involved in drift processes.

However in case of a linear drift

$$\frac{df}{f_0} = at \tag{10}$$

which is valid at least for short periods of time, the corresponding $\sigma_y(\tau)$ in time domain is

$$\sigma_y(\tau) = \frac{a}{\sqrt{2}} \tau \tag{11}$$

which is just proportional to τ.

The relative contributions of these different noise sources are summarized in Fig. 14 and 15 respectively for frequency and time domains.

In Fig. 16 are presented the one-sided power spectral densities of the phase noise of four oscillators utilizing a 1.2 GHz SAW delay line [24], a

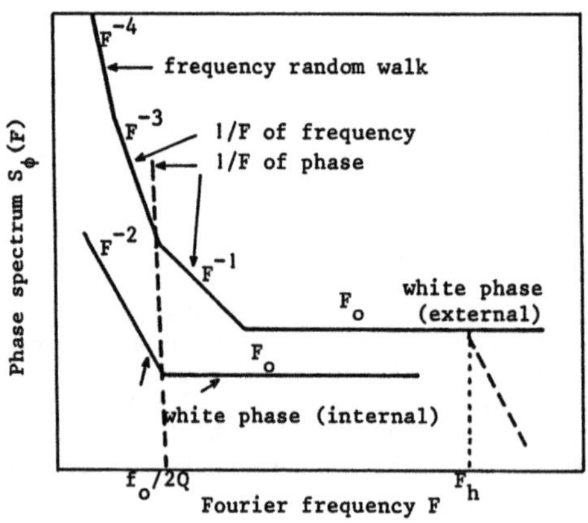

Fig. 14. Noise contributions in frequency domain

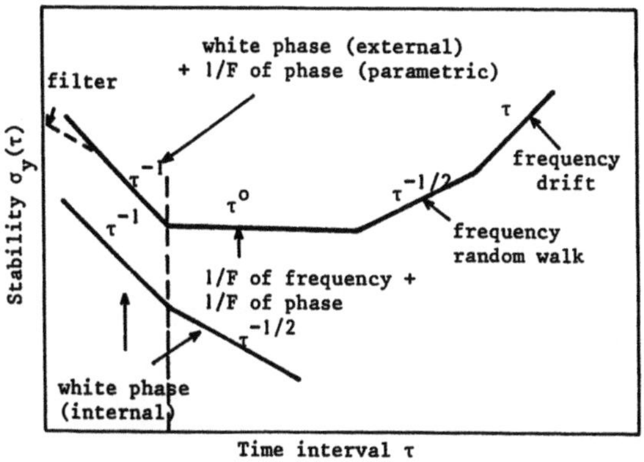

Fig. 15. Noise contributions in time domain

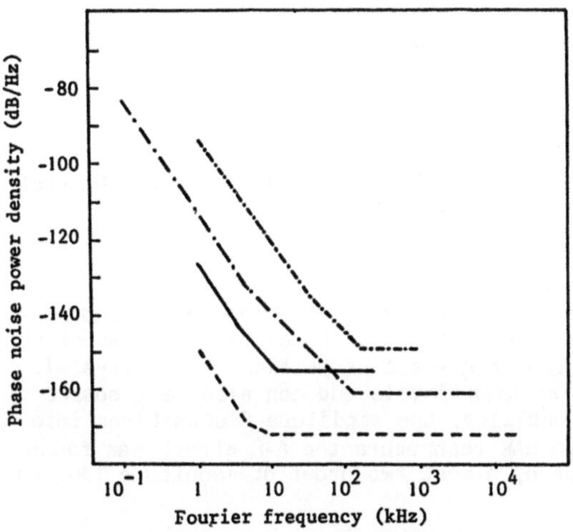

Fig. 16. Comparison of the phase spectral density of SAW and BAW oscillators
--- 250 MHz SAW resonator (ref. 26) —— 489 MHz BAW resonator (ref. 27)
—·— 401 MHz SAW delay-line (ref. 25) -.-. 1.2 GHz SAW delay-line (ref. 24)

401 MHz SAW delay line [25], a 250 MHz SAW resonator [26] and a 489 MHz BAW resonator [27]. The 1.2 Hz oscillator exhibits essentially internal additive noise characterized by $1/F^2$ and white spectra which in principle intersects at the equivalent half bandwidth of the delay line, defined from the group delay. (The higher noise floor which is observed in the present case can be explained by additional noise due to the measurement system). The three other oscillators present also internal additive noise, and in addition a $1/F^3$ spectrum, which is due to parametric noise, either to 1/F phase (electronics) or to 1/F frequency (delay line or resonator). The level of phase fluctuations due to internal additive noise depends on Q-factor, following relation (3) ; lower Q's enhance the $1/F^2$ spectrum. On the other hand it has been shown that the 1/F frequency fluctuations of a resonator are related to its unloaded Q-factor, following a $1/Q^4$ dependence [28].

The stability in time domain is illustrated by the curve of Fig. 17, which corresponds to a 160 MHz SAW-resonator oscillator [29]. This short term

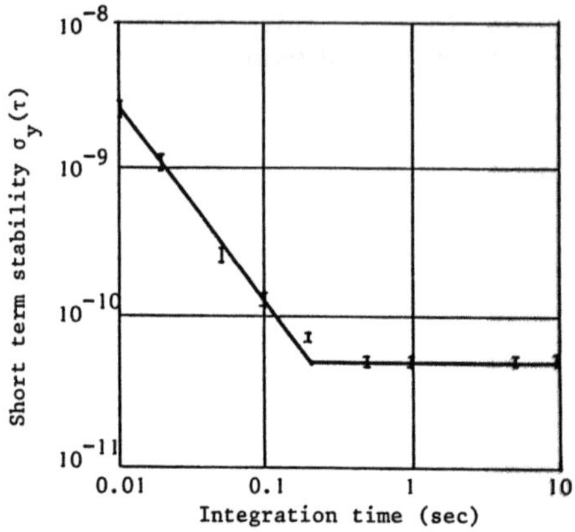

Fig. 17. Short term stability of a 160 MHz SAW resonator or oscillator, from ref. 29.

stability is characterized by $\sigma_y(\tau) \approx 10^{-11}/\tau$ (for $\tau < 0.3$ sec) and by a flicker floor at 5×10^{-11} (for $\tau > 0.3$ sec). These results are among the best ones which have been achieved with SAW oscillators.

e) Amplitude-frequency effect

The frequency of a resonant or oscillating system is not completely independent of the driving or oscillating level. This amplitude-frequency dependence is due, in a resonator, to the nonlinear elastic properties of the crystal. This effect can prevent operation at high levels and can also be a source of noise since it transforms, in the oscillor, the amplitude fluctuations into frequency fluctuations. Measured on SAW resonators the A-F effect was found to be smaller than in BAW resonator by almost two order of magnitude [30] as shown in table II.

Table II - Amplitude-Frequency effect $\Delta f/f_o = kI^2$ (k : A-F coefficient, I : driving current)

	cut	frequency	Q-factor	A-F coefficient
SAW	ST	110 MHz	26 000	$1.1\times10^{-3}/A^2$
BAW	AT	100 MHz	63 000	$2.5\times10^{-1}/A^2$
BAW	AT	5 MHz	2×10^6	$2\times10^{-1}/A^2$

This property offers the possibility of operating the oscillator at rather high level, therefore with a better signal/noise ratio and enables higher spectral purity.

f) Long term aging

The frequency stability of the oscillator over large periods of time, i.e. days, months or years, is limited by aging phenomena of the components and mainly of the resonator. There is a long experience of that problem in BAW resonators, and certainly SAW can still learn from BAW [31]. The first aging measurements were made without the necessary care in cleaning, vacuum, enclosure, mounting, etc and were not so relevant. More recently [32,33,34, 35,36] the aging of delay lines and resonators was measured over long periods of time (two years) with devices under vacuum, controlled in temperature and with the minimum of possible pollutions. The different data, which can be collected, indicate aging of the order of a few ppm in average and exceptionally lower than 1 ppm [37]. This is illustrated in Table III. The data on SAW resonators were obtained from ref. 37. In addition are given data on BAW resonators at 488 MHz and at 5 MHz.

Table III - Long term aging of resonators

Devices	Power [dBm]	Aging (averaged) [ppm]	Ref.
SAW resonators	10	$1.3\ 10^{-6}$ / 37 weeks	37
416 MHz	0	3.10^{-7} / 37 weeks	
BAW resonators 488 MHz	-3	5.10^{-7} / year	27
BAW resonators 5 MHz	-30	10^{-8} - 10^{-10}/year	

5. Sensitivity to the environment

The previous spectral purities and stabilities were measured in laboratory conditions, by keeping the oscillators at temperatures as constant as possible, and by protecting them from mechanical perturbations.

In real conditions the oscillators can be submitted to drastic temperature variations, to accelerations, vibrations, forces, pressures, etc, which induce frequency shifts in the resonator. This sensitivity comes from the nonlinear properties of the crystal and in particular from the fourth order elastic constants [38,39,40,41,42]. The anisotropy of the crystal gives some possibilities for reducing these effects, by choosing particular crystallographic orientations and propagation directions.

Temperature

When the substrate is submitted to temperature variations its elastic properties are modified. There is a direct change of elastic constants characterized by their fundamental temperature coefficients, and a bias is induced by thermal expansion which also modifies the properties on account of the nonlinear effects. Both contributions are of the same order of magnitude. The second one is neglected with satisfactory approximation only if the crystal is free to expand and if the Bechmann's phenomenological elastic constant T.C.'s are used [43].

Calculations of SAW temperature sensitivity were performed for all crystal configurations defined by the two cut angles (Φ and θ) and propagation direction Ψ. ST-cut is today the most widely used. It corresponds to a zero-first-

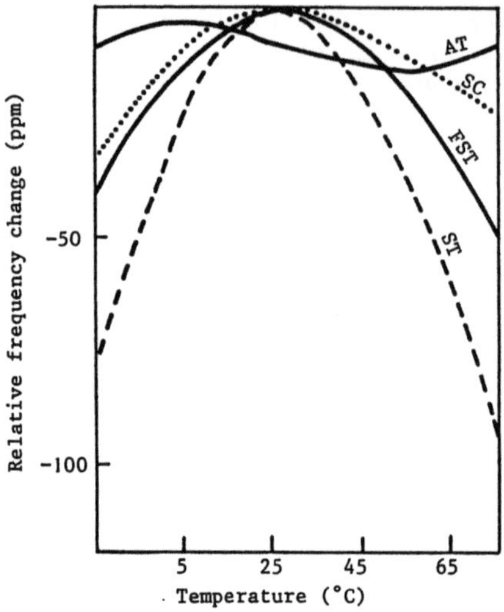

Fig. 18. Comparison of BAW and SAW frequency-temperature characteristics

order temperature coefficient, and presents a parabolic frequency-temperature characteristics. In spite of the efforts it was not possible to find a cut comparable to the AT or SC ones of bulk waves, with cubic characteristics. Only a reduction of the second order temperature coefficient, by a factor 2, was possible with FST cut [44]. Comparison of BAW and SAW cuts is made in Fig. 18.

Thermal properties described above correspond to a static thermal behavior, i.e., the temperature distribution within the substrate is homogeneous. When the crystal is submitted to fast temperature fluctuations, temperature gradients take place and induce thermal stresses and strains. This is known as the dynamic thermal behavior. A phenomenological model of the thermal dynamic effect was proposed by Ballato [45]. This model corresponds to frequency variations directly proportional to the time derivative of the external temperature : $df/f_0 = \tilde{a}\ dT/dt$.

This behavior is illustrated in Fig. 19, where are given the frequency variations of a SAW delay-line oscillator as a function of temperature for

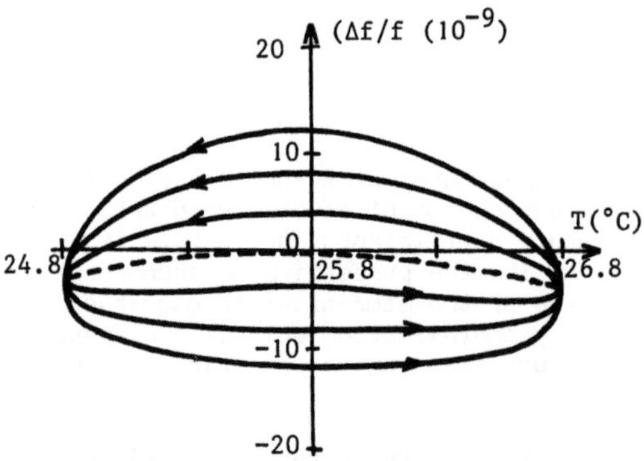

Fig. 19. Dynamic frequency-temperature characteristics of ST-cut delay line oscillator. Temperature variation rate : (a) 1.8 K/mn ; (b) 3.6 K/mn ; (c) 5.4 K/mn. Dotted line : static characteristic

different temperature variation rates. Large frequency variations can be induced, which do not follow anymore the static characteristics. The $\tilde{a}$ coefficient of ST-cut was found to be of the order of -10^{-7} sec/K. As a comparison the value of the BAW AT-cut is -10^{-5} sec/K and of BAW SC-cut is 10^{-7} sec/K. In fact, SC-cut was calculated for compensating this effect.

Axial and bending forces

The sensitivity of rectangular plates (delay-lines) clamped at one edge and submitted to uniformly distributed axial and bending forces at the other edge is given in Table IV. Such effects are often induced in the mounting of the device and the sensitivities are large enough to create spurious frequency shifts. Circular plates were also studied, and as shown in Fig. 20, for SAW and BAW the effect is minimized for two directions of force applications.

Table IV - Sensitivity (Hz/N) of a rectangular quartz plate (length 17 mm, thickness 2 mm, width 15 mm) submitted to axial and bending forces. The surface wave is propagating along Oa_1 at 100 MHz

	Axial force		bending force	
	ST,X	Y,X	ST,X	Y,X
measured	58	90	-1140	-740
calculated	54	45	-1200	-930

Acceleration fields

The first measurements of the G-sensitivity of a SAW oscillator show a sensitivity of the order of 10^{-7}/G, which is larger by two orders of magnitude than that of BAW resonators [46]. In fact it strongly depends on the mounting structure. Different types of mounting were tested, as shown in Fig. 21.

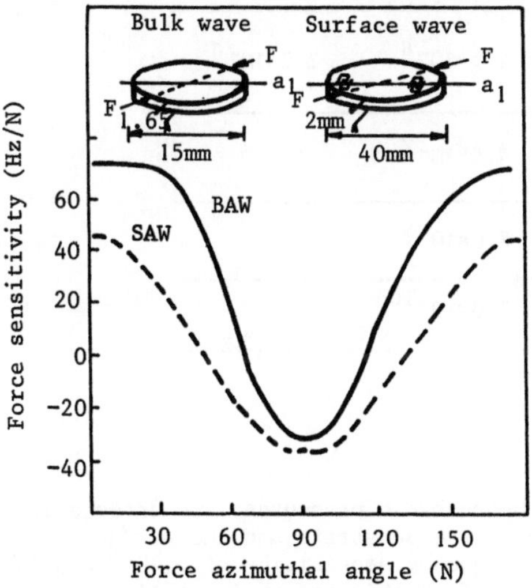

Fig. 20. Comparison between bulk and surface waves submitted to diametrically applied forces, from ref. 49

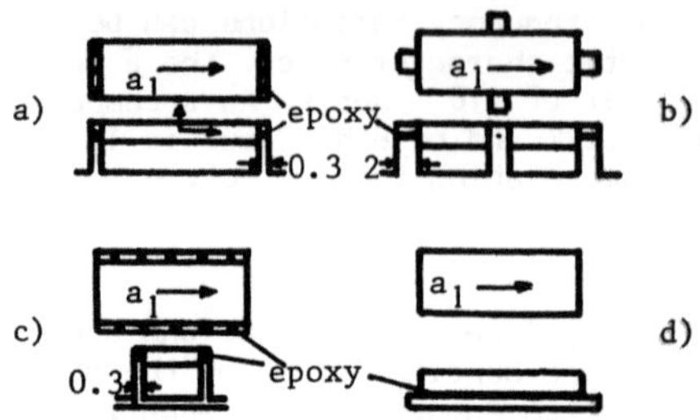

Fig. 21. Four different configurations for SAW acceleration field sensitivity. 16a : L = 33 mm W = 13 mm 2h = 2 mm. 16b,c,d : L = 40 mm W = 10 mm 2h = 2 mm, normal acceleration field Γ, from ref. 42

The corresponding G-sensitivities are given in Table V, in addition with the sensitivities of 5 MHz BAW resonators. It appears that SAW oscillators still have a large sensitivity to accelerations, and even when reduced it remains much larger than that of BAW oscillators.

Table V - Calculated and measured acceleration sensitivities (at 105 MHz, propagation along Oa_1) as a function of the mounting configuration

	mounting	crystal cut	G-sensitivities for 1 G: measured	G-sensitivities for 1 G: calculated
SAW 105 MHz	2 supports along Z_3 (fig. 21a)	AT	1.8×10^{-7}	2.0×10^{-7}
	4 supports (fig. 21b)	ST	7.3×10^{-8}	8.2×10^{-8}
	2 supports along Z_1 (fig. 21c)	ST	4.5×10^{-8}	2.2×10^{-8}
	Epoxy (fig. 21d)	ST	4.5×10^{-7}	
BAW 5 MHz	2 supports	AT	2.0×10^{-9}	
	BVA	SC	3.0×10^{-10}	

6. Hydrostatic Pressure

Rayleigh waves are also sensitive to hydrostatic pressures. Two effects are combining, the mechanical deformation of the substrate and the air/gas loading effect. This gives a fractional frequency sensitivity of 3.10^{-8}/kPa [47] (measured on a 100 MHz delay line). Therefore this effect is small, and is generally neglegible and especially if the device is in an enclosure.

7. Coriolis Force

Since a Rayleigh wave has an elliptical polarization, it can experience Coriolis acceleration when placed in rotation [48]. The fractional frequency shift is proportional to the rotation angular frequency, and inversely proportional to the SAW frequency. For ST-cut quartz at 100 MHz the sensitivity is of the order of 2×10^{-9} for a rotation of 1 rad/sec.

8. Sensor applications

Since it is possible to make the Rayleigh wave selectively sensitive to a given physical quantity, it can be used as a transducer which transforms the variations of that physical quantity into frequency variations. Following this approach, temperature, force, pressure sensors and accelerometers have been studied [49]. The main characteristics are summarized in the following tables. In table VI are given the sensitivities of temperature sensors. SAW and BAW have similar fractional sensitivities, but on account of the different configurations of these waves and therefore of the mounting of the devices, the time constant of the SAW sensor is as small as 0.3 sec (this corresponds to the conduction time constant measured in a liquid).

The accelerometer is in fact a force sensor ; the acceleration is transformed into a force by means of an auxiliary inertial mass.

In this example the sensitive direction corresponds to F_2. The directions F_1 (compression) and F_2 (flexure) correspond to transverse sensitivities, which must be reduced as far as possible.

Table VI - Temperature sensors * at 100 MHz ** at 28 MHz.

quartz cuts	orientation Φ	θ	Ψ	sensitivity	time constant
LST	11.4 °	59.4 °	35.25 °	33 ppm/K*	0.3 sec
JCL	0	42.1 °	39 °	22 ppm/K*	
LC (BAW)	11.1 °	9.4 °		30 ppm/K**	6 sec

Table VII - Accelerometer sensitivities along the three directions

sensitivity [Hz/G]	cuts Y,X	AT,X	ST,X
Compression $\vec{F}_1$	6.34	7.2	8
Flexure $\vec{F}_2$	1 877	1 352	1 387
Flexure $\vec{F}_3$	2.97	3.47×10^{-2}	3.3×10^{-2}

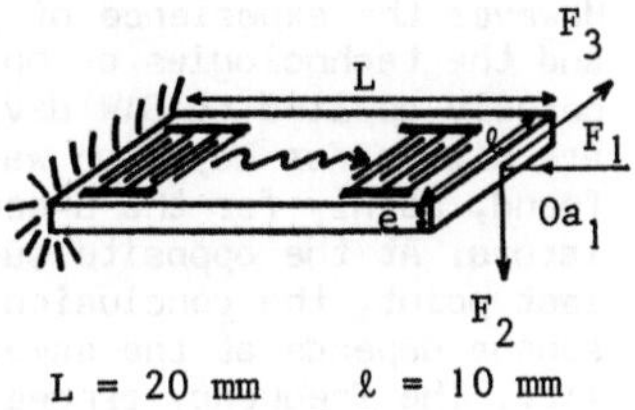

L = 20 mm ℓ = 10 mm
e = 0,5 mm

The sensitivity to hydrostatic pressure, which was given in the previous section is too low for sensor application. A solution was used, which consists in applying the pressure on a thin diaphragm in order to induce a flexure deformation [49,50]. The sensitivity depends on the location of the SAW delay line on the diaphragm, as shown in Fig. 22, and it is possible to inverse its sign. This enables operation with two delay-lines, or resonators, on the same substrate in a differentiel way, with a global sensitivity of 400 Hz/kPa at 100 MHz. A structure similar to a diaphragm, loaded at the center by an additional mass was also used as an accelerometer [51].

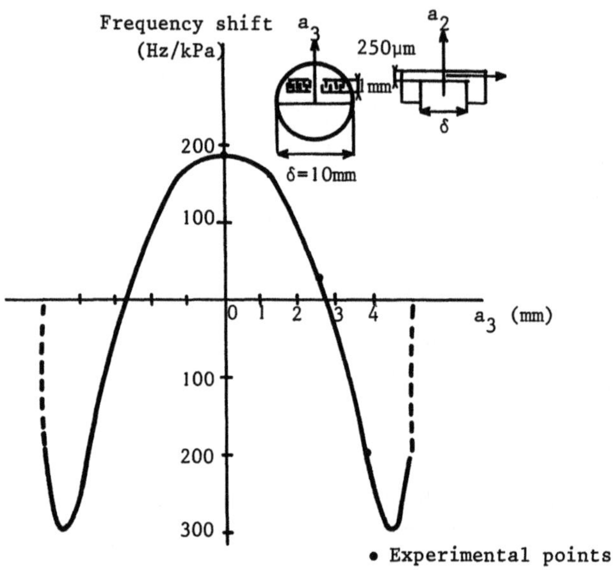

Fig. 22. Pressure sensor (Y,X)cut quartz crystal at 100 MHz, from ref. 42

9. Conclusion

During this brief survey the main characteristics of Rayleigh wave resonators, delay-lines, and oscillators have been reviewed. When comparing bulk waves and Rayleigh waves it appears that both have their advantages and disadvantages, and often are complementary, either for high stability oscillator applications, or for sensor applications. Since nonlinear effects, like amplitude-frequency effects, are smaller, more power can be applied to Rayleigh waves, and, as a consequence, oscillators with the best spectral purities can be achieved. But this is at the disadvantage of long term aging. However the experience of Rayleigh waves is still rather short in that domain, and the technologies of bulk waves used for many years now could be advantageously applied to SAW devices. The sensitivities to external perturbations are higher for Rayleigh waves and more efficient compensations are to be found, mainly for the G-sensitivity, to improve the stabilities of the oscillators. At the opposite such sensitivities are in favour of sensors. For this last point, the conclusion is not definitive, since the resolution of a sensor depends at the same time on the sensitivity and the intrinsic stability. The frequency ranges covered by Rayleigh and bulk waves are now more overlapping, this could induce an interesting emulation for the future.

10. References

1 E.A. Ash : "Surface wave Grating Reflectors and Resonators", IEEE Symp. Microwave Theory and Techniques, Newport Beach, Calif., (1970).

2 E.J. Staples : "UHF Surface Acoustic wave Resonators", Proc. 28th Ann. Freq. Cont. Symp., Fort Monmouth (1974), NTIS*.

3 P. Hartemann : "Acoustic Surface wave Resonator using ion-implanted Gratings", Proc. Ultrasonics Symp. IEEE Pub 75CH0994-4SU (1975).

4 R.V. Schmidt : "Acoustic Surface wave Velocity Perturbation in $LiNbO_3$ by Diffusion of Metals", Appl. Phys. Lett., 27, n ° 1 (1975).

5 R.C.M. Li, J.A. Alusow, R.C. Williamson : "Surface wave Resonators using Grooved Reflectors", Proc. 29th Ann. Freq. Cont. Symp., Fort Monmouth (1975).

6 DeLamar T. Bell, R.C.M. Li : "Surface-acoustic-wave Resonators", Proc. IEEE, 64, n ° 5 (1976).

7 G.L. Matthaei, B.P. O'Shaughnessy, F. Barman : "Relations for analysis and design of surface-wave resonators", IEEE Trans. Sonics Ultrasonics, SU-23 (1976).

8 P.S. Cross, "Reflective Arrays for SAW Resonators", Proc. Ultrasonics Symp. IEEE Pub 75CH0994-4SU (1975).

9 F.G. Marshall : "Surface Acoustic wave Resonators Constructed of Aluminum on ST quartz for use in high Stability Feedback Oscillators", Proc. Ultrasonics Symp. IEEE Pub 75CH0994-4SU (1975).

10 J.S. Schoenwald, W.R. Schreve, R.C. Rosenfeld : "Surface Acoustic wave Resonator Development", Proc. 29th Ann. Freq. Cont. Symp., Fort Monmouth (1975).

11 P.S. Cross, S.S. Elliott : "Surface-acoustic-wave Resonators", Hewlett Packard Journal (déc. 1981).

12 W.R. Schreve : "Surface-wave two-port Resonator Equivalent Circuit", Proc. Ultrasonics Symp. IEEE Pub 75CH0994-4SU (1975).

13 G.L. Matthaei, F. Barman, E.B. Savage, B.P. O'Shaughnessy : "A study of the Properties and Potential Application of Acoustic Surface-wave Resonators", Proc. Ultrasonics Symp. IEEE Pub 75CH0994-4SU (1975).

14 T.R. Joseph, K.L. Lakin : "Equivalent Circuit and Properties of Surface-wave Planar Resonators", Proc. 29th Ann. Freq. Cont. Symp., Fort Monmouth (1975).

15 R.C.M. Li, J.A. Alusow, R.C. Williamson : "Experimental Exploration of the Limits of Achievable Q of Grooved Surface-wave Resonators", Proc. Ultrasonics Symp. IEEE Pub 75CH0994-4SU (1975).

* Proceedings of the Annual Frequency Control Symposium are available from National Technical Information Service - Sills Building 5285 Port Royal Road, Springfield VA 22161

16 A.J. Slobodnik Jr. : "Attenuation of Microwave Acoustic Surface waves due to gas loading", J. Appl. Phys., 43 (1972).

17 R.C.M. Li, R.C. Williamson, D.C. Flanders, J.A. Alusow : "On the Performance and Limitations of the Surface-wave Resonators using Grooved Reflectors", Proc. Ultrasonics Symp. IEEE Pub 74CH0896-1SU (1974).

18 E.J. Staples, R.C. Smythe : "Surface Acoustic wave Resonators on ST quartz", Proc. Ultrasonics Symp. IEEE Pub 75CH0994-4SU (1975).

19 C. Dunrowicz, F. Sandy, T. Parker : "Reflection of Surface waves from Periodic Discontinuities", Proc. Ultrasonics Symp. IEEE Pub 76CH1120-5SU (1976).

20 W.A. Haydl, P. Hiesinger, B. Dischler, R. Smith : "Design of Quartz and Lithium Niobate Resonators using Aluminum Metalization", Proc. 30th Ann. Freq. Cont. Symp., Fort Monmouth (1976).

21 W.T. Tanski : "Surface Acoustic wave Resonators on Quartz", IEEE Trans. Sonics Ultrasonics, SU-26, n ° 2 (1979).

22 L.S. Cutler, C.L. Searle : "Some Aspects of the Theory and Measurement of Frequency Fluctuations in Frequency Standards", Proc. IEEE, 54, n ° 2 (1966).

23 R. Brendel : "Internal Noise of a Quartz Crystal Oscillator", Proc. 29th Ann. Freq. Cont. Symp., Fort Monmouth (1975).

24 R.D. Weglein : "1.2 GHz Temperature-stable SAW Oscillator", Proc. 31st Ann. Freq. Cont. Symp., Fort Monmouth (1977).

25 T.E. Parker, D.L. Lee : "Stability of Phase Shift on Quartz SAW Devices", Proc. 32nd Ann. Freq. Cont. Symp., Fort Monmouth (1978).

26 C. Lardat, H. Gautier, Thomson-CSF, Division des Activités Sous-Marines, DTAS, Valbonne, France, (1985) Private Communication.

27 L. Bidart, J. Chauvin : "Direct Frequency Crystal Oscillators", Proc. 35th Ann. Freq. Cont. Symp., Fort Monmouth (1981).

28 T.E. Parker : "1/f Noise in Rayleigh wave Resonators and Delay Lines", These Proceedings.

29 C.A. Adams, J.A. Kusters : "Deeply Etched SAW Resonators", Proc. 31st Ann. Freq. Cont. Symp., Fort Monmouth (1977).

30 M. Planat, D. Hauden, J. Groslambert, J.J. Gagnepain : "Nonlinear Propagation of Surface Acoustic Waves on Quartz", Proc. 34th Ann. Freq. Cont. Symp., Fort Monmouth (1980).

31 T. Lukaszek, A. Ballato : "What SAW can learn from BAW : Implications for Future Frequency Control, Selection and Signal Processing", Proc. IEEE Ultrasonics Symp. IEEE Pub 80CH1602-2 (1980).

32 T.E. Parker : "Analysis of aging data on SAW oscillators", Proc. 34th Ann. Freq. Cont. Symp., Fort Monmouth (1980).

33 M. Gilden, G.K. Montress, R.A. Wagner : "Long term aging and Mechanical Stability of 1.4 GHz SAW Oscillators", Proc. IEEE Ultrasonics Symp. IEEE Pub 80CH1602-2 (1980).

34 J.S. Schoenwald, J. Wise, E.J. Staples : "Absolute and Differential aging of SAW Resonator Pairs", Proc. 35th Ann. Freq. Cont. Symp., Fort Monmouth (1981).

35 C. Bennett : "SAW Resonator Stabilized Oscillator", Proc. 37th Ann. Freq. Cont. Symp., Fort Monmouth (1983).

36 R.B. Stokes, K.H. Yen, K.F. Lan : "SBAW versus SAW Oscillator aging", Proc. 38th Ann. Freq. Cont. Symp., Fort Monmouth (1984).

37 T.E. Parker : "Random and Systematic Contributions to long term Frequency Stability in SAW Oscillators, Proc. IEEE Ultrasonics Symp. IEEE Pub 83CH1947-1 (1983).

38 C. Maerfeld : "Rayleigh wave Nonlinear Components", These Proceedings.

39 H.F. Tiersten : "On the Nonlinear Equations of Thermo-electroelasticity", Int. J. Engng. Sci., 9 (Pergamon Press) (1971).

40 D.F. Nelson : "Theory of Nonlinear Electroacoustics of Dielectric, Piezo-electric and Pyroelectric Crystals", J. Acoust. Soc. Am. 63 (1978).

41 M. Planat, G. Théobald, J.J. Gagnepain : "Propagation non linéaire d'ondes élastiques dans un solide anisotrope, I-Ondes de volume, Onde Electrique, 60, n ° 8-9 (1980), II-Ondes de surface, Onde Electrique 60, n ° 11 (1980).

42 G. Théobald, D. Hauden, J.J. Gagnepain : "Propagation des ondes élastiques en milieu perturbé, I-Théorie générale, Onde Electrique, 65, n ° 2 (1985), II-Applications aux oscillateurs et aux capteurs, Onde Electrique, 65, n ° 3 (1985).

43 R. Bechmann, A.D. Ballato, T.S. Lukaszek : "Higher Order Temperature Coefficients of the Elastic Stiffness and Compliances of Alpha-quartz Proc. IRE, 50 (1962).

44 D. Hauden, M. Michel, J.J. Gagnepain : "Higher Order Temperature Coefficients of quartz SAW Oscillators", Proc. 32nd Ann. Freq. Cont. Symp., Fort Monmouth (1978).

45 A. Ballato, J.R. Vig : "Static and Dynamic Frequency-temperature Behavior of Singly and Doubly Rotated Oven-controlled Quartz Resonators", Proc. 32nd Ann. Freq. Cont. Symp., Fort Monmouth (1978).

46 P. Levesque, M. Valdois, D. Hauden, J.J. Gagnepain, P. Hartemann, J. Uebersfeld : "Theoretical and Experimental Analysis of SAW Quartz Oscillator Acceleration Sensitivity", Proc. IEEE Ultrasonics Symp. IEEE Pub 79CH1482-9SU (1979).

47 J.F. Dias, H.E. Karrer : "Stress Effects in Acoustic Surface wave Circuits and Applications to Pressure and Force Transducers", IEEE Int. Conf. Solid-state Circuits (1974).

48 H.F. Tiersten, D.S. Stevens, P.K. Das : "Acoustic Surface wave Accelerometer and Rotation rate Sensor", Proc. IEEE Ultrasonics Symp. IEEE Pub 80CH1602-2 (1980).

49 D. Hauden, S. Rousseau, J.J. Gagnepain : "Sensitivities of SAW oscillators to Temperature, Forces and Pressure : Application to Sensors", Proc. 34th Ann. Freq. Cont. Symp., Fort Monmouth (1980).

50 T.M. Reeder, D.E. Cullen, M. Gilden : "SAW Oscillator Pressure Sensors", Proc. IEEE Ultrasonics Symp. IEEE Pub 75CH0994-4SU (1975).

51 P. Hartemann, P.L. Meunier : "Tensioned or Flexured SAW Accelerometers", Proc. IEEE Ultrasonics Symp. IEEE Pub 83CH1947-1 (1983).

1/f Noise in Rayleigh Wave Resonators and Delay Lines[1]

T.E. Parker

Raytheon Research Division, 131 Spring Street,
Lexington, MA 02173, USA

1.0 Introduction

Flicker or 1/f noise has been observed in many different physical systems, and Rayleigh wave devices are no exception. Flicker noise has been reported for both Rayleigh wave delay lines and resonators [1,2] as well as bulk acoustic wave (BAW) devices [3,4]. Though the ultimate source of the 1/f noise is not yet understood, some less fundamental questions are now beginning to be answered. For resonators, these questions involve the influence of Q, device frequency and one port versus two port design on 1/f noise levels. For delay lines the parameters under question are delay time and device frequency. In both cases fabrication and enviromental parameters may also influence the 1/f noise levels.

This paper will discuss the results of 1/f noise measurements on over 110 resonators and delay lines. As a result of these measurements, empirical relations have been derived for estimating the 1/f noise levels on Rayleigh wave resonators and delay lines. Since there is a large amount of scatter among "identical" devices, these relations are only estimates with an accuracy of approximately ± 10 dB.

2.0 1/f Noise in Rayleigh Wave Resonators

A 1/f noise power spectral density of frequency fluctuations is known to exist in Rayleigh wave resonators over the offset frequency range of approximately 10^{-2} Hz to 10^{4} Hz [5]. Recently this noise level was measured on 56 vacuum sealed Rayleigh wave (SAW) resonators (fabricated on 40 degree rotated Y

[1]This work supported in part by U.S. Army Electronics Technology and Devices Laboratory Contract No. DAAK20-84-0401.

cut quartz) covering the frequency range of 187 MHz to 984 MHz. This corresponds to values of unloaded Q (no electrical loading) from over 40,000 to approximately 7,000. Nearly all of these devices exhibited values of Q_u that were greater than 75% of the material limit. The noise measurements were made by operating the resonators in oscillators and measuring the phase noise of the oscillators over the frequency range of 1 Hz to 300 Hz. Figure 1 shows a typical plot of single sideband phase noise, $\mathcal{L}(f)$, on a SAW oscillator. (Above ~300 Hz the noise from the reference synthesizer dominates.) From this data an average value for the level at 1 Hz was determined. The value of the power spectral density of frequency fluctuations, $S_{\Delta F}(f=1\ Hz)$, was calculated from $\mathcal{L}(f=1\ Hz)$ [in dBc/Hz] by using the relation

$$S_{\Delta F}(f) = (2\times 10^{\mathcal{L}(f)/10})/f^2. \qquad (1)$$

Figure 2 shows the measured values of $S_{\Delta F}(f)$ at 1 Hz plotted as a function of unloaded Q, Q_u, for the 56 SAW resonators. The values of $S_{\Delta F}(f=1\ Hz)$ serve as a measure of the magnitude

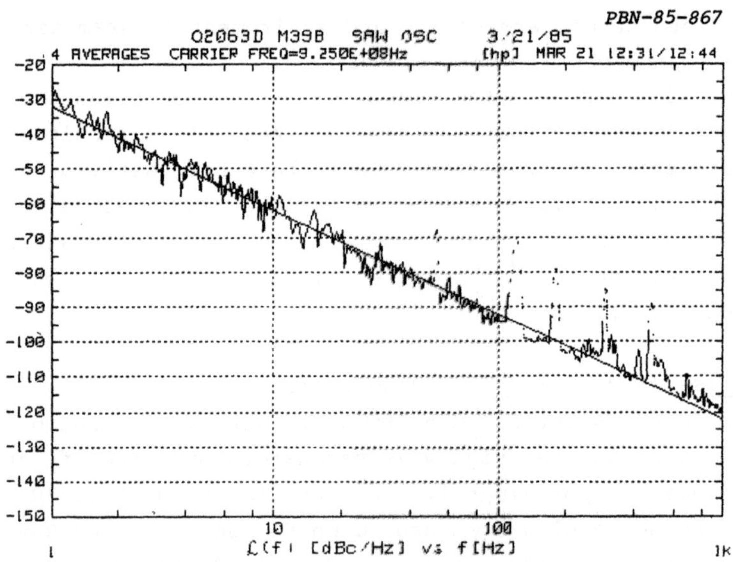

Figure 1. Typical plot of single sideband phase noise as a function of offset frequency for a Rayleigh wave resonator oscillator

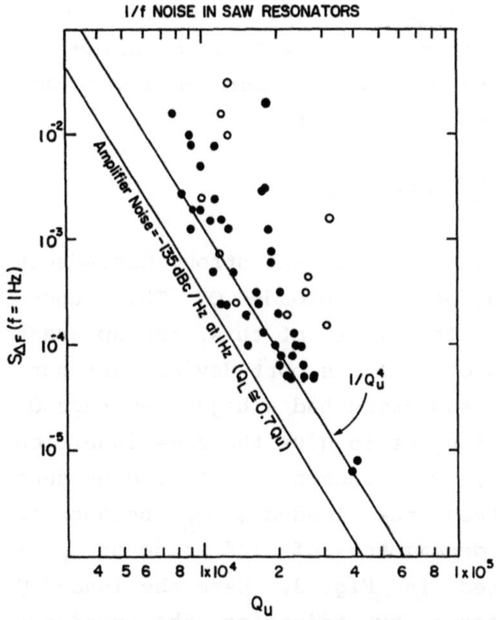

Figure 2. 1/f noise in Rayleigh wave (SAW) resonator oscillators as a function of unloaded Q. Open data points represent one port resonators and solid data points represent two port resonators

of the 1/f noise. The solid data points represent two port resonators and the open data points represent one port resonators.

2.1 One Port vs Two Port

The data in Fig. 2 shows that there is approximately ± 10 dB of scatter for devices with the same unloaded Q. This is similar to what was observed by Elliott and Bray [2]. Though there is considerable scatter, it is clear that there is not much difference in the noise levels of the one port devices as compared to the two port devices. In fact, the one port resonators have a slightly higher (~4 dB) average noise level. The electronic circuits used for the one port resonators were somewhat different from those used for the two port devices so it was desirable to verify that the circuits were not influencing the observed noise levels. This was easy to do since two port devices could be operated in the one port circuits (but not vice versa). Such measurements with a low noise two port device confirmed that the one port measurements were not being affected by the oscillator electronics.

2.2 Dependence on Loaded and Unloaded Q

Another feature of the data in Fig. 2 is the strong dependence of 1/f frequency fluctuations on the unloaded Q. This dependence is close to $1/Q_u^4$. In light of this, the apparent slightly higher noise levels of the one port devices are surprising since the one port resonators had a higher average Q, due to the need for fewer fingers to give the same insertion loss. The unloaded Q, Q_u, was chosen for the independent variable in Fig. 2 rather than the loaded Q, Q_L, because it has been observed that the dependence of 1/f noise on Q_L is weak [6]. This is illustrated in Fig. 3. Here the loaded Q was varied on four resonators by adjusting the impedance matching circuits. The accompanying change in device insertion loss was compensated for by changing the gain of the loop amplifier by the addition or removal of attenuation. As can be seen in Fig. 3, the dependence of flicker noise on loaded Q

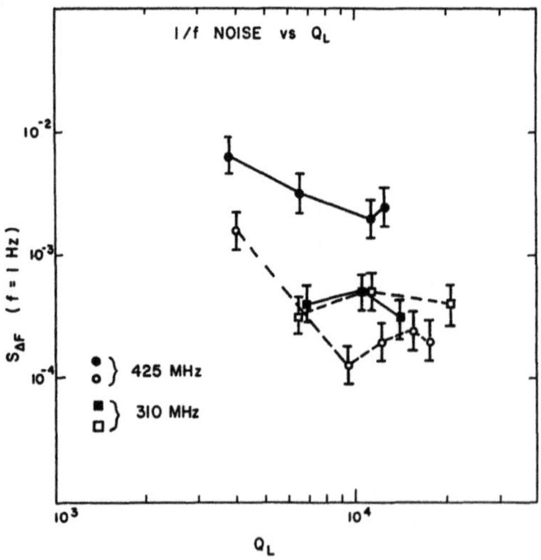

Figure 3. 1/f noise level at 1 Hz of four Rayleigh wave resonators as a function of unloaded Q

in SAW resonators is much weaker then the dependence on unloaded Q.

It should be pointed out that, because the devices in Fig. 2 are operating near the material limit for unloaded Q, there is a strong relationship between resonator frequency, F_0, and Q_u. This relationship occurs because the $Q_u x F_0$ product for material limited acoustic resonators is a constant. (Approximately $1x10^{13}$ for ST cut quartz.) The variable $1/F_0$ could be substituted for Q_u in Fig. 2 and the data would look essentially unchanged except for a scale factor. To test the dependence on Q_u, flicker noise measurements were also made on 27 ~400 MHz SAW resonators in which a metal pad of either aluminum or gold was added to the acoustic path to lower the value of Q_u [7]. The results of these measurements seemed to confirm that Q_u was the important parameter rather than F_0, since a strong dependence on Q_u was also observed.

However, some recent results with air loading have cast some doubts on this conclusion. In these recent experiments, the 1/f noise was measured on a 425 MHz Rayleigh wave resonator that was contained in a chamber in which the temperature and pressure could be varied. By varying the pressure of a nitrogen atmosphere from 10^{-6} Torr to 35 PSI the value of Q_u could be changed by a factor of two. The data from this experiment, and the metal loading results, are shown in Fig. 4. The strong dependence of the 1/f noise on Q_u for the metal loaded resonators is clearly seen, but the dependence on Q_u for the air loaded device is weak, and may be even less than $1/Q_u^2$. Therefore, it must be concluded that for devices operating well below the material limit, the means by which Q_u is decreased is important to the level of 1/f noise. However, for material limited devices, the 1/f frequency fluctuations can be thought of as varying as $1/Q_u^4$ or F_0^4. Though the Q_u dependence is not as simple as first thought, the data in Fig. 4 clearly demonstrates that metal films play a major role in determining 1/f noise levels.

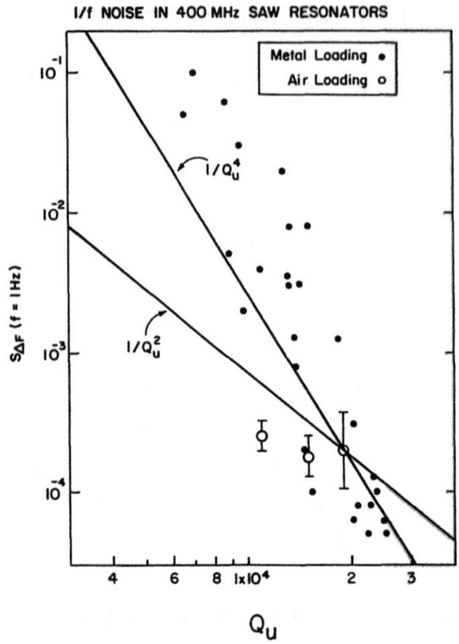

Figure 4. 1/f noise in 400 MHz Rayleigh wave resonator oscillators as a function of unloaded Q

2.3 Other Environmental Factors

In addition to the pressure dependence of the 1/f noise on the 425 MHz resonator, two other environmental factors were investigated. The flicker noise was measured at -5°C, 40°C, and 90°C, and was found to be constant to within ± 1 dB. Also, it was observed that exposure to UV light (350 to 380 nm) at 62 mW/cm^2 and white light at 20 mW/cm^2 had no effect on the noise level.

2.4 Correlation With SEM Data

As part of this investigation of 1/f noise in SAW devices, some of the resonators for which the noise had been measured were examined with a scanning electron microscope (SEM) to determine if there was any correlation between the physical appearance of the surface of the SAW device and the observed 1/f noise level. At this time 15 devices have been examined. These include ten 416 MHz resonators that have just completed a two year aging test and five two port resonators in the range of 650 MHz to 800 MHz. The bare quartz surface is essentially featureless and has provided very little information. However, the transducer is quite visible and can have considerable structure.

At this time no correlation has been observed between 1/f noise and the number of open fingers or other defects in the metallization. Also there is no apparent correlation between 1/f noise and the quality of the finger definition. This is illustrated in Fig. 5 which shows transducer fingers from two 416 MHz resonators. For the device in Fig. 5a the lift-off was poor and the finger edges were very ragged. In Fig. 5b a very clean lift-off was achieved. The 1/f noise levels for these two devices are also shown in Fig. 5, and as can be seen, they are very similar.

Another parameter that has been examined for correlation with 1/f noise is the width to space ratio of the transducer

(a)

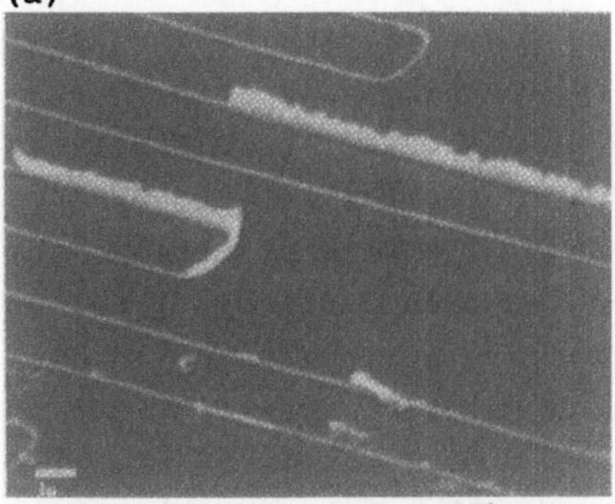

$S_{\Delta F}$(f=1 Hz) = 2×10^{-4}

(b) PBN-85-533

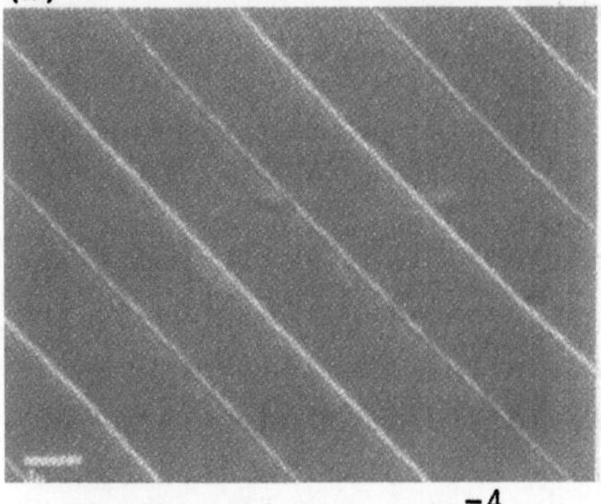

$S_{\Delta F}$(f=1 Hz) = 1×10^{-4}

Figure 5.
(a) Ragged finger edges and (b) clean finger edges

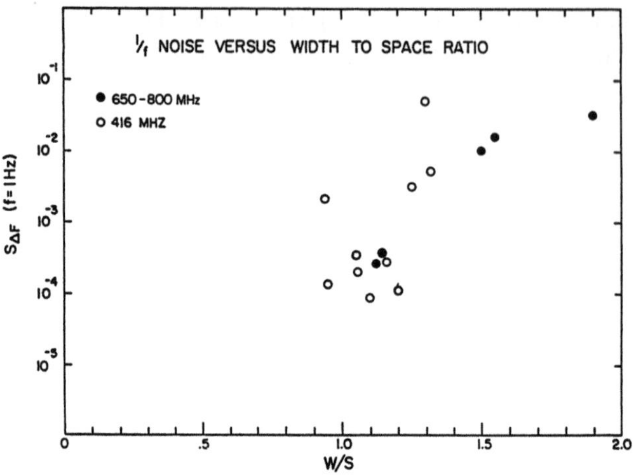

Figure 6. 1/f noise levels as a function of width to space ratio

metallization. The width to space ratio is defined as the ratio of the finger width to the distance between fingers. For an ideal transducer this ratio should be one (i.e., finger width equals finger space), but in practice fabrication variables often result in fat fingers, particularly at high frequencies. Figure 6 shows a plot of observed 1/f noise levels as a function of width to space ratio for the fifteen devices that have been examined with the SEM. The data points for the ten 416 MHz resonators are plotted as open circles, while the data from the other five resonators are plotted as solid circles. There does appear to be a correlation between width to space ratio and 1/f noise, with the larger values of w/s giving higher noise levels. The data in Fig. 6 is very intriguing, but results from more devices will be required before a convincing case can be made. These measurements are now in progress.

2.5 1/f Noise in Amplifiers

When making noise measurements on oscillators it is important to be sure which components are the dominant source of the excess noise. The amplifiers used in the oscillator circuits for these measurements were commercial wide band silicon bipolar devices which have a very low 1/f noise level. By interchanging SAW devices and amplifiers it was easy to confirm that nearly all of the Rayleigh wave resonators in Fig. 2 were the dominant source of the flicker noise. However, the 1/f noise levels of several typical amplifiers were also measured directly. These measurements were done both with the amplifiers terminated in 50 ohms and also with the amplifiers terminated with an impedance similar to what they would see in the oscillator loop. In both cases the <u>open loop</u> single sideband phase noise at 1 Hz, $\mathcal{L}'(f=1Hz)$, was ~ -135 dBc/Hz over the full frequency range of the amplifiers.

By using the definition of group delay

$$\tau = \Delta\phi/\Delta\omega_0 \tag{2}$$

and the relation between loaded Q and group delay, τ,

$$Q_L = \omega_0\tau/2 \tag{3}$$

it can be shown that, in an oscillator, the power spectral density of frequency fluctuations, $S_{\Delta F}(f)$, caused by open loop phase fluctuations with power spectral density $S_\phi'(f)$ in the loop amplifier is

$$S_{\Delta F}(f) = (F_0/2Q_L)^2 S_\phi'(f). \tag{4}$$

In these expressions $\omega_0 = 2\pi F_0$, where F_0 is the resonant frequency and ϕ is the phase shift through the SAW device. Note that

$$\mathcal{L}'(f) = 10\ \mathrm{Log}[S_\phi'(f)/2]. \tag{5}$$

For a constant value of $S_\phi'(f)$ [or $\mathcal{L}'(f)$], (4) predicts a $1/Q_L^2$ dependence for $S_{\Delta F}(f)$, which clearly was not present in Fig. 3. This is further evidence that 1/f amplifier phase noise is not dominant in these SAW oscillators.

Using the relation

$$Q_u \times F_0 = C \tag{6}$$

and assuming

$$Q_L \cong 0.7\ Q_u \tag{7}$$

(4) becomes

$$S_{\Delta F}(f) \cong (c^2/2Q_u{}^4)S_\phi{}'(f). \tag{8}$$

We now see that a constant $S_\phi{}'(f)$ will give a $1/Q_u{}^4$ dependence to the 1/f frequency fluctuations of acoustic oscillators operating near the material limit for Q_u. However, when the measured value of the amplifier phase noise [$\mathcal{L}'(f=1Hz) = -135$ dBc] is used in (5) and (8), the calculated oscillator noise levels are found to be too low. This is illustrated in Fig. 2 by the line repesenting the amplifier noise. It is now clear that the SAW resonators are the dominant source of the 1/f noise in nearly all cases. Furthermore, (8) and the $1/Q_u{}^4$ dependence of the data in Fig. 2 also show that the average <u>open loop</u> phase noise of the SAW resonators, $S_\phi{}'(f)$, [or $\mathcal{L}'(f)$] is a constant value which is independent of F_0 or Q_u. From the data in Fig. 2, the average value of $\mathcal{L}'(f=1$ Hz) for the SAW resonators is calculated to be -124 dBc/Hz. This value is 11 dB higher than the amplifier noise and is in close agreement with the results of Elliott and Bray [2].

It should be noted that not all amplifiers will exhibit the same low 1/f noise level that was present in the amplifiers used in this investigation. In particular, GaAs FET amplifiers have a 1/f noise level that can be as much as 20 dB higher than silicon bipolar amplifiers.

2.6 <u>Possible Causes of 1/f Noise in Rayleigh Wave Resonators</u>

There are two potential fundamental types of noise in a Rayleigh wave resonator that can result in frequency fluctuations in an oscillator. One is a variation of the resonant frequency of the acoustic device, and the second is fluctuations in the phase relationship between the electrical signal (at the pins of the package) and the acoustic energy stored in the resonator. In an open loop measurement, both will result in fluctuations of the transmission phase shift. However, the observed phase fluctuations due to changes in the resonant frequency will depend on the loaded Q, which determines the phase slope. Conversely, fluctuations in the phase relationship may very well be largely independent of the loaded Q. In a closed loop, the situation is reversed. Here the oscillator frequency will track the resonator frequency independent of loaded Q, but fluctuations in the phase relationship will result in oscillator frequency variations that are dependent on loaded Q, as given in (4). At this time the experimental evidence favors fluctuations in resonant frequency because of the weak dependence of oscillator noise on loaded Q as shown in Fig. 3.

For cavity resonators, the resonant frequency is determined by the equation

$$F_0 = NV/L \tag{9}$$

where N is a large integer or half integer (~390 for these resonators), V is the average SAW velocity, and L is the effective cavity length. Any parameter that affects V or L can cause a change in resonant frequency. Table 1 is a list of possible sources of frequency fluctuations. These sources are divided into two basic catagories, (1) those that affect velocity and, (2) those that affect cavity length. The ones that affect velocity are subdivided into those that affect the quartz directly, and those that affect the transducer. The transducer parameters are divided into mechanical and electrical phenomena.

There are many possible sources of 1/f noise, but comments can be made about some of them. First of all, any fluctuation that influences the entire substrate in the same fashion as static or dynamic temperture stability, or vibration sensitivity is not a likely source since the fourth power dependence on Q_u (or F_0) of 1/f noise cannot be explained. Temperature and vibration sensitivity are constant with respect to fractional frequency changes. Therefore, temperature and stress

Table 1
Possible Sources of Frequency Fluctuations

ACOUSTIC VELOCITY			EFFECTIVE CAVITY LENGTH
QUARTZ	TRANSDUCER		
	MECHANICAL	ELECTRICAL	
Temperature, Temperature gradients	Temperature, Temperature gradients	Electric field strength and distribution	Temperature, Temperature gradients
Mechanical stress	Mechanical stress	Finger resistance	Mechanical strain
Defects	Defects	Inter electrode resistance	Defects
Damage	Acoustic attenuation	Capacitance	
Acoustic attenuation	Adsorption, desorption of contaminants		
Piezoelectric effect			
Adsorption, desorption of contaminants			
Surface preparation			

related phenomena that affect the entire substrate are probably ruled out. This also includes fluctuations due to time dependent changes in bulk defects or damage. Fluctuations in acoustic attenuation can also cause frequency fluctuations [4], but the predicted Q dependence for this process is $1/Q_u^4$ for fractional frequency fluctuations, $S_y(f)$, rather than absolute frequency fluctuations, $S_{\Delta F}(f)$, as has been observed here. Surface related phenomena such as surface damage, surface temperature fluctuations, or adsorption and desorption of contaminants would of course be frequency dependent, but again it is not clear how a Q to the fourth power dependence could be obtained. Furthermore, the 1/f noise is not strongly affected by surface perturbations caused by illumination or pressure changes.

The transducer metallization can also influence the average SAW velocity through both it's mechanical and electrical properties. Though the transducer makes only a small perturbation on the velocity, the fractional frequency fluctuations that are observed are also very small (~3 to 10 parts in

10^{11}). Also, the physical properties of the transducer metal are not nearly as stable as the quartz substrate. The strong dependence of the 1/f noise on the amount of metal added to reduce the unloaded Q, as seen in Fig. 4, is an indication that the transducer metal plays an important role in 1/f noise. This is further supported by the preliminary data in Fig. 6 which shows a dependence on width to space ratio.

Changes in the physical length of the cavity would also cause frequency fluctuations. Though this seems unlikely to be a source of noise, it must be noted that a change in frequency of 3×10^{-11} corresponds to a change in physical length for a typical SAW resonator of only 0.0004 Å. It is not hard to believe that one or two interstitial atoms could cause this much perturbation.

After looking at the list in Table 1, one begins to wonder if the difficulty in understanding 1/f noise does not stem from the possibility that there may be many simultaneous sources of noise.

2.7 1/f Noise in Rayleigh and BAW Resonators

Measured levels of 1/f noise in quartz BAW resonators are now available in the literature [3,4,8,12], and when this data is combined with the Rayleigh wave data, the $1/Q_u^4$ dependence becomes even more dramatic. This is shown in Fig. 7. All of the BAW data are taken from the literature as indicated in the figure with the exception of the data point for the 466 MHz device. This device was measured in the author's laboratory using the same type of amplifier as was used with the SAW devices. The frequencies of the BAW devices at or above 100 MHz are indicated beside each data point. The BAW data points with no indicated frequencies fall in the range of 1 to 25 MHz. The presence of error bars indicates that the value of Q_u was not available and had to be estimated.

A least mean square fit to the data in Fig. 7 gives a slope of $Q_u^{-3.85} \pm 0.11$. The standard deviation of the data is ± a factor of 8, or ± 9 dB. The data clearly indicates that within ± 10 dB, the 1/f noise process is the same in SAW and BAW resonators.

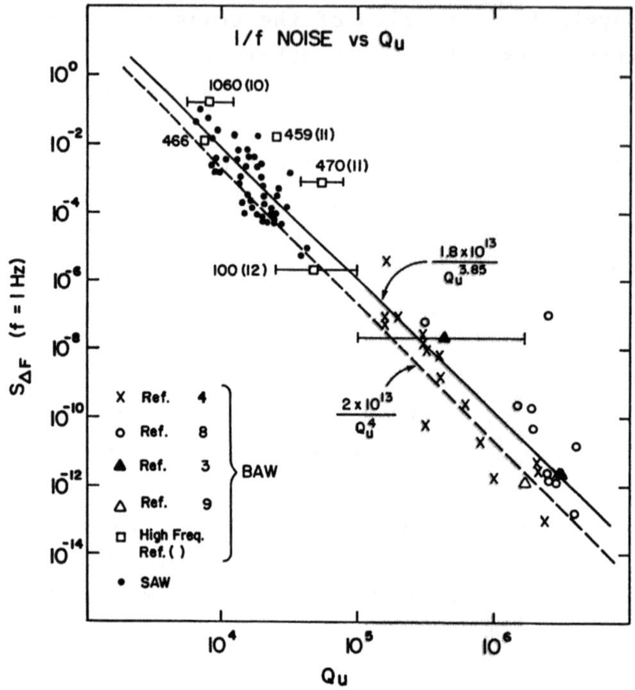

Figure 7. 1/f noise levels of quartz Rayleigh wave and bulk acoustic wave (BAW) resonators as a function of unloaded Q

An empirical relation for estimating the level of 1/f noise in material limited SAW and BAW oscillators can be obtained from the data in Fig. 7.

$$S_{\Delta F}(f) \cong (2\times10^{13}/Q_u{}^4)(1/f). \qquad (10)$$

Equation (10) gives a value that is near the low end of the range of data in Fig. 7 and therefore, represents an approximate lower limit for 1/f frequency fluctuations in quartz acoustic resonators. In general it is not unusual to obtain devices with relatively high noise levels, but devices that are more than 10 dB lower than the value calculated from Eq. (10) are very rare. Equation (10) can also be written in terms of F_0 (in MHz).

$$S_{\Delta F}(f) \cong [2\times10^{-15}(F_0{}^4)](1/f) \qquad (11)$$

or can be expressed as fractional frequency fluctuations, $S_y(f)$,

$$S_y(f) \cong (2\text{x}10^{-13}/Q_u^{\ 2})(1/f). \tag{12}$$

The $1/Q_u^{\ 4}$ (or $F_0^{\ 4}$) dependence of the 1/f noise in quartz acoustic resonators means that when a low frequency quartz oscillator is multiplied to a higher frequency, the degradation in close-to-carrier noise is less than that obtained by operating a quartz oscillator directly at the higher frequency. When multiplying to a higher frequency the phase noise is increased by the square of the multiplication factor. By operating directly at the higher frequency, the close-to-carrier noise is increased by the frequency ratio to the fourth power. This explains why the close-to-carrier noise of a high quality frequency synthesizer is lower than that of a free running high frequency SAW or BAW oscillator.

3.0 1/f Noise in Rayleigh Wave Delay Lines

Flicker, or 1/f noise is known to exist in quartz Rayleigh wave delay lines over the offset frequency range of 10^{-2} to 10^5 Hz [5]. The 1/f noise level has been measured on over 40 delay lines (on rotated Y cut quartz) operating at frequencies from 340 MHz to over 1400 MHz using essentially the same measurement technique as was used with the resonators. The observed noise levels in delay line oscillators are 20 to 40 dB higher than the 1/f noise levels on resonators at the same frequency. However, a spread of ± 10 dB is also present on the delay lines, which is similar to that observed on resonators. Since the delay time of the delay lines is significantly less than the group delay of resonators (at the same frequency), one might naturally expect the delay line oscillators to have higher noise levels than resonator oscillators. However, there is considerable evidence that the noise process in delay lines is fundamentally different from (and higher than) that in resonators.

Specifically, it has been observed that the noise on delay lines tends to decrease when the substrate surface gets contaminated, but on resonators the noise either does not change at all, or it increases. In fact, the noise on delay lines can be reduced by 10 dB or more by surface treatment with certain silicone materials [1]. However, these surface treatments cause no noise reduction on resonators, and generally tend to increase the noise. As mentioned in Sec. 2.2, the addition of extra metal (gold or aluminum) in the acoustic path increases the noise level of resonators, but similar metal pads have no affect on delay line noise levels.

Further evidence that the noise process is fundamentally different in delay lines and resonators comes from the observation that the 1/f noise in delay lines varies as $1/\tau^2$, [1] where τ is the delay time, and also that the noise is independent of device frequency (for equivalent delay times). In practice, the delay time of delay lines for oscillator applications is limited to approximately 1000 wavelengths or less, which means that the delay time for most delay line oscillators varies roughly as the inverse of the device frequency. Therefore, the 1/f noise level for practical delay line oscillators typically increases as $F_0^{\,2}$. This is in contrast to the $F_0^{\,4}$ dependence for resonators as expressed in (11).

By using the data gathered from the noise measurements on numerous delay line oscillators, an empirical equation for estimating the average 1/f noise level of a delay line oscillator has been derived. This expression is given in (13).

$$S_{\Delta F}(f) = [2\times10^{-13}/\tau^2 f]. \tag{13}$$

Equation (13) gives the approximate 1/f noise level (to an accuracy of about ± 10 dB) for an untreated delay line oscillator (for τ in seconds). If the Rayleigh wave delay line has been silicone treated, the average noise level will be approximately 10 dB lower.

The average <u>open loop</u> noise level, $\mathcal{L}$ (f=1 Hz), for an untreated delay line is independent of device frequency and delay time [13] and is equal to approximately -114 dBc/Hz, with a ± 10 dB uncertainty. Silicone treated delay lines are roughly 10 dB better. As long as the associated amplifier 1/f noise is lower than the contribution from the delay line, the oscillator noise level will be controlled by the delay line.

4.0 Acknowledgements

The author would like to thank Gordon Jackson for valuable assistance in obtaining much of the SAW data and J. Lang, E. Sabatino and J. Columbus for fabricating the devices. The author would also like to thank Gary Montress and Joe Callerame for valuable discussions and encouragement.

5.0 References

1. T.E. Parker, "1/f Phase Noise in Quartz Delay Lines and Resonators", Proc. 1979 IEEE Ultrasonics Symp., p. 878 (1979).

2. S.S. Elliott and R.C. Bray, "Direct Phase Noise Measurements of SAW Resonators", Proc. 1984 IEEE Ultrasonics Symp., p. 180 (1984).

3. F.L. Walls and A.E. Wainwright, "Measurement of the Short Term Stability of Quartz Crystal Resonators and the Implications for Crystal Oscillator Design and Applications", IEEE Trans. Instrum. and Meas., IM-24, p. 15 (1975).

4. J.J. Gagnepain, J. Ubersfeld, G. Goujon and P. Handel, "Relation Between 1/f Noise and Q Factor in Quartz Resonators at Room and Low Temperatures, First Theoretical Interpretation", Proc. 35th Annual Symp. on Frequency Control, p. 476 (1981).

5. T.E. Parker, "Random and Systematic Contributions to Long Term Frequency Stability In SAW Oscillators", Proc. 1983 IEEE Ultrasonics Symp., p. 257 (1983).

6. T.E. Parker, "1/f Frequency Fluctuations in Quartz Acoustic Resonators", Applied Physics Letters, 46, p. 246 (1985).

7. J. Day, G. Jackson and T.E. Parker, "Bias Controlled Frequency Trimming Of SAW Devices in a DC O_2 Plasma", Proc. 38th Annual Symp. on Frequency Control, p. 310 (1984).

8. J.J. Gagnepain, M. Oliver and F.L. Walls, "Excess Noise in Quartz Crystal Resonators", Proc. 37th Annual Symp. on Frequency Control, p. 218 (1983).

9. Guaranteed noise performance of Model FTS 1050A/004 Quartz Frequency Standard (5 MHz) manufactured by Frequency and Time Systems Inc., 34 Tozer Rd., Beverly, MA 01915.

10. J.P. Aubry, "Quartz and $LiTaO_3$ VHF Resonators for Direct Frequency Generation in the GHz Range", Proc. 1983 IEEE Ultrasonics Symp., p. 487 (1983).

11. L. Bidart and J. Chauvin, "Direct Frequency Crystal Oscillators", Proc. 35th Annual Symp. on Frequency Control, p. 365 (1981).

12. Guaranteed noise performance of Model 1830025 High Frequency Crystal Oscillator (100 MHz) manufactured by Piezo Systems, 100 K St., Carlisle, PA 17013.

13. R.L. Baer, D.M. Hoover, D. Molinari and E.C. Herleikson, "Phase Noise in SAW Filters," Proc. 1984 IEEE Ultrasonics Symp., p. 30 (1984).

Rayleigh Wave Non-Linear Components

C. Maerfeld

THOMSON-CSF DASM/DTAS, BP 38, F-06561 Valbonne Cedex, France

The analysis of non-linear phenomena has never been ignored by physicists or engineers in acoustics, but for most of them non-linear effects become a way of extending linear phenomena beyond the strict limits of the linear theory itself. In this vein, large signal properties, harmonic generators and, as in the previous paper, asynchronism in oscillators have been studied for a long time and undoubtedly will be the subject of yet more work. In cases such as these, components are basically linear and non-linear effects appear largely as a limitation of the field of application.

On the other hand, there are devices based directly on non-linear effects to obtain new interactions and new properties. The non-linear interactions of concern here are those called parametric effects and the only non-linear component to be considered are those using such effects.

Initially, we will look only at the second order parametric interactions. As will be developed later, in most cases, the non-linear components can be considered as bilinear.

The scope of this paper is a review of the properties and performance of Rayleigh wave non-linear components in the sense of the above definition, i.e. using parametric effects as the basis of operation. The parametric effect here is a non-linear interaction between two waves, or pseudo waves, giving rise to a third one, thus it can be called a 3-wave interaction.

The parametric amplification of acoustic waves as studied in the 60's did not have the success observed in the parametric amplification of electromagnetic waves. Few applications emerged, and only one, the Rayleigh wave convolver demonstrated enough potential to warrant the allocation of a lecture at this Rayleigh Wave Centenary Symposium. Thus, this paper will be devoted primarily to convolvers.

It is important in a presentation such as this, to give an historical review of the birth and development of the subject. Unfortunately the author's talents as an historian are mediocre, moreover he did not have a "dating apparatus for ideas" a device as efficient as those giving the age of ancient parchments. Restricted to a careful bibliographical search for earlier work he quickly became lost in the maze of references, footnotes and quotations from yet earlier papers thus completely lost and limited by time, he must now confess to having given up an organized historical review. Therefore, the historical flashbacks are limited and must be considered for what they are, a simple sentimental recall of some facts which have impressed me over the years.

1. THEORY of NON-LINEAR ACOUSTICS in SOLIDS

Before describing the properties of the non-linear components to be considered, it is useful to look at the origin of the non-linearity in solids. It is important to point out *that acoustic waves are basically non-linear*, but that we usually use linearised expressions.

To get a feeling of the origin of this non-linearity, let us first consider a macroscopic presentation.

In all the following development tensorial notation will be used. It has two advantages:

- it gives an exact formulation of the phenomenon for those who are familiar with the notation,

- it gives a phenomenological expression for those who are not familiar with the notation, because they can read the expressions and ignore the indices.

In this notation each repeated index is supposed to represent a sum on this index (Einstein's convention) and δ_{ij} is the Kronecker-delta symbol.

1.1 MACROSCOPIC OBSERVATION

Let us look at the classical mechanical equation of motion of a solid

$$\rho \frac{\partial^2 u_i}{\partial t^2} = \frac{\partial T_{ij}}{\partial x_j} \tag{1}$$

ρ is the density, u_i a component of the mechanical displacement, T_{ij} the stress tensor, x_j the reference axes.

This is clearly not a linear expression.

a) ρ depends on the extension or compression of the solid and therefore on $\partial u_i/\partial x_i$

b) T_{ij} is a ratio of force to a surface element $dx_i dx_j$ which is deformed by the acoustic motion.

c) At a free surface, $x_3 = 0$, the normal stress T_3 of a surface wave should be zero. But the motion distorts the surface and $T_3 \neq 0$ at the surface $x_3 = 0$ which is no longer the fixed surface.

Is there a specific non-linearity for Rayleigh waves?

d) There is a well known phenomenon existing in a linear dielectric which relates the stress and the electric field E. It is the electrostatic force $T = \varepsilon E$.

In a very short time we have seen that acoustic non-linearities do not come only from the non-linear relationship between strain and stress.

1.2 ELASTIC NON-LINEAR THEORY

We give here the results of the theory elaborated by [1,2,3] and others . . .

Two reference frames are considered, the first one $\{\vec{a}_1, \vec{a}_2, \vec{a}_3\}$ corresponds to the solid at rest and is called the material coordinates.

In it, a material point M is characterised by its coordinates at rest $\{a_i\}$. After deformation the position of M is changed and becomes M $\{x_1, x_2, x_3\}$ with $x_i = f(a_1, a_2, a_3, t)$.

The second frame is the reference frame at the final state $\{\vec{x}_1, \vec{x}_2, \vec{x}_3\}$ after deformation.

It is called the spatial reference frame.

Some effects are easily described in the material system of coordinates $\{\vec{a}_i\}$, others such as the stress-strain relation are more easily described in the $\{\vec{x}_i\}$ system.

Thus displacement u_i is $u_i = x_i - a_i$ (2)

1.3 STRAIN

Strain is defined by the relative change of length between two material points due to the deformation. More precisely, for a small length ds_o which becomes ds after deformation, we consider

$$\lim_{ds \to ds_o} \left(\frac{1}{2} \frac{ds^2 - ds_o^2}{ds_o^2} \right) \tag{3}$$

Developing ds and ds_o in the $\{\vec{a}_i\}$ frame gives

$$ds^2 - ds_o^2 = dx_i\, dx_i - da_j da_j = 2\, S_{ij}\, da_i da_j \tag{4}$$

$$S_{ij} = \frac{1}{2} \left(\frac{\partial u_j}{\partial a_i} + \frac{\partial u_i}{\partial a_j} + \frac{\partial u_k}{\partial a_i} \frac{\partial u_k}{\partial a_j}\right) \tag{5}$$

S_{ij} is the strain and it is seen to contain non-linear terms.

1.4 MOMENTUM CONSERVATION

The conservation of the momentum expressed in the final reference frame $\{\vec{x}_i\}$ leads to

$$\rho \frac{\partial u_j}{\partial t^2} = \frac{\partial T_{ij}}{\partial x_i} \tag{6}$$

With the following boundary condition at a surface whose normal components are n_i

$$[n_i\, T_{ij}] = 0$$

The brackets [ ] mean the discontinuity of the expression within the brackets across the surface.

To get (6) we have to consider a stress T_{ij} related to the thermodynamic function χ, the free enthalpy of motion, by

$$T_{ij} = \rho \frac{\partial x_i}{\partial a_m} \frac{\partial x_j}{\partial a_k} \frac{\partial \chi}{\partial S_{mk}} + T_{ij}^{ES} \tag{7}$$

$$T_{ij}^{ES} = + \varepsilon_o E_i E_j - \tfrac{1}{2} \varepsilon_o E_k E_k \delta_{ij} \tag{8}$$

T_{ij}^{ES} is the electrostatic stress tensor.

In the electrical domain the polarisation vector P_i is given by

$$P_i = -\rho \frac{\partial x_i}{\partial a_m} \frac{\partial \chi}{\partial W_m} \tag{9}$$

The W_m are the electric field components in the material frame $\{\vec{a}_i\}$

1.5 TENSOR NON-LINEARITY

The general extension of χ as a function of the variables S_{ij} and W_k is

$$\chi = \chi_L + \chi_{NL} \tag{10}$$

$$\rho_o \chi_L = \tfrac{1}{2} c_{ijkl} S_{ij} S_{kl} - e_{m.ij} W_m S_{ij} - \tfrac{1}{2} \varepsilon_{mn} W_m W_n \tag{11}$$

It is a second order tensor while c_{ijkl}, $e_{m.ij}$, ε_{mn} are the linear elastic, piezoelectric and dielectric tensors respectively. Whereas the material non-linearity comes from the further extension:

$$\rho_o \chi_{NL} = 1/6\, c_{ijklmn} S_{ij} S_{kl} S_{mn} - \tfrac{1}{2} e_{m.ijkl} W_m S_{ij} S_{kl}$$

$$- \tfrac{1}{2} l_{mn.ij} W_m W_n S_{ij} - 1/6\, \varepsilon_{mnp} W_m W_n W_p \tag{12}$$

We have restricted ourself to the third-order extension.

c_{ijklmn} is the third-order elastic tensor

e_{mijkl} is the third-order piezoelectric tensor

l_{mnij} is the electrostrictive tensor

ε_{mnp} is the tensor or the third-order dielectric components.

These tensors represent the non-linearity of the material itself

ρ_o is the density of the material at rest.

1.6 GENERAL EXPRESSIONS in the MATERIAL FRAME

The material reference frame is the most interesting to use since it is the one for the solid rest. Using the previous notation various authors [1,2,3] have developed the following expressions

$$\begin{cases} \rho_o \dfrac{\partial^2 u_i}{\partial t^2} = \dfrac{\partial P_{ij}}{\partial a_j} \\ \dfrac{\partial D_i}{\partial a_i} = 0 \end{cases} \tag{13}$$

With the boundary condition

$$[N_i P_{ij}] = 0, \quad [N_i P_i] = \sigma \tag{14}$$

Where P_{ij} is the stress and P is the electric induction Piola Kirchoff Tensor and σ the surface density of charge.

We have

$$P_{ij} = P_{ij}^{(L)} + P_{ij}^{(NL)}; \quad D_i = D_i^{(L)} + D_i^{(NL)} \tag{15}$$

a) $P_{ij}^{(L)} = c_{ijkl}\, u_{k,l} + e_{m.ij}\, \phi,_m$

b) $P_{ij}^{(NL)} = c'_{ijklmn}\, u_{k,l}\, u_{m,n} + e'_{m.ijkl}\, \phi_{,m}\, u_{k.l} - \frac{1}{2}\, l'_{mn.ij}\, \phi_{,m}\, \phi_{,n}$

(16)

a) $D_m^{(L)} = e_{m.ij}\, u_{i,j} - \varepsilon_{mn}\, \phi_{,n}$

b) $D_m^{(NL)} = \frac{1}{2}\, e'_{m.ijkl}\, u_{i,j}\, u_{k,l} - \frac{1}{2}\, l'_{mnij}\, \phi_{,n}\, u_{i,j} + \varepsilon_{mnp}\, \phi_{,n}\, \phi_{,p}$

(17)

with
$$\begin{cases} c'_{ijklmn} = c_{ijklmn} + 2c_{ijkl}\, \delta_{jm} + c_{ijnl}\, \delta_{km} \\ e'_{m.ijkl} = e_{m.ijkl} + e_{m.il}\, \delta_{kj} \\ l'_{mn.ij} = l_{mn.ij} - \varepsilon_o(\delta_{im}\, \delta_{jn} + \delta_{in}\, \delta_{jm} - \delta_{ij}\, \delta_{mn}) \end{cases} \tag{18}$$

In these expressions symbols such as $u_{k,l}$ mean $\partial u_k/\partial a_l$ and ϕ is the electrical potential.

1.7 FINAL REMARKS

Expressions (13) to (18) are of great interest, and despite their complicated appearance are worth presenting in a review paper.

The main advantage of the formulation in (13) is to have all the different non-linearity terms in the Piola Kirchoff tensor. The left-hand side of (13) is linear. By examining (18) the two different origins of non-linearity can be recognized. The tensors c_{ijklmn}, l_{mijkl} and l_{mnij} contain non-linearity of the relationship between strain and stress, and are non-linearities intrinsic to the material. The other term in (18) with the linear tensors (c_{ijkl}, e_{mil}) and functions producing effects coming from the non-linearity of acoustic waves, we will call geometrical non-linearities such as density modulation, length variation and surface ripples.

The general formulation, (13), applies for bulk waves as well as for surface waves and there is no non-linearity specific to a surface wave.

It should be pointed out that the "primed tensors" such as C'ijklmn have lost the symmetry properties which exist in the c_{ijklmn} tensor.

The two separate mechanisms can be quantified. Let us consider first the non-linear coefficient C'_{111} where we have limited the number of indices, reducing a pair ij into one index k which takes on a value 1 to 6 in the usual way. Equation (16) becomes

$$C'_{111111} = C_{111} + 3C_{11} \qquad (19)$$

where in $LiNbO_3$ $C_{111} = - 510 \pm 200$ GPa where $C_{11} = 203$ GPa

in Quartz $C_{111} = - 210$ GPa and $C_{11} = 85$ GPa

It is seen that the geometrical non-linearity has the same order of magnitude as the tensor non-linearity and therefore can never be neglected.

Furthermore, for convolvers, which use both non-linearity and piezoelectricity, we can predict that generally speaking, the larger the piezoelectric coupling of the crystal the larger will be the convolution merit factor described below since due to geometrical effects there will always be a sufficient non-linear interaction to be detected in a high piezoelectric material. Table 1 shows this to be true.

2. PARAMETRIC EFFECT with RAYLEIGH WAVES, an OVERVIEW

2.1 PARAMETRIC INTERACTION

Studies on the parametric interaction of acoustic waves started in the 60's just after the success of electromagnetic parametric amplifications. In principle, the method of analysis is always the same.

First, (13) to (16) are resolved in a linear way by neglecting all non-linear terms. Wave solutions are found in the form of:

$$U_1 = A_1 \cos(\omega_1 t - \vec{k}_1\vec{r}) \text{ and } U_2 = A_2 \cos(\omega_2 t - \vec{k}_2\vec{r})$$

Table 1. Magnitude of the linear and non-linear coupling constants for various crystals.

Materials	Plane	Direction	M (10^{-1}) (V m/W)	ΔV/V	ΔV/V Refs.
PZT-8	Basal	All	4.95×10^{-1}	For PZT-4 242×10^{-1}	a
$LiNbO_3$	Y cut	Z	1.21×10^{-1}	246×10^{-1}	b
$Bi_{12}GeO_{20}$	(001)	(110)	1.02×10^{-1}	66×10^{-1}	c
ZnO	Basal	All	2.46×10^{-5}	44×10^{-1}	d
Quartz	Y cut	X	$<3.3\times10^{-6}$	11×10^{-1}	b

[a] H.Thomann: Electron. Letters 5, 25, 652 (1969).

[b] W.R. Smith, H.M. Gerrad, J.H. Collins, T.M. Reeder, and H.J. Shaw: IEEE Trans. Microwave Theory Tech. MTT-17, 856 (1969).

[c] E.A. Kraut, B.R. Tittmann, L.J. Graham, and T.C. Lim: Appl. Phys. Letters 17, 271 (1970).

[d] J.J. Campbell and W.J. Jones: IEEE Trans. Son. Ultrason. SU-15, 209 (1968).

where k_1 and k_2 are the wave vectors. (For electromagnetic waves we can apply the same expressions putting in the quasi statistic approximation $\vec{k} = 0$)

Then, (13) is written in the form:

$$\text{a)}\ \rho_0 \frac{\partial^2 u_i^{(L)}}{\partial t^2} - \frac{\partial P_{ij}^{(L)}}{\partial a_j} = \frac{\partial P_{ij}^{(NL)}}{\partial a_j} \qquad \text{c)}\ \frac{\partial D_i^{(L)}}{\partial a_i} = \frac{\partial D_i^{(NL)}}{\partial a_i}$$

$$\text{b)}\ [N_i P_{ij}^{(L)}] = [N_i P_{ij}^{(NL)}] \qquad \text{d)}\ [N_i D_i^{(L)}] = [N_i D_i^{(NL)}] \tag{20}$$

This is a linear inhomogeneous set of equations where the non-linear terms on the right-hand side of the equations appear, as electrical and acoustic sources; bulk sources in (20)a and c; surface sources in (20)b and d.

Such sources have the form of $U_1 U_2 + U_1^2 + U_2^2$

Parametric effects are interactions of the U_1 and U_2 type. The result is 2 pseudo waves

$A_1A_2/2 \cos[(\omega_1 \pm \omega_2)t - (\vec{k}_1 \pm \vec{k}_2).\vec{r}]$

Those are genuine waves if $\vec{k}_1 \pm \vec{k}_2$ is a possible wave vector at the frequency $\omega_1 \pm \omega_2$. If one wave at $\omega_1 \pm \omega_2$ exists prior to the interaction there is parametric amplification of this wave.

2.2 PARAMETRIC ANTENNA

The first demonstrated parametric effect with acoustic waves which had a clear technological impact was the parametric antenna proposed by *WESTERVELT* [4] in underwater acoustics. He considered a transducer array working in the 30 kHz range, launching two waves at frequencies f_1 and f_2. As long as the two other waves propagate in the water, they generate, due to the parametric interaction, two other waves at frequencies f_1 and f_2 which propagate at the same velocity. But the f_1 and f_2 waves are far less attenuated than others and, at a distance of a few hundred meters, they remain practically alone. An antenna, tens of meters long has been realised. A beam is formed with a directivity which would have required, with a linear transducer, a gigantic antenna (60 wavelengths i.e. 90 cm at 1 kHz). Nowadays, one of the most striking applications of such an antenna is for sounding underwater sediments. Only low frequency waves (of the order of kHz) can propogate into sediments. But at this frequency the low achievable directivity of linear antennas does not allow good image resolution. Therefore a parametric antenna is used. The pump frequency is in the 100 kHz range and the narrow beam formed in the 5 kHz range.

It is not surprising that the first published paper on parametric interaction using Rayleigh waves [5] is an exact transposition of the antenna technique above. But it is not very attractive in solids since the medium of propagation is limited by the side of the crystal.

2.3 EARLY DEMONSTRATION of CONVOLVERS

Svaasand's configuration [6] is much more innovative. The parametric interaction takes place on a piezoelectric substrate between two contradirected waves at practically the same frequency ω. Moreover the resultant is not a wave. It is an electric field at 2ω and $k = 0$, i.e. with no spatial periodicity, which is detected by a simple electrode.

But the idea of the convolver, the first recognition that Svaasand's configuration can be used to compute the convolution of two signals launched at the two acoustic inputs, was published by *QUATE* [7]. It is said, however, that the idea was proposed by *WANG* [8]. As a consolation W.C. Wang first published how to enhance non-linear effects by using a semiconductor piezoelectric substrate [9].

Figure 1 reproduces the drawing of [7] and its photographs of two signals $s_1(t)$ and $s_2(t)$ at a frequency ω. At a given point x, parametric interaction produces surface charges (see (18)b) in the form.

$$\sigma_{NL}(x) = S_1(t - \frac{L}{2v} - \frac{x}{v}) \; . \; s_2(t - \frac{L}{2v} + \frac{x}{v}) \cdot \cos 2\,\omega t \qquad (21)$$

L is the interaction length, v the acoustic wave velocity

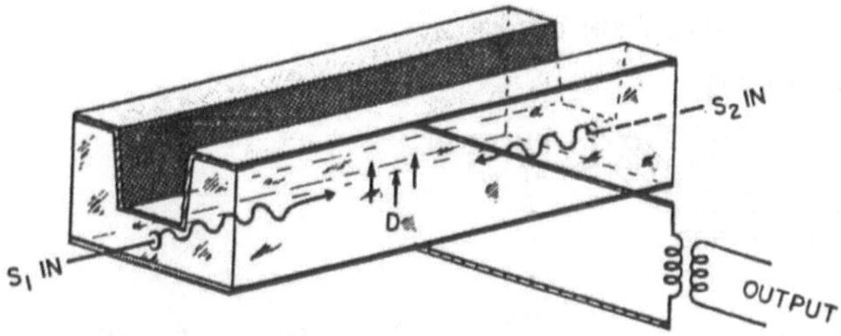

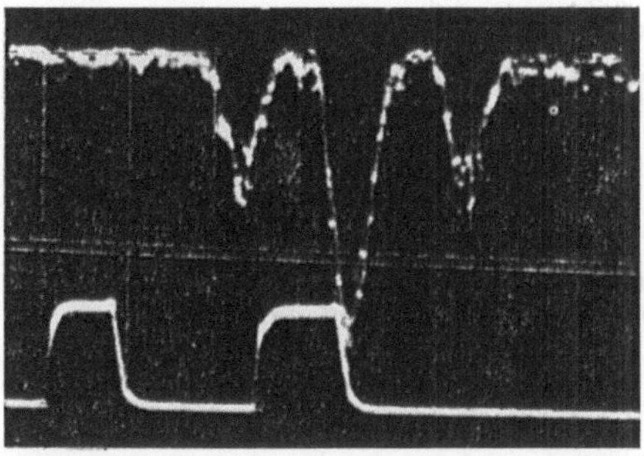

Fig. 1

Illustration of the convolution signal (upper trace) at 2700 GHz as generated by two similar but opposite sound waves at 1350 GHz. The two sound waves are amplitude-modulated by the double pulse envelope illustrated in the lower trace. Horizontal time scale is 2 μ sec/cm.

The output voltage is

$$V_{NL}(t) = \cos 2\,\omega t \int_{L/2}^{L/2} s_1\left(t - \frac{Td}{2} - \frac{x}{v}\right)\ s_2\left(t - \frac{Td}{2} + \frac{x}{v}\right)dx \qquad (22)$$

which is the convolution product of s_1 and s_2, compressed in time by a factor 2, if the duration of the two signals are shorter than $Td = L/v$, the propogation time.

As can be seen, bulk waves were used by [7] and the working frequency was in the GHz range.

LUKKALA and *KINO* [10] came back to Rayleigh wave interactions with a generalisation of the required configuration of the output electrode when the frequencies ω_1 and ω_2 of the two input waves are different.

Then the parametric interaction gives pseudo waves of frequency $\omega_1 \pm \omega_2$ and wave number $k_1 \pm k_2$.

To collect the output an interdigital transducer of periodicity ρ is needed with

$$2\,\pi/p = k_1 \pm k_2$$

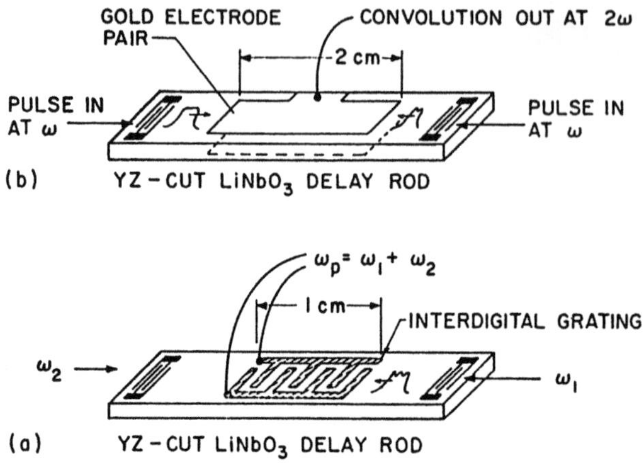

Fig.2: - *C.F. QUATES and al* [7] demonstration
- The two different electrode configuration as seen by [10]

This is the non-degenerate case, while for the degenerate case:

$$\omega_1 = \omega_2 = \omega; \quad k_1 = -k_2 = -k;$$

Figure 2 coming from [10] recalls the two definitions.

It should be noted that one of the reasons for the success of the device comes from the fact that the output configuration is only matched to one wave or pseudo-wave. All the others for which the spatial periodicity is not matched are filtered out.

However some residue of two incoming waves which have very large amplitude compared to the parametric counterpart may be received at the output. An electrical filter is thus required to select the appropriate frequency at the output.

2.4 CONVOLVER MERIT FACTOR

To define the merit factor of a convolver let us consider two plane contradirective waves of the form

$$U_i^{(1)} = A_1 U_i (kZ) \cos(\omega t - kx) \quad \text{and}$$

$$U_i^{(2)} = A_2 U_i (-kZ) \cos(\omega t + kx)$$

A_1 and A_2 are the amplitudes of the waves, Z is the transversal direction U_i (kZ) the transversal dependance of the wave for unit amplitude.

The convolver output voltage has the form:

$$V_a = A_1 A_2 \int C_{NL} (kZ) \, |U_i (kZ)|^2 dZ \tag{23}$$

C_{NL} (kZ) comes from all the non-limear terms in (16 b) and (17 b) and we have also used the fact that $U_i(-kZ)$ is the complex conjugate of $U_i(kZ)$.

(23) is very similar to the power flux P per unit width W in the x direction which has the form:

$$\frac{P}{W} = A_i^{\,2} \int C_L(kZ)|U_i(kZ)|^2 dz \tag{24}$$

By comparing (23) and (24) we can write

$$\left.\begin{aligned} V_o &= M\,\frac{P_1 P_2}{W} \\ M &= \frac{C_{NL}(kZ)|U_i(kZ)|^2 dZ}{C_L(kZ)|U_i(kZ)|^2 dZ} \end{aligned}\right\} \tag{25}$$

M is the factor of Merit of the substrate.

It can be seen from (25)b that M is independant of the frequency ω in a degenerate convolver. In a degenerate case, M will depend on $(\omega_1 - \omega_2)$ but not on $(\omega_1 + \omega_2)$.

Table 1 taken from [12] gives the M value of different materials. It can be seen as said earlier, that roughly speaking, the higher the piezoelectric coupling factor, the higher the value of M.

However, at equal values of $\Delta V/V$, PZT-8 is much more non-linear than $LiNbO_3$-YZ.

It is straightforward, starting from (25)a, to deduce the output available power P_o as a function of the two input electrical powers, G and P_2, taking into account the two transducer losses α_1 and α_2, the propagation loss α_o. The output has capacity C and loss given by Q.

a) $P_{out} = F\, P_1 P_2$

b) $F = \alpha_o \alpha_1 \alpha_2 \dfrac{QC\omega}{4} \dfrac{M^2}{W^2}$ (26)

Here F is the convolver factor of merit, usually expressed in - dBm, when P_{out}, P_1 and P_2 are expressed in dBm. To give an order of magnitude, in the experiments quoted earlier F was - 90 to - 100 dBm.

2.5 GUIDED RAYLEIGH WAVE CONVOLVER

In the description above, Rayleigh waves have no particular advantage over bulk waves. The merit factors are the same. But in fact it is the ability to guide Rayleigh waves which allows the surface configuration to be more

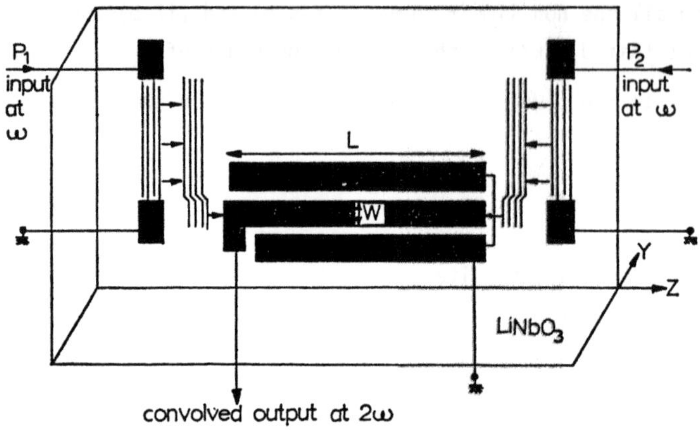

Figure 3. Diagram of b.w.c. convolver [5].

effective [13,14]. We have given [15] a clear demonstration of the efficiency of such a guided structure. A guide 5 wavelengths wide was used on $LiNbO_3$ and excited through a multistrip beamwidth compressor which allowed for impedance matching of the transducer. Furthermore, the ground electrodes were placed beside the convolution electrode, giving a full planar configuration and keeping the output capacitance at a reasonable value independant of W (fig. 3). Therefore, since the beamwidth was reduced by 10, the F factor was increased by 20 dB according to (26)b. Indeed, the F merit factor was - 73 dBm. The device was able to process a signal bandwidth of 50 MHz with an interaction time of 12 μs. Pulse compression experiments performed with this device demonstrated the ability to provide efficient processing.

Since then, newly developed Rayleigh wave convolvers have had the same structure:

- guided waves in the interaction region
- planar configuration
- guide excitation is done, not only by multistrip beamwidth compressors, [15], but also by means of horns, lenses and focussing transducers [13].

3. RAYLEIGH WAVE CONVOLVERS - STATE of the ART

3.1 GENERAL DESCRIPTION of BILINEAR COMPONENTS

Stimulated by important applications in Spread Spectrum Communication [16] researchers have produced convolvers with a very high degree of sophistication and performance. But now, we need an accurate way of measuring such performance and we have to come back to earlier definitions. For this Morgan's analysis is of particualr interest [17].

3.2 DEFINITION of BILINEAR PROCESSOR

Let us consider a processor with two inputs and one output. It will be called a bilinear processor if the output is a linear function of a signal applied to either of the inputs when the second is maintained constant.

3.3 BILINEAR PROCESSOR PROPERTIES

Let $s_1(t)$ and $s_2(t)$ be the two signals applied at the inputs, $S_1(\omega_1)$ and $S_2(\omega_2)$ their Fourier Transforms. It is easily demonstrated, following Morgan's suggestion, that the output signal V(t) can be written in the form (as long as the bilinearity assumption is still valid)

$$V_o(t) = \frac{1}{4\pi^2} \iint_{-\infty}^{+\infty} H(\omega_1,\omega_2)\, S_1(\omega_1)\, S_2(\omega_2)\, e^{j(\omega_1 + \omega_2)t}\, d\omega_1 d\omega_2 + \text{c.c.} \quad (27)$$

$H(\omega_1,\omega_2)$ is the bilinear transfer function. The knowledge of $H(\omega_1,\omega_2)$ is necessary and sufficient to describe completely the performance of the bilinear processor.

H has the following properties:

$$\begin{cases} H(\omega_1,\omega_2) = 0 \quad \text{if } \omega_1 \times \omega_2 < 0 \\ H(-\omega_1,-\omega_2) = H^*(\omega_1,\omega_2) \end{cases} \quad (28)$$

H is the two dimensional Fourier Transform of the bilinear impulse response $h(t_1,t_2)$. $h(t_1,t_2)$ is real because of (28).

$$h(t_1,t_2) = \frac{1}{4\pi^2} \iint_{-\infty}^{+\infty} H(\omega_1,\omega_2)\, e^{j(\omega_1 t_1 + \omega_2 t_2)}\, d\omega_1 d\omega_2 \quad (29)$$

Note that in (28) the frequency $\omega_1 + \omega_2$ has been taken deliberately to fit the degenerate convolver case. An equivalent analysis using $\omega_1 - \omega_2$ could have been made depending on the proper choice of the central frequency of the output filter.

We can also use the two-dimensional convolution theorem

$$V_o(t) = \iint_{-\infty}^{+\infty} h(t_1,t_2)\, s_1(t-t_1)\, s_2(t-t_1)\, dt_1 dt_2 \quad (30)$$

A multiple operator is such that

$$\begin{cases} H(\omega_1,\omega_2) = m_o \\ h(t_1,t_2) = m_o\, \delta(t_1)\, \delta(t_2) \end{cases} \quad (31)$$

δ is the delta function

An ideal convolver is such that

$$\begin{cases} h(t_1,t_2) = \sqrt{F}\ \delta(Td - t_1) - t_2) \\ H(\omega_1,\omega_2) = \sqrt{F}\ e^{-j\omega_1 Td}\ \delta(\omega_1-\omega_2) \end{cases} \tag{32}$$

It is easy to verify that, in putting (32) into (27) (or (30)), we obtain (22).

In setting

$$T_1 = \frac{Td}{2} + \frac{x}{v} \quad \text{(and therefore } t_2 = \frac{T_2}{2} - \frac{x}{v}\text{)}$$

t_1 and t_2 are related to the position of the interaction:

$$\frac{2x}{v} = t_1 - t_2 \tag{33}$$

which gives a physical meaning to their definition.

For a perfect convolver with finite plate length, T_0 being the propagation time along it, the transfer function is

$$\begin{cases} H(\omega_1,\omega_2) = \sqrt{F}\ e^{-i(\omega_1+\omega_2)\frac{Td}{2}}\ \dfrac{\sin(\omega_1-\omega_2)T_p/2}{(\omega_1-\omega_2)T_p/2} \\ h(t_1,t_2) = \sqrt{F}/T_p\ \ \mathrm{rect}(T_p-t_1-t_2).\ \delta(Td-t_1-t_2) \end{cases} \tag{34}$$

Figure 4 taken from [17] shows a representation of (34).

When the signal durations are short enough compared to Tp, (34) is equivalent to (32).

Note that F has the same meaning as in (25)a

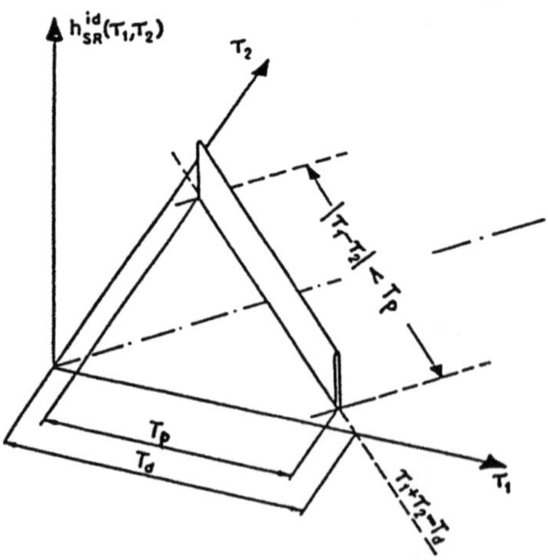

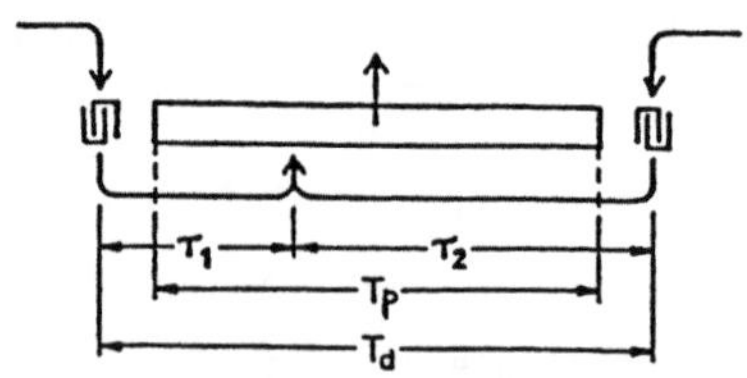

Figure 4. Bilinear impulse response of ideal convolver after [17]

3.4 ACTUAL CONVOLVER

It will be shown that actual convolvers do not have transfer functions exactly as in (34). Theoretically, a complete characterisation of the convolver requires the measurement of $H(\omega_1,\omega_2)$ over the whole frequency domain ω_1 and ω_2.

Looking at (34) it appears that $\omega_1+\omega_2$, $\omega_1-\omega_2$, t_1+t_2, t_1-t_2 are better choices of variables. Furthermore, in most cases, the knowledge of two special cuts in the frequency domain or in the time domain are sufficient for characterisation.

- The $\omega_1 - \omega_2 = 0$ cut - i.e. $H(\omega_1 + \omega_2, \omega_1 - \omega_2 = 0)$

This is the *degenerate transfer function*, often called the *output transfer function.*

- The $t_1 + t_2 - Td = 0$ cut, i.e. $h(t_1 + t_2 = Td, t_1 - t_2)$ which gives an idea of the uniformity of the convolution efficiency versus $\frac{x}{2v} = t_1 - t_2$

In fact, the uniformity function is the one-dimensional Fourier Transform of $h(t_1 + t_2, t_1 - t_2)$:

$$\text{a)}\quad H(\omega_1+\omega_2,t_1-t_2) = \frac{1}{2\pi}\int M(\omega_1,\omega_2)\, e^{j(\omega_1-\omega_2)\,(t_1-t_2)}\, d(\omega_1-\omega_2)$$

$$\text{b)}\quad = \int_{-\infty}^{+\infty} h(t_1,t_2)\, e^{j(\omega_1+\omega_2)\,(t_2+t_1)}\, d(t_1+t_2) \qquad (35)$$

The equivalence between (35)a and (35)b comes from Fourier Transform properties.

3.5 DETAILED ANALYSIS of MODERN CONVOLVERS

An actual convolver will differ from the ideal presentation (32) in two ways.

a) defects which give non-bilinear behaviour

b) defects which deteriorate the convolver response but converse the bilinear property.

3.5.1. NON-LINEAR BEHAVIOUR DEFECTS

The main non-bilinear behaviour is *autoconvolution*. The convolution of one signal on itself after reflection at the 2nd transducer which is similar to the partial convolution between s(t) and s(t-Td). This spurious signal is large if $s_1(t)$ is a long and symmetrical signal.

Unfortunately, this is the case of a CW signal emitted by a jammer. By definition, the autoconvolution suppression is the ratio of the autoconvolution of a CW signal S_1 applied to one part to the convolution of two CW signals of the same frequency and the same amplitude S_1.

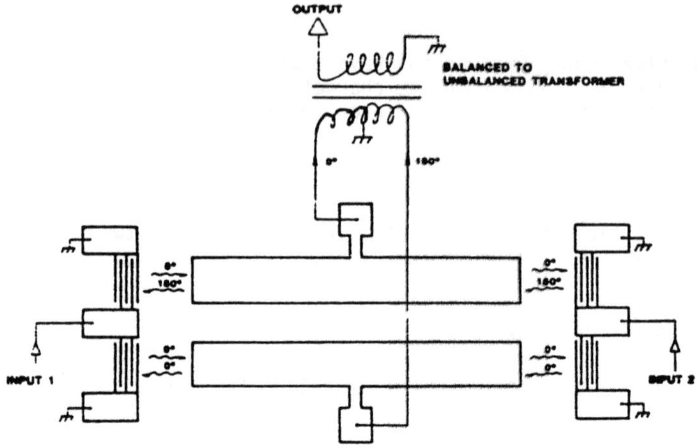

Figure 5. Schematic diagram of the dual-track convolver structure for self-convolution suppression, after [19]

In the single configurations presented previously autoconvolution is about -25 dB or -30 dB, which is not sufficient. In order to improve the device a dual configuration is proposed by MOTZ [18] and Fig. 5, taken from [19], explains how it works.

By using this technique, autoconvolution is reduced down to -40 or -45 dB.

3.5.2 SATURATION

The dynamic range of a convolver is limited at low signal levels, by thermal noise and at the high levels by saturation. A definition of dynamic range has been given in [20]. Usually, the -1 dB saturation point is achieved at input levels of the order of 30 dBm which is high enough for many applications.

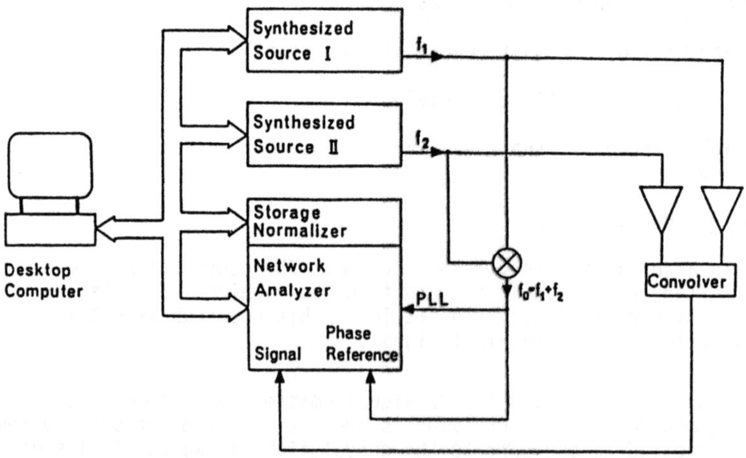

Figure 6. Schematic of set up for measurement of principal two-dimensional frequency response [21]

3.5.4 GUIDE DISPERSION and PHASE CORRECTION

Guided Rayleigh wave propagation is by nature dispersive. It can be shown from (22) or (30) that the result of guide dispersion is equivalent to the response of an ideal convolver followed by a dispersive delay line whose dispersion is the same as that of a guide of the same width and length.

The transfer function becomes

$$H(\omega_1,\omega_2) = e^{j\omega_1 T_d(\omega_1)} \delta(\omega_1 - \omega_2)$$

Now, we can analyse the different aspects.

3.5.3 BILINEAR DEFECTS

The first question is how to measure the bilinearity? The block diagram of a measurement set up is given in Fig. 6 taken from [21]. It has also been proposed by [22]. It provides precise measurement of $H(\omega_1,\omega_2)$. Other characteristics such as the uniformity function are obtained by computing the appropriate Fourier Transform.

Again by analysing (27), it is seen clearly that this phase distortion can be compensated by using the proper filter at the input. [23] and [24] propose the use of two almost identical dispersive transducers at the input. The slight difference on the dispersion of the two transducers compensates the guide dispersion.

3.5.4 MULTIMODES

Higher modes can perturb the convolver behaviour as bulk waves modify the transfer functions of SAW filters.

To simplify we will suppose that the only two modes which can exist at a given frequency in the guide are the S and the A modes. S (for symmetrical) is the main one. Then the $s_1(t)$ input launches two waves A+ and S+ propagating at two different velocities V_A and V_S which interact with the two equivalent contradirected waves A- and S-. The parametric interaction gives four products

- two terms which we will call the autoconvolutions A+ A- and S+ S-
- two terms which are the cross convolutions A+ S- and A- S+

In the later case the parametric wave number should be zero in order that a signal be collected by the plate. This condition occurs for

$$\omega_1/VA = \omega_2/VS \quad \text{and} \quad \omega_1/VS = \omega_2/VA$$

Mathematically the impulse response and transfer function are:

$$h(t_1,t_2)=F_S\delta(t_1+t_2-T_s)+F_A\delta(t_1+t_2-T_A)+F_{AS}[\delta(t_1+\frac{V_A}{V_S}t_2-T_S)+\delta(t_2+\frac{V_A}{V_S}t_1-T_A)] \tag{36}$$

$$H(\omega_1,\omega_2)=(F_S e^{-j\omega_1 T_S}+F_A e^{-j\omega_1 T_A})\delta(\omega_1-\omega_2)+F_{AS}e^{-j\omega T}[\delta(\omega_1-\frac{V_S}{V_A}\omega_2)+\delta(\omega_1-\frac{V_A}{V_S}\omega_2)]$$

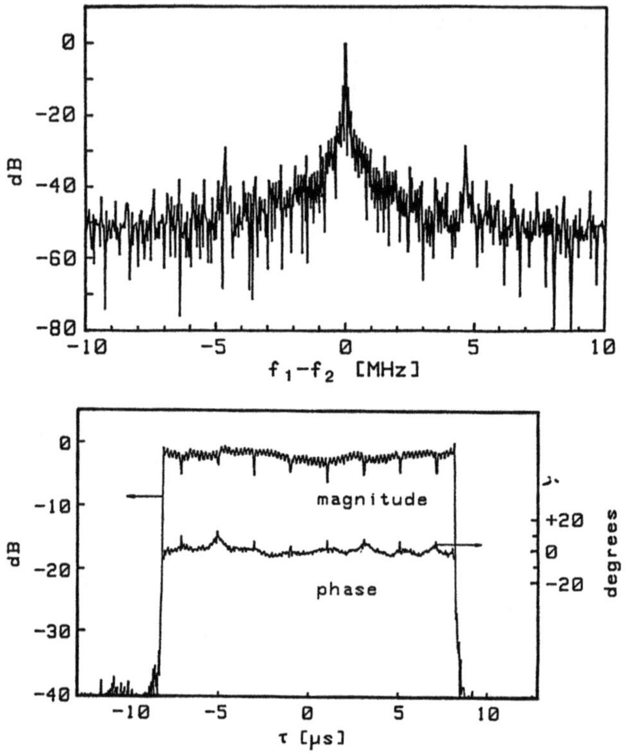

Figure 7. Measurement of convolver uniformity [20]

a: in the frequency domain: $H(2\omega_0,\omega_1 - \omega_2)$

b: in time domain $h(2\omega_0,t_1 - t_2)$

Figure 7 taken from [20] gives a clear demonstration of the existence of the cross convolution for the cut $\omega_1 + \omega_2 = 2\omega_0$

It is clear that a multimode structure should be avoided when designing a convolver.

3.5.5 CHOICE of the GUIDE WIDTH

To avoid multimoding effects, it is necessary to reduce the width of a guide in such a way that only the first symmetrical mode exists. Since we know how to compensate the dispersion, the guide width can be quite small. In fact, the restriction comes from the efficiency. *PLANAT* and al [25] showed that the M factor for a 3λ guide is reduced to 0.6 10^{-4} V/m (instead of 1.2 10^{-4} V/m for a large guide).

If we take into account the difficulties in exciting the guide mode, the 3λ guide on YZ $LiNbO_3$ seems the best compromise, the electrode thickness being kept as thin as possible.

Fig. 8

3.5.6 PLATE UNIFORMITY

To get (22) we assumed that amplitudes of all the parametric charges collected at a position x are added in phase. But errors are introduced because:

a) Due to plate resistivity, loss appears along the distance to the output tap.

b) On long plates, electromagnetic propagation leads to phase rotation, preventing correct addition of local parametric charges.

The only way of overcoming both difficulties is to use a multi-tap correcting structure along the convolution plate. Each tap should be connected to electromagnetic lines of exactly the same delay as proposed by [26]. A small loss efficiency appears at the position of each connection as is clearly visible in Fig. 7. The photograph of Fig. 8 shows details of this structure.

3.5.7 GUIDE EXCITATION

Many different ways of exciting the guided structure have been used. A simple transducer having the same width as the guide has a very high impedance which gives high losses when operating with large bandwidth signals. Beam-Width Compressors (BWC), focussed transducers, and chirp transducers have been proposed for exciting the guide.

3.5.8 BEAMWIDTH COMPRESSOR

The Multistrip BWC has already been mentioned [15]. A very attractive design has been made by *RONNEKLEIV* [27] which excites only the first symmetrical mode. By taking into account diffraction effects, the Multistrip BWC [27] shapes the output beam profile to match that of the symmetrical guided mode.

Hornes have also been widely studied. (See for example [19]).

3.5.9 CHIRP TRANSDUCERS

A long chirp transducer also has a small input impedance even if its width is no more than 3 wavelengths. An efficient guide exciter has been demonstrated, [28]. Here the two chirp transducers, at each input, must be identical.

With this type of excitation we can also compensate for the dispersion in the guide as mentioned earlier, however, diffraction along the transducer must be controlled.

3.5.10 FOCUSSED TRANSDUCER

The focussed transducer was used by [29] while [30] used a focussed chirp transducer. The latter has the advantage of controlling diffraction along the transducer as well as optimizing the guide excitation over all the bandwidth, (Fig. 9).

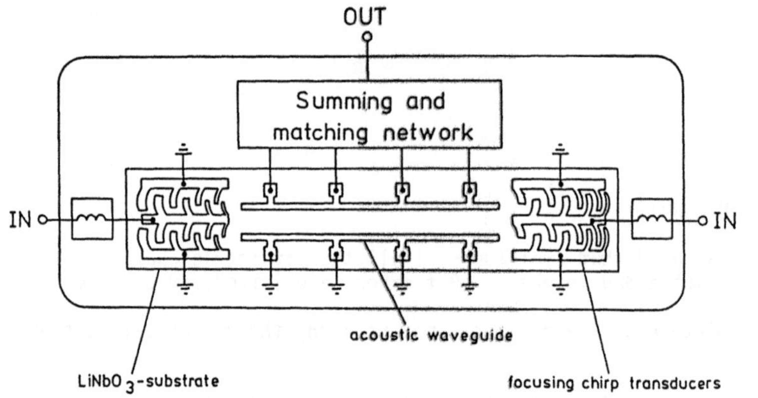

Figure 9. Dual track convolver using chirp focusing transducer [30]

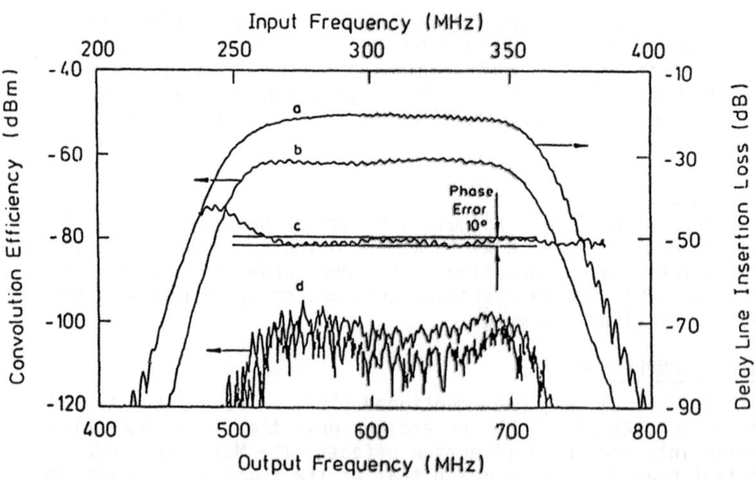

Figure 10. Its degenerate transfer function

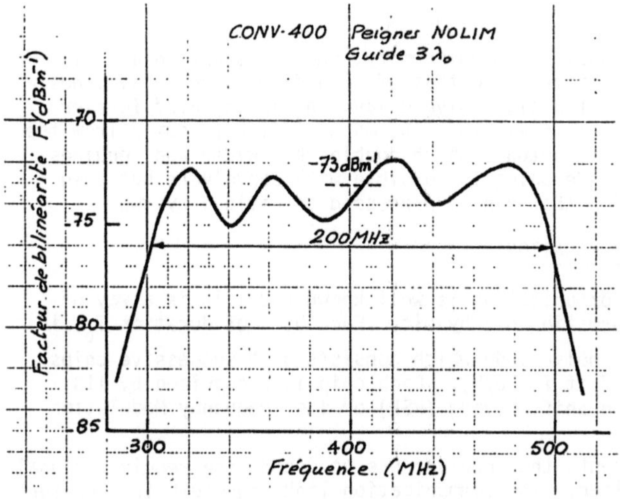

Figure 11. Degenerate transfer function of 200 MHz input bandwidth convolver [31]

3.6 PERFORMANCE

All these new improvements have lead to high performance devices with high efficiency and low distortion.

As an illustration we will consider the device mentioned earlier in (27) using a dual-track convolver, and focussed dispersive transducers. Figure 10 shows its degenerate transfer function. The very flat response and small phase errors are demonstrated. The device handles signals of 95 MHz bandwidth and 16 μs duration.

A 200 MHz input bandwidth convolver has been demonstrated [31] and its degenerate function transfer is illustrated in Fig. 11.
Table 2 gives some examples of the state of the art performance.

Table 2. Example of convolver performance

Bandwidth (MHz)	20	95	20	200
Interaction Time (μs)	32	16	30	12.5
Factor (+ dBm)	-60	-61	-63	-73
Amplitude Ripple (dB)	dB	dB	dB	+1 dB
Phase Ripple (^{o})	±10°	±15°	±10°	N.C
Spatial Uniformity	1 dB	1 dB	dB	±0.5
Autoconvolution (dB)	-35	-45	-45	-30
Guide Excitation type	MBWC	Dispersive & Focussing	Horns	MBWC + Dispersive Transducer
Tracks	Double	Double	Double	Single
Origin	TH CSF	Siemens	RSRE (29)	TH CSF (28)

4. APPLICATIONS

The performance of the convolver in uniformity, phase distortion and BT product has become very comparable to that of good dispersive delay lines. But the transfer function of a dispersive delay line is engraved forever whereas the convolver is more flexible. The main difficulty comes from the time compression by 2 of the outputs which doubles bandwidth and requires special care. Synchronous decoding is possible with convolvers but then the largest signal duration that can be processed is divided by two.

4.1 SPREAD SPECTRUM COMMUNICATION

The principle of spread communication is well known [32], it is a way of transmitting digital information bit by bit. Each bit, of duration T_b, is in fact coded by a pseudo noise code which consists of N successive chips (therefore the duration of it is Tb/N). Each chip is recognized by its phase modulation for Phase Shift Keying (PSK) or its Frequency Modulation (FSK).

To recognize the bit sent, the receiver has to decode the received signal with the code matched filter. The communication link bandwidth is therefore spread by N. Thanks to the processing gain, the receiver can detect the signal below thermal noise. Therefore, someone who did not have the correct decoder would not only be unable to recognize the message but also would not recognize its existence. For more secure links the codes are changed very often (for example after each word). As a consequence, a programmable filter is necessary.

With the convolvers of table 2, chip rates could reach 100 MHz, thus each code could have a BT product as high as 1028.

Figure 12 shows the results of the decoding by a convolver of a PN Gold's code of 512 chips of 12.5 ns each.

The autocorrelation peak is so short that it is not clearly visible on the photograph. The small arrow marks its location. For a signal-to-noise ratio (S/B) of 10 dB the decoded signal is close to the theoretical autocorrelation function. For S/B = 0 dB, the figure is similar. For S/B = -20 dB, the peak is still there well above threshold.

4.2 ASYNCHRONOUS MODE

In spread spectrum communication, the message has a preamble in order to measure the Time of Arrival (TOA) [33]. After the preamble the receiver is synchronised properly and the information of the message is decoded in a synchronous way. By definition, detection of the preamble is asynchronous. But even then, despite the compression in time of the output, the convolver could measure TOA, for example as shown in Fig. 13 [34]. The convolver used is supposed to have an interaction time of 2T and it possesses a preamble of duration T. The reference signal is launched every 2T at port 2, at time t = 0, 2T, 4T . . . It is clear that, during the time intervals (T, 2T), (3T, 4T) a reference signal is entirely in the interaction region. Then any signal arriving from port 1 would be intercepted.

At the output port 3 during these time intervals, we observe the exact convolution of the signal but compressed in time by two. Apart from the time compression, it is the same signal that would be observed at the output of a matched Rayleigh interdigital filter during the time of arrival

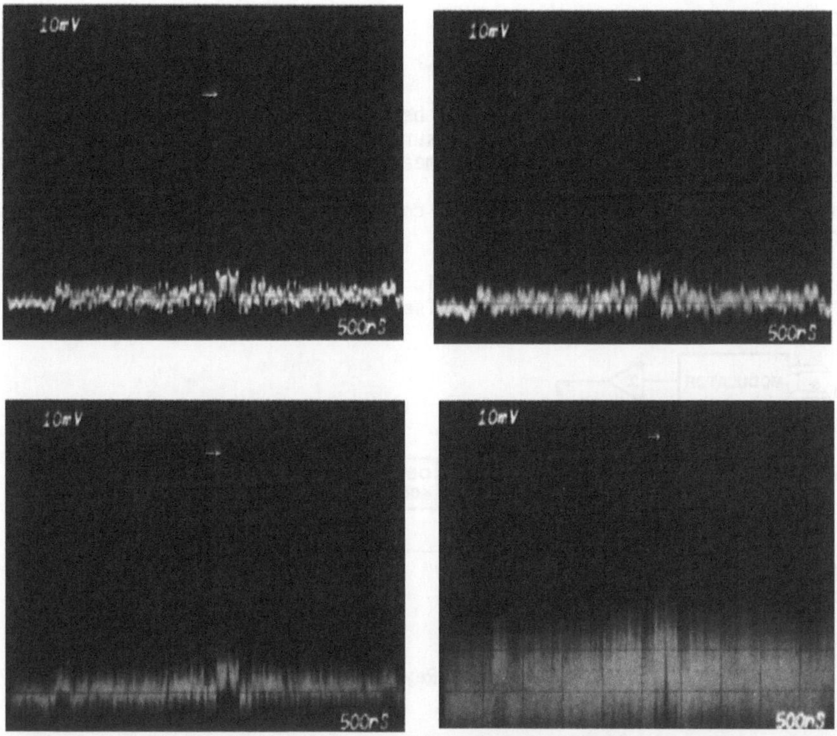

Figure 12. Demonstration of convolution of two matched PN gold's codes of 512 chips of 12.5 ns

a : S/B = 10 dB   c : S/B = -10 dB
b : S/B = 0 dB   d : S/B = -20 dB

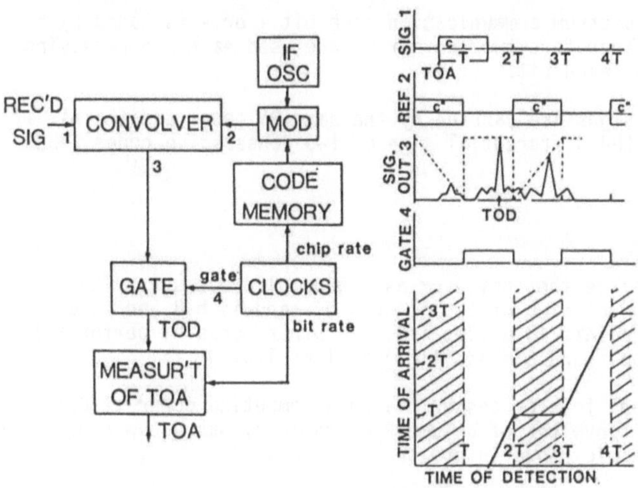

Figure 13. Simplified block diagram of a spread spectrum system employing one convolver operating in an asynchronous mode to detect and measure T.O.A. of incoming signals [34]

$$\left|\frac{T}{2} \quad \frac{5T}{2}\right| \quad \left|\frac{5T}{2} \quad \frac{7T}{2}\right|.$$

Therefore there is a unique relation between the time of arrival measured at port 1 and the time of detection measured at port 3. This relation is shown in Fig. 14. The TOA is properly measured.

During the period (0,T) (2T, 3T) the collected signal of port 3 has no significance and should be gated out.

This mode of operation is not specific to PSK codes. The same techniques could be used to detect the expanded pulse of a pulse compression radar.

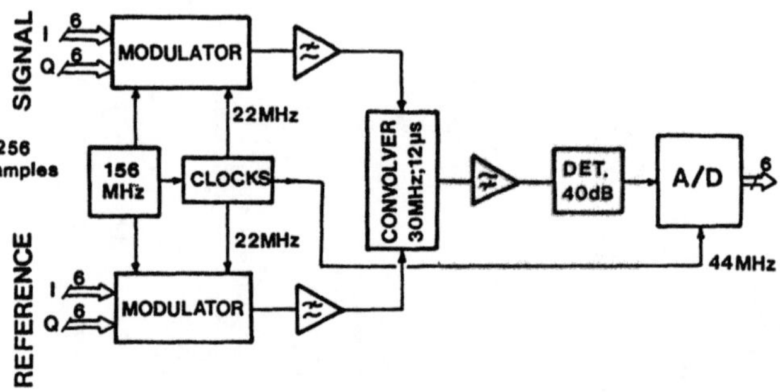

Figure 14. Digital correlator using a Rayleigh wave convolver [35]

4.3 SYNCHRONOUS MODE

Obviously, detection in the synchronous mode is easier, because the TOA is known, the whole interaction time can be used.

Usually in spread spectrum communication each bit + or - is coded by two orthogonal PN codes. Therefore two convolvers are used each one receiving the reference signal of each bit.

In other cases bit signs are carried by the same PN code, and the bit + or - is recognized by the differential sign of two consecutive codes [33] (System DPSK).

4.4 DIGITAL CONVOLVER

Let us try to characterise the convolver as a general processor. The convolver performs the convolution of two signals of bandwidth B and duration T, i.e. BT information points in a time 1/B. In other words it performs BT analogue multiplications (and one addition) in time 1/B.

It is thus equivalent to a processor having a computing power of BeT FLOPS. Therefore the convolver of column 2 of table 2, has a power of 150 G FLOPS! The dynamic range is greater than 60 dB.

Since digital multipliers are bulky and power consuming, why not use an analogue convolver as an ultra fast digital processor?

This is the idea of [35] for which the schematic of the subsystem is presented in Fig. 14. The 256 samples of the input signal at a clock rate of 22 MHz, (6 bits for the imaginary part 6 for the real part) are converted to analogue. The same for the reference. The convolution is performed in a convolver with 30 MHz bandwidth 12 μs interaction time, and is then connected back to digital form at a clock frequency of 44 MHz. This subsystem which consumes 40 W is used in a side-looking radar to resolve the lateral range cell.

4.5 FUTURE OF CONVOLVERS - CONVOLVERS VS DISPERSIVE DELAY LINES

Today, convolvers have reached a high degree of sophistication and performance and I am a bit disappointed that no large scale application has appeared. Spread Spectrum Communication Convolvers are still in a prototype stage.

However, it is reasonable to use convolvers in pulse radar detection for two reasons.

a) The transfer function is as well controlled as in a dispersive delay line.

b) More and more often the expanded pulse is generated digitally.

Then, why not use the same signal generator to send periodically the matched code to the second port of the convolver in the asynchronous mode described in Fig. 13?

The advantages of such a subsystem are clear:

i) FLEXIBILITY

a) Possibility to change the code after each radar pulse

b) Same component can be used whatever the distance range of the radar

c) Reference signal can be modified to take Doppler into account

ii) THERMAL INSENSIBILITY

In convolvers the time delay variation with temperature does not deteriorate the convolution. Only the time of detection is slightly modified.

5. OTHER RAYLEIGH WAVE NON-LINEAR COMPONENTS

There are many other proposed non-linear Rayleigh Wave Components. Most of them take advantage of acoustoelectronic effects.

There is a recent component proposed by *COLVIN* [36] which should be mentioned, namely the Rayleigh Wave Mixer. Two Rayleigh waves at frequencies f_2 and f_3 are launched in the same direction. The parametric interaction generates two waves at frequencies $f_2 - f_3$ and $f_2 + f_3$. The output transducer is matched to one of these frequencies.

COLVIN [36] found one main advantage in this SAW mixer; it is a true bilinear device, i.e. the linearity in both inputs is larger than 60 dB. It is a true analogue multiplier. A double balanced mixer is not a bilinear component: linearity with respect to one input is only achieved if the signal applied to the other input is high and consistent.

A secondary advantage lies in the fact that on the same crystal one can put a mixer plus an interdigital filter, i.e. an entire IF receiver, and this could be a genuine advantage.

6. CONCLUSION

To summarize, it appears that parametric Rayleigh Wave Interaction gives rise to the convolver. A component which has no counterpart in other areas of Physics. In principle the convolver requires two low-velocity waves propagating in opposite directions and one very fast wave to collect the parametric result.

Rayleigh Wave Components appear to be very good bilinear processors. In other words the main characteristics of Rayleigh wave non-linear components is *their ability to implement high performance, analogue multiplications.*

High performance means:

a) True bilinear properties

b) Ultra-high speed (more than 100 G FLOPS) in a very small volume.

I still believe that this very attractive performance should be incorporated in digital processors where multiplication is a difficult, bulky and power consuming operation especially if one has in mind high-speed processors.

I expect the growth of high-speed digital processors will boost the need for Rayleigh Wave Components, especially non-linear ones.

7. REFERENCES

1 R.N. Thurston: "Waves in Solids", In. Handbuch der Physik, Vol. VI, a.4 (1974)

2. D.F. Nelson: "Theory of Non-Linear, Electroacoustics of Dielectric, Piezoelectric and Pyroelectric Crystals", J.A.S.A. 63, p. 1738 (1978)

3 H.F. Tiersten: "On the Non-Linear Equation of Thermo-Electro-Elasticity", Int. J. Eng. Sci. 9, pp. 587-604 (1971)

4 P.J. Westervelt: "Parametric Acoustic Array", J.A.S.A. 35, 4 (1963)

5 E.G.H. Lean, C.G. Powell & L. Kumn: "Acoustic Surface Wave Mixing on X Quartz", A.P.L. Vol. 15, No. 1, p. 10 (July 1969)

6 L.O. Svaasand: "Interaction between Elastic Surface Waves in Piezoelectric Materials", A.P.L. Vol. 15, No. 9 (Nov. 1969)

7 C.F. Quate & R.B. Thompson: "Convolution and Correlation in Real Time with Non-Linear Acoustics", A.P.L. Vol. 16, No. 2 (June 1970)

8 W.C. Wang: Joint Services Technical Advisory Committee Meeting, Polytechnic Institute of Brooklyn, Brooklyn N.Y. (1966), unpublished

9 W.C. Wang: "Signal Generation Via Non-Linear Interaction with Oppositely Directed Sonic Waves in Piezoelectric Semiconductors", Appl. Phys. Letters 18 , p. 337 (1971)

10 M. Lukkala & G.S. Kino: "Convolution and Time Inversion using Parametric Interaction of Acoustic Surface Waves", A.P.L. Vol. 18, No. 9 (May 1971)

11 G.S. Kino, Sludvier, H.J. Shan, W.R. Shreve, J.M. White & D.K. Winslow: "Signal Processing by Parametric Interaction in Delay Lines Devices", I.E.E.E. Trans. on MTT 25, pp. 244-255 (1973)

12 T.C. Lim, E.A. Kraut & R.B. Thomson: "Non-Linear Materials for Acoustic Surface Wave Convolver", A.P.L. Vol. 20, No. 3 (Feb. 1972)

13 P.E. Lagasse, I.M. Mason & E.A. Ash: "Acoustic Surface Waveguides Analysis and Assessments", I.E.E.E. Trans. on MTT 21, no. 4 (April 1973)

14 R.A. Zakarevicius, E.A. Ash & I.M. Mason: "Convolution and Imaging in Acoustic Surface Wave Thin Film Guides", Elect. Letters, Vol. 9, no. 16 (April 1973).

15 P.H. Defranould & C. Maerfeld: "A SAW Planar Piezoelectric Convolver", Proceedings of the I.E.E.E., Vol. 64, no. 5 (May 1976)

16 R.D. Haggarty, E.A. Palo & W.M. Bridge: "SAW Processors vs. Computing Technologies for Spread Spectrum Data Link Applications", Proc. of I.E.E.E. Ultrasonics Symposium, pp. 567-572 (1978)

17 D.P. Morgan: "General Analysis of Bilinear SAW Convolvers", Elect. Letters, Vol. 17, No. 7 (April 1981)

18 M.D. Motz, J. Chambers & I.M. Mason: "Suppression of Spurious Signal in Degenerate SAW Convolvers", Proc. of 1973 Ultrasonic Symposium, pp. 152-154

19 I. Yao: "High Performance Elastic Convolver with Parabolic Horns", Proc. of 1980 Ultrasonic Symposium

20 H. Gautier & C. Maerfeld: "Wideband Elastic Convolvers", Proc. of 1980 Ultrasonic Symposium

21 H.P. Grassl & R. Veith: "High Performance SAW Convolvers Characterised by Two-Dimensional Frequency Domain Measurements", Proc. of 1982 Ultrasonic Symposium

22 D.R. Selviah, D.H. Warne & D.P. Morgan: "Spatial Uniformity Measurement of SAW Convolvers", Proc. of the I.E.E.E. 1980 Ultrasonic Symposium

23 E.L. Adler & J.H. Cafarella: "Design Criteria for Surface Acoustic Wave Convolvers", Proc. of the I.E.E.E., Vol. 69, No. 3 (March 1981)

24 W.R. Smith: "Broadband, Amplitude and Phase Compensating Transducers for Monolithic Convolvers", Proc. of the I.E.E.E. 1980 Ultrasonic Symposium

25 M. Planat, G. Vanderborck, H. Gautier & C. Maerfeld: "A finite Element Analysis of the Piezoelectric Waveguide Convolver", to be published

26 H. Gautier & C. Maerfeld: "Patents Fr. 80, 11, 225, May 80, U.S. 4, 388, 599"

27 A. Ronnekleiv: "Diffraction Compensated SAW Multistrip Beam Compressors", Proc. of the I.E.E.E. 1983 Symposium

28 D.P. Morgan, D.H. Warne, P.N. Naish & D.R. Selviah: "Monolithic SAW Convolvers using Chirp Transducers", Proc. of the I.E.E.E. 1981 Symposium

29 J.B. Green & G.S. Kino: "SAW Convolvers using Focussed Interdigital Transducers", I.E.E.E. Trans. on Sonics and Ulstrasonics, Vol. 30 No. 1 (Jan. 1983)

30 H.P. Grassl & H. Eugan: "Small Aperture Focussing Chirp Transducers vs Diffraction Compensated Beam Compressors in Elastic Convolvers", accepted for publication in I.E.E.E. Trans. on Sonics and Ultrasonics

31 H. Gautier: TH-CSF Internal Report

32 R.C. Dixon: "Spread Spectrum System", Willey (1976)

33 D. Brodtkorb & J.E. Laynor: "Fast Synchronisation in a Spread Spectrum System Based on Acoustoelectric Convolvers", Proc. of the I.E.E.E. 1978 Ultrasonic Symposium

34 H. Gautier: "Application of SAW Convolvers and Correlations", Proc. of the 36th Annual Frequency Control Symposium (1982)

35 H. Gautier & P. Tournois: "Very Fast Signal Processors as a Result of the Coupling of SAW and Digital Technologies", I.E.E.E. Trans. Microwaves T and T - MTT 29 (1981)

36 R.D. Colvin: "Surface Acoustic Wave Non-Linear Interactions in Lithium Niobate", Rome Air Development Centre in House Report No. RADCTR 83-9, (Jan. 1983)

Acoustoelectronic Rayleigh Wave Devices*

E. Stern

M.I.T. Lincoln Laboratory P.O. Box 73, Lexington, MA 02173, USA

I. Introduction

Electrons or charged particles can be made to interact with the electric fields of Rayleigh waves in piezoelectric crystal surfaces. The charged particles can be in vacuum, in semiconductors or bound to traps in solids or on surfaces. During the past two decades a large variety of acoustoelectronic amplifiers, signal processors and image processing devices were demonstrated. I shall discuss in some detail the more successful versions of these devices by quoting or reproducing device characteristics that appeared primarily in the 1972 to 1984 Ultrasonics Symposium Proceedings. The reader interested in the full range of concepts and devices could refer to the references contained in the articles from which this material is drawn.

II. Amplifiers

a. Purpose

The intended purpose for Rayleigh wave amplifiers is the compensation of acoustic losses in delay lines. The relevant power levels, translated to the interior of delay lines, are shown in Figure 1. The acoustic input levels are shown on the left. The maximum power level is set either by the limitations of the acoustic medium or by the input amplifier. This maximum signal decreases exponentially as it propagates towards the output by the loss factor L. Delay lines typically require a substantial linear dynamic range, here shown as the range between the maximum and minimum signal at the input. The minimum signal is attenuated by L, and if the loss is large, then LS_{min} can fall below the noise floor kTB within the delay line or the output amplifier $kTBF/L_t$, where F is the noise figure of the amplifier and L_t is the output transducer loss factor. It is the purpose of the acoustic amplifier to raise the minimum signal above the noise.

When an amplifying section is inserted, the noise of the amplifier raises the internal noise to a level of $kTBF_A$ at the input to the amplifier, where F_A is the acoustic amplifier noise figure. The noise is increased by the gain factor G and then decreased by the acoustic loss factor L_A as the noise propagates to the output. To this is added the noise from the output amplifier and delay line for a combined approximate level of $kTB\{F/L_t + F_AGL_A\}$.

The minimum output signal LS_{min} is below the noise floor, and an amplifying section with gain G is inserted to raise the output signal (LS_{min}) to

---

*Supported by the U.S. Air Force and Army

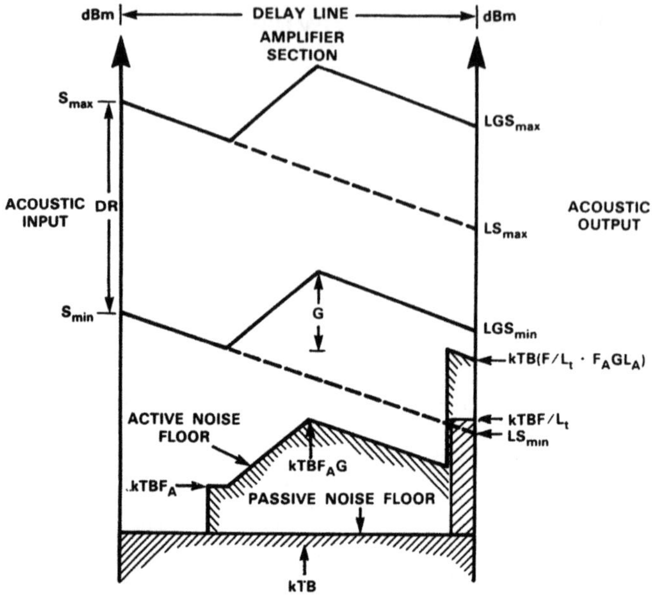

Fig. 1 Relative power and noise levels of delay lines with acoustic amplification

well above the noise both inside and at the output of the delay line. The amplifier is positioned along the delay line to keep peak power levels in the linear range in order to preserve the full dynamic range of the signal.

During the late 1960's it was believed that wideband delay lines with large time-bandwidth products would require internal amplification. Two basic structures[1] shown in Figure 2 were investigated. Figure 2a is a straightforward adaptation of the bulk acoustic wave amplifiers that had been perfected in the early 1960s. For example, WHITE[2] used a semi-insulating block of CdS on which were deposited metallic Rayleigh wave transducers and drift field electrodes. The region between the contacts was illuminated with sufficient light intensity to create photocarriers for a low resistivity region at the surface of the crystal. The piezoelectric field of a propagating Rayleigh wave drags along these charges, contributing an additional resistive loss. This loss diminishes as the drift field increases, and substantial gain is achieved when the drift velocity of the carriers is approximately twice the propagation velocity of the Rayleigh wave. For wideband, large time-bandwidth (TB) product delay lines it is desirable to have substrates with large piezoelectric coefficients to provide efficient, large bandwidth transducers and strong interaction with carriers. Large bandwidths imply high carrier frequencies, therefore it is important that the acoustic loss coefficient is low. The power dissipated in the conductive region should be kept low by selecting materials with large carrier mobility. Neither CdS nor any other piezoelectric semiconductor material posses these combined properties.

Lakin and Shaw conceived and demonstrated the composite structure in Figure 2b that permits the independent selection of piezoelectric and semiconducting materials. They selected a lithium niobate substrate over

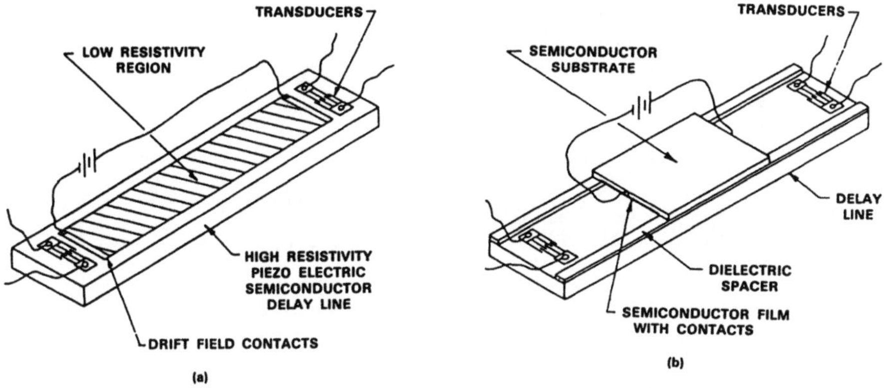

Fig. 2 Acoustoelectric Amplifier Structures, from LAKIN and SHAW[1]

which is suspended a thin silicon layer on a sapphire substrate. A gap separation of a small fraction of the acoustic wavelength is established by spacer rails. The piezoelectric field extends across the gap to interact with drifting carriers within the Si film. The film is a fraction of a μm thick to minimize parasitic heating of the semiconductor by carriers not participating in the transfer of energy to the acoustic wave.

Several laboratories produced practical amplifiers. One reproducible example[3] is shown in Figure 3. Here the gap between $LiNbO_3$ and Si is maintained with spacer posts etched into the surface of $LiNbO_3$. The Si film is forced to conform to the posts through the use of gas pressure above the sapphire substrate. A photograph of the interaction region is in Figure 3b. Four amplifying sections, each kept short to reduce the drift potential below 100 V, are visible. The gain and noise figure obtained with a 1 -cmlong CW amplifier are in Figure 3c. These results conform closely to the theories of KINO-REEDER[4] and KINO-COLDREN[5]. The values of gain and noise figure are fully compatible with acoustic amplifier requirements for large TB-product delay lines.

Many other amplifier configurations were explored, particularly planar structures that did not require the mechanical complexities of uniform gap structures. One of the most successful configurations[6] is in Figure 4. A 550-Å film of InSb is deposited on $LiNbO_3$ substrate that is coated with a 300 Å passivating SiO layer. The Rayleigh wave propagates through the films. Carriers, caused to drift in InSb film, provide the energy for gain. The difference in acoustic velocity and impedence in films with respect to the substrate causes undesirable dispersion, especially at high frequencies. COLDREN and KINO[7] subsequently solved this problem by reducing the width of InSb so that the dispersion introduced by the waveguiding action of the strip is equal to but opposite in sign to dispersion caused by film.

The amplifiers described here could be applied to those situations where acoustic gain is necessary. However the need for amplifiers did not materialize because of the successful development of wideband, efficient transducers on piezoelectric substrates with relatively low acoustic loss for frequencies up to 1 GHz. In the past decade, Rayleigh wave devices with large TB product typically operated below 1 GHz. The increasing emphasis on

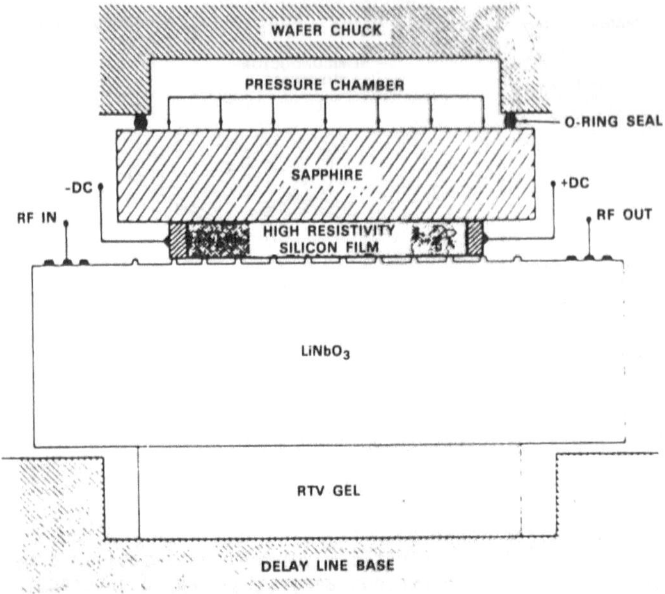

Fig. 3a

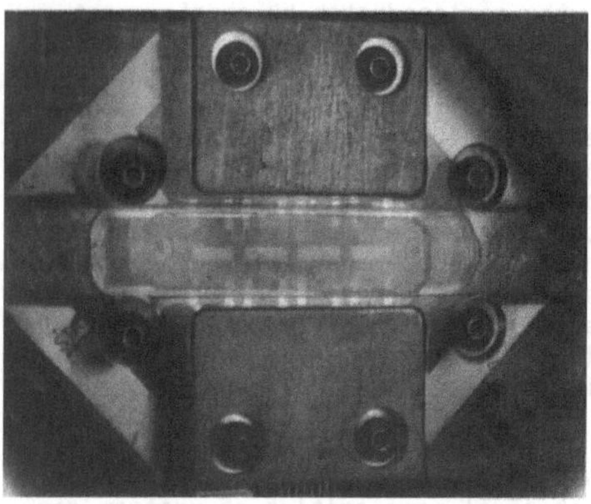

Fig. 3b

imaging capability for radars and the use of optical carriers for communication may well result in a need for extremely wideband devices operating in the 1 to 10 GHz range. Such devices very likely will require acoustic amplification. The basic technology is available should this need arise.

Amplifiers were the first successful acoustoelectronic Rayleigh wave devices. Although practical applications were not needed, the work nevertheless resulted in basic device configurations that provide the essential piezoelectric coupling of Rayleigh waves to carriers in semiconductors across minuscule gaps, or through the use of metallic coupling strips.

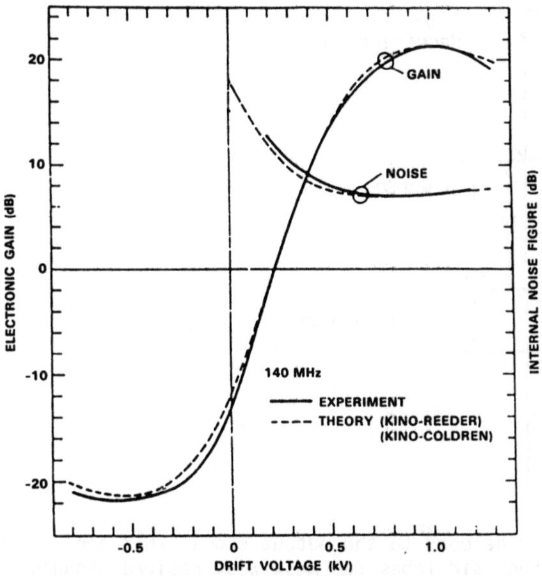

Fig. 3a, 3b, 3c Gap-Coupled Silicon-on-Sapphire to $LiNbO_3$ Acoustoelectric Amplifier, from RALSTON[3]

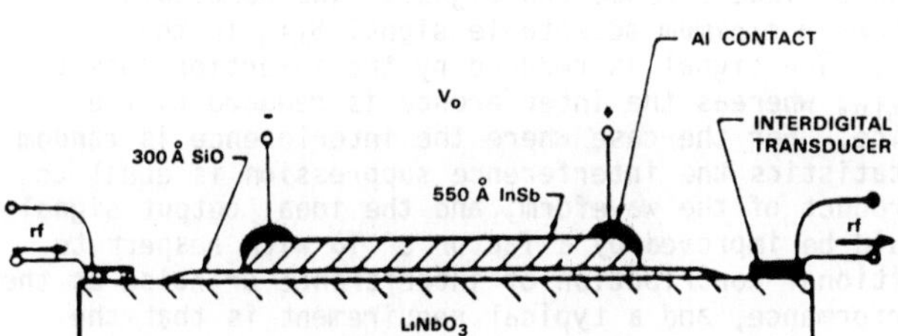

Fig. 4 Monolithic Acoustic Surface-Wave Amplifier, from COLDREN[6]

III. Signal Processing Devices

A. Purpose

The purpose of signal processing devices is the suppression of interference or noise with respect to the signal. The input and output power levels are shown in Fig. 5 along the left and right vertical axes. Underlying the signal is the noise floor that appears at the output of the device in the absence of the signal, but with all other pump and reference signals present. The noise floor is a combination of noise or interference contributed by the output amplifier, by the interactions of reference or pump energy with the structure, including fixed-pattern noise, bulk-wave and other spurious signals, intermodulation signals, signals due to reflections, spurious coupling or pickup, etc. The performance of most acoustoelectronic devices is limited by this signal-independent noise, and considerable engineering effort has been concerned with its suppression. Some of these

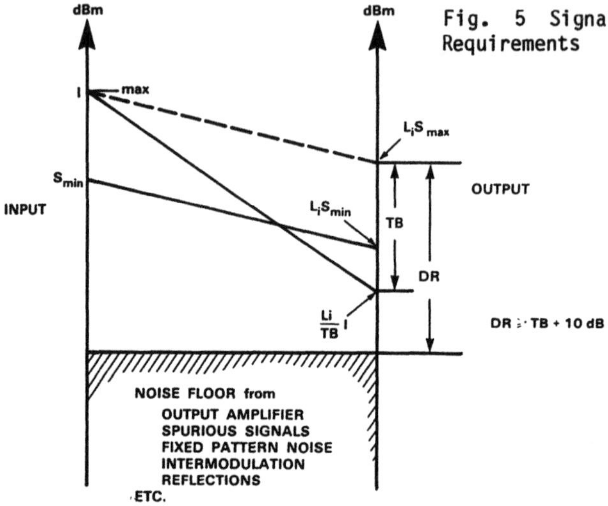

Fig. 5 Signal Processor Performance Requirements

effects can also cause spurious sidelobes to the output signal that are proportional to signal level. Such sidelobes can mask weak desired signals and considerable effort is required to suppress spurious sidelobes to an acceptable level. Typical signal processors have at the input a fixed power level that is a combination of interference and signal. The particular situation in the figure shows a minimum acceptable signal S_{min} in the presence of interference I. The signal is reduced by the insertion loss L_i to an output level of L_iS_{min}, whereas the interference is reduced by the the signal processing device. For the case where the interference is random in nature with Gaussian statistics the interference suppression is equal to the time-bandwidth (TB) product of the waveform, and the ideal output signal-to-interference ratio should be improved by a factor of TB with respect to the input ratio. Any additional contribution of interference or noise at the output degrades device performance, and a typical requirement is that the noise floor be at least 10 dB below the maximum interference appearing at the output. Another way to express this requirement is by specifying the dynamic range (DR), defined as the ratio of the maximum possible output signal L_iS_{max} to the noise floor, should exceed TB by 10 dB or more.

We are living in an era where the pervasive, ubiquitous technology is based on intergrated digital circuits. This technology is usually preferred over other, less accessible components, and acoustoelectronic Rayleigh wave devices are likely to be used only if digital circuits cannot provide the desired performance. In Fig. 6 is shown the performance range of several signal processing technologies in terms of the bandwidth (vertical axis) and time duration (horizontal axis) of the waveform to be processed. The diagonal line represents a constant TB product of 1000. Currently digital circuits are limited to bandwidths of about 20 MHz, but that limit is likely to rise to 50 MHz in the next decade. Rayleigh wave acoustoelectronic devices have achieved bandwidths of 200 MHz, and they therefore are likely to be chosen for applications where the bandwidth exceeds 50 MHz and TB exceeds a few hundred. Devices which fall outside of this niche would be used only in highly specialized situations when the advantages of lower volume or power consumption are compelling justification for their use. Therefore the discussion will concentrate primarily on acoustoelectronic devices capable of functioning above 50 MHz. These devices include convolvers, memory correlators and programmable filters.

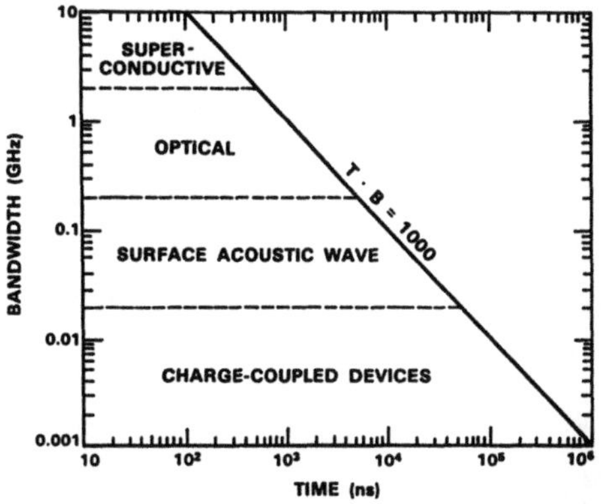

Fig. 6 Comparison of Analog Signal Processing Technologies

B. Convolvers

Convolution in an acoustoelectronic device was first demonstrated by QUATE and THOMPSON[8] in GaAs with bulk acoustic waves. WANG[9] proposed a Rayleigh wave version, shown in Fig. 7, that combines the piezoelectric properties of $LiNbO_3$ with the semiconducting properties of Si. The piezoelectric fields couple across a gap to the semiconductor surface. Typically the Si surface is stripped of oxide in order to expose a high density of trapping states and to deplete the surface of carriers. Stray dc fields caused by pyroelectric effects in the substrate terminate on traps. However, the relatively slow relaxation times of traps permit signal fields in the hundred MHz range to penetrate the surface and attract or repel free carriers. This action, equivalent to a distributed veractor, provides the desired non-linearity. A signal at frequency ω is entered at the left transducer, a reference identical to the signal but reversed in time is entered at the right, and the veractor action produces mixing at every point beneath the Si where signal and reference overlap. The mixing products are summed at the output electrode, which rejects all products except sum frequency terms. This output signal at frequency 2ω is proportional to the complex convolution of signal with reference. Most spurious signals at ω are rejected with a high-pass filter in the output. Only those signals caused by the self-convolution of the reference are a problem, and techniques for their suppression have been developed[10]. A photograph of a practical device[11] is in Fig. 8. It consists of two silicon strips that are pressed against spacer rails on $LiNbO_3$. The transducers on $LiNbO_3$ are connected to microstrip immpedence transformers near each end of the $LiNbO_3$ crystal. Also visible is the ground plane on the underside of $LiNbO_3$. The near-ideal

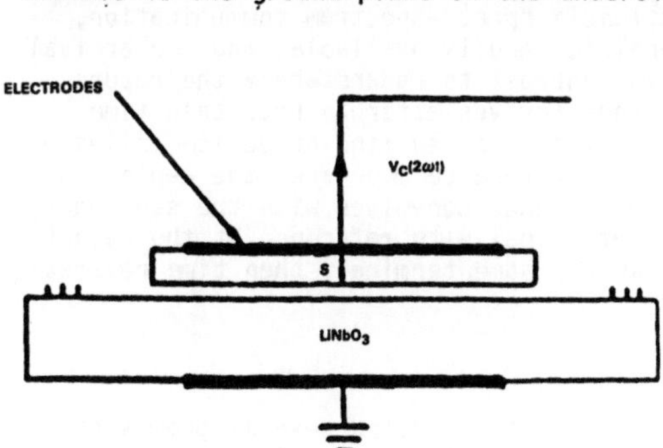

Fig. 7 Gap-Coupled Si-$LiNbO_3$ Convolver, from WANG[9]

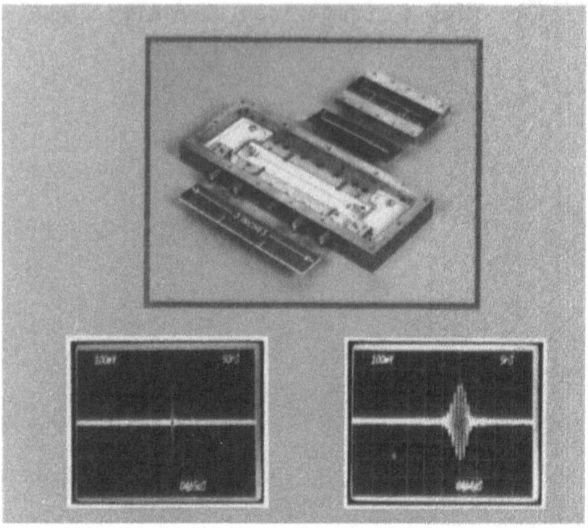

Fig. 8 Wideband Si-$LiNbO_3$ Convolver, from YAO and REIBLE[11]

output waveforms in the figure demonstrate the perfection of the device. It can accept any waveform with a bandwidth of up to 200 MHz and a time-duration of up to 12 µs. Although acoustoelectronic convolvers work exceptionally well, they are being replaced by the simpler and less expensive elastic convolvers featured in Maerfeld's article within this volume. Nevertheless the work on acoustoelectronic convolvers yielded the analytical tools and a gap technology capable of achieving performance parameters well beyond the limits of digital techniques.

Many other less successful versions of the convolver were demonstrated as well. For example, ADKINS[12] used a piezolectric AlN film adjacent to Si film on a sapphire substrate. Coupling between semiconductor and piezoelectric was achieved with lateral metallic strips. MORGAN[13] proposed the use of back-biased diodes connected to the terminals of a tapped delay line where the mixing products from discrete diodes are summed to produce the convolution signal. KHURI-YAKUB and KINO[14] used a thin film of piezoelectric ZnO on Si. These methods are limited either in bandwidth, or time-bandwidth product, or both, and have not yet produced devices that lie outside the range of digital methods.

Convolvers require the simultaneous presence of signal and time-reversed reference. They have found application in spread-spectrum communication, where a time-reversed reference signal is readily available, and the arrival time of the signal is known. This in contrast to radars where the return signal from a target at an unknown range arrives after an uncertain time delay. STERN and WILLIAMSON[15] conceived of a new kind of device called a memory-correlator, in which a stationary reference modulates the depletion depth of a semiconductor. A subsequent signal convolves with the stationary reference to produce the correlation of signal with reference at the output. If signal and reference are entered at the same terminal, then time reversal of the reference is not required.

C. Memory Correlator

The search for a suitable signal storage method yielded several promising techniques. The first[16] shown in Fig. 9, consists of a piezoelectric delay

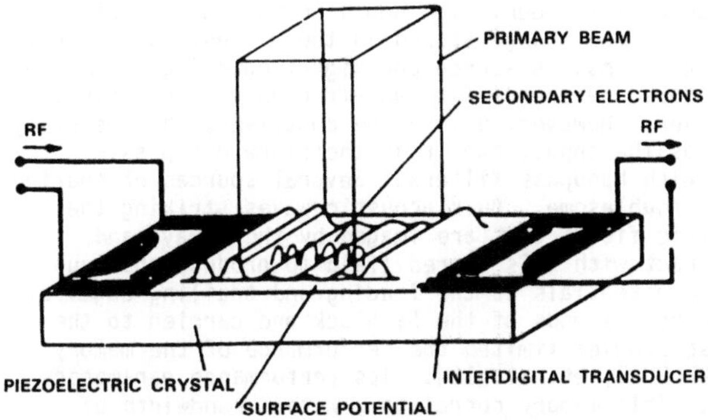

Fig. 9 Electrostatic Storage of Rayleigh Wave Signal, from BERT, et.al.[16]

line and a source of high energy electrons. A Rayleigh wave is entered on the delay line, a sharp burst of electrons strikes the surface, low energy secondary electrons are emitted which follow the piezoelectric field to regions of positive polarization where they are trapped. The Rayleigh wave continues onward, but a surface-charge snapshot of the wave, taken at the instant of the electron impulse, is retained. The surface charge distribution causes a stressed pattern within the surface that correspond to the stored signal. A subsequent Rayleigh wave matched to the stored signal is strongly reflected. The use of free electrons is difficult to implement, and several laboratories searched for other, all-solid-state methods for achieving signal storage. The preferred method at Lincoln Laboratory is the use of an array of Schottky diodes [17] shown schematically in Fig. 10. The diode array, located in the surface of Si, is adjacent to the $LiNbO_3$ delay line. A Rayleigh wave reference signal enters from the left and propagates beneath the Si. A sharp forward-bias impulse is applied to the diode array, charge flows into diode contact partly in response to the impulse, partly in response to piezoelectric fields. At cessation of impulse all diodes are

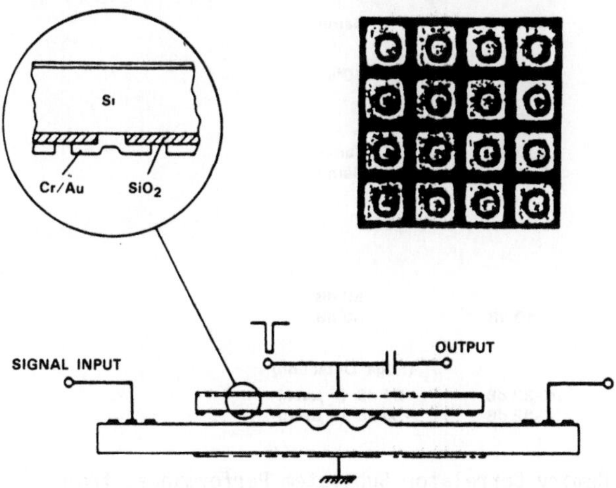

Fig. 10 Schottkey-Diode Memory Correlator, from INGEBRIGTSEN, et.al.[17]

strongly back-biased, which causes a depletion depth at the surface of the semiconductor in proportion to the charge stored in the diodes. The charge is retained for several tens of ms. A subsequent signal entering from the left correlates with the stored reference with an efficiency comparable to the acoustoelectric convolver. However, unlike the convolver, the frequency at the output is the same as the input, and it is therefore not possible to suppress spurious signals with bandpass filters. Several sources of spurious signals were particularly troublesome. Bulk acoustic waves striking the surface produce piezoelectric fields that are imaged by the array, and subsequent bulk waves interact with this stored field to produce spurious outputs. The piezoelectric potentials at the leading and trailing edges of signal pulses are detected by the ends of the Si block and carried to the output terminal. This last problem limited the performance of the memory correlator developed by HURLBURT, et. al.[18]. Its performance parameters are summarized in Fig. 11. This memory correlator, with a bandwidth of 60 MHz and time duration of 5 µs, met every goal except the suppression of spurious levels. The complex structure of Fig. 12 was developed to suppress spurious signals to an acceptable level. Bulk waves are blocked by a notch cut in front of the transducer. The Rayleigh wave is coupled transversely with a multistrip coupler to the interaction region. Two channels, driven in anti-phase, are used in order to provide common-mode rejection of edge detection. This was supplemented by additional grounded precurser strips and shaped metallic screens.

Other memory-correlator structures were demonstrated at Thompson-CSF[19] with pn diode arrays, and at Stanford[20] with ZnO overlay of pn diodes. A monolithic GaAs version was developed at Texas Instruments, but spurious levels associated with this device resulted in a recommendation for a lateral combination[21] of GaAs and $LiNbO_3$ as shown in Fig. 13. Here $LiNbO_3$ and GaAs are bonded together, and metallic interconnections are deposited over the bond as shown. The technique might provide monolithic performance with composite structures without the need of gaps or of dispersive or lossy films.

	GOALS	PERFORMANCE
CENTER FREQUENCY	150 MHz	Same
BANDWIDTH	60 MHz	Same
FRACTIONAL BANDWIDTH	40%	40%
PULSE DURATION TIME		
MAXIMUM	5 µsec	Same
MINIMUM	2 µsec	Same
DYNAMIC RANGE (above kTB × N.F. with 23 dBm reference)		
5 µsec PULSE		50 dB
2µsec	> 40 dB	46 dB
SPURIOUS LEVELS		(Edge Detection)
NEAR-IN	<-32 dB	-24 dB (5 µsec)
RMS	<-32 dB	-20 dB (2 µsec)

Fig. 11 Schottky-Diode Memory Correlator Subsystem Performance, from HURLBURT, et.al.[18]

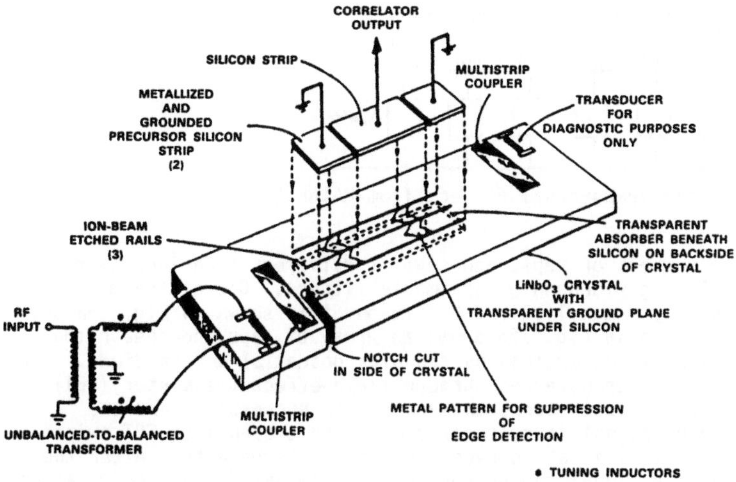

Fig. 12 Schottky-Diode Memory Correlator Structure, from HURLBURT, et.al.[18]

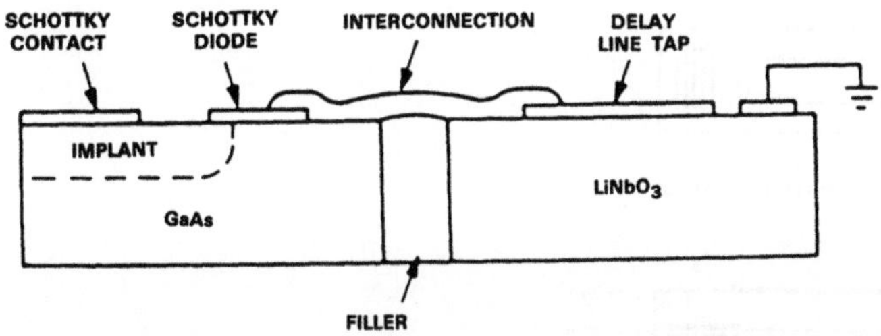

Fig. 13 Hybrid GaAs/$LiNbO_3$ SAW Memory Correlator, from WAGERS, et.al.[21]

In summary, memory correlators have been perfected for signals with bandwidths to 100 MHz and TB-products to 1000 for applications should the need arise. This is likely to happen in the coming decade as the bandwidth of radars is increased to hundreds of MHz for imaging applications.

Memory correlators are matched to the radar problem where both reference and signal occur at wide bandwidth. There are many situations where this is not the case, for example in communications sytems where the reference code is generated at low data rate with conventional MOS circuitry. Here a wideband matched filter programmable at a low data rate would be desirable.

D. Programmable Matched Filter

The programmable matched filter concept[22] is shown schematically in Fig. 14. It consists of an input transducer and an array of taps, each of which is connected to an output bus with programmable complex tap weights. These weights are typically provided by integrated circuits external to the delay line to which wideband connections are made. The bandwidth increases the density of taps so that at bandwidths above 50 MHz the tap spacing is no more than several tens of µm. It is not feasible to achieve this in a hybrid

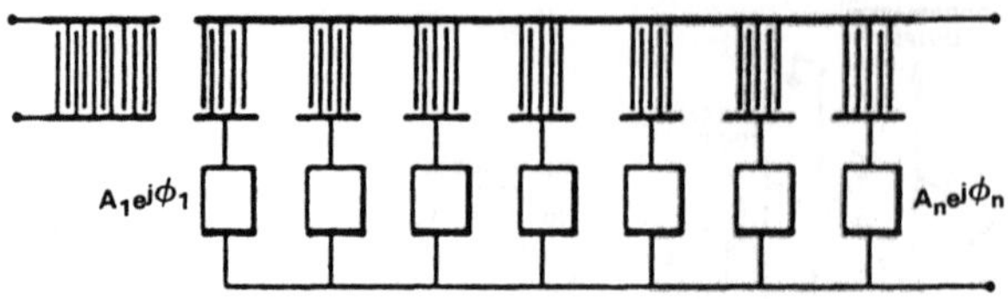

Fig. 14 Programmable Transversal Filter, from MAINES[22]

fashion for large numbers of taps, and several different techniques were developed for realizing wideband programmable filters. One of the more effective methods[23], a monolithic ZnO film on Si is shown in Fig. 15a and b. The circuit diagram in Fig. 15b shows an array of taps underneath the ZnO film. Each tap weight is determined by the bias potential on its diode. The bias from terminal 1 is transferred through field-effect transistor (FET) switches. The shift register turns on each FET in succession, and the filter-function input signal is synchronized to the sequence of FET switch closings, which occurs at a rate convenient for MOS components. After the tap weights are set, a signal can be entered at terminal 2 and a correlation signal appears at terminal 3. The reverse process, with input at 3 and output at 2, works as well. The use of ZnO introduces dispersion and excess

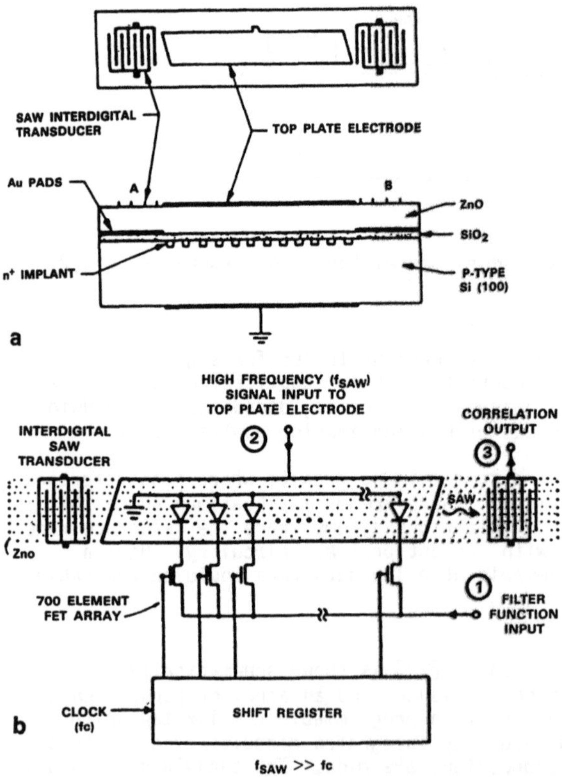

Fig. 15a, 15b The SAW/FET: A new programmable SAW transversal filter, from GREEN, et.al.[23]

loss, and the technology is therefore limited to a relatively narrow bandwidth. The same technique, implemented in gap-coupled Si-$LiNbO_3$ structure[24], shown schematically in Fig. 16 overcomes this constraint. Here edge-bonded transducers provide octave bandwidth performance. Dual tracks are used to provide common mode suppression of spurious signals. The circuit components on the Si surface include shift registers, FET switches, and output summing electrodes. A photograph of the device in Fig. 17 shows the component parts as well as a micrograph of the Si integrated circuit. Note the relative complexity of the circuit.

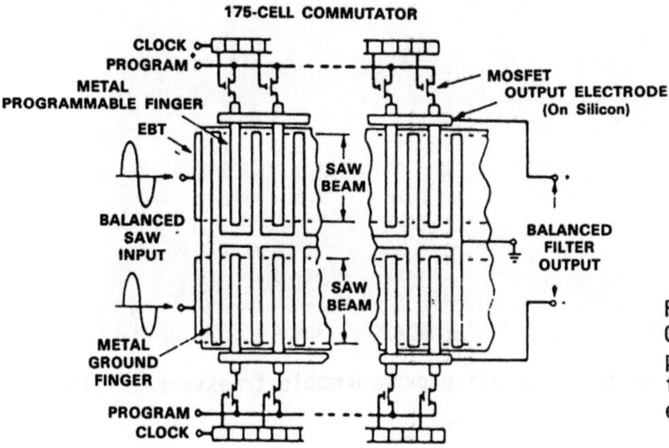

Fig. 16 Si-$LiNbO_3$ Gap-Coupled SAW/FET programmable filter, from OATES, et.al.[24]

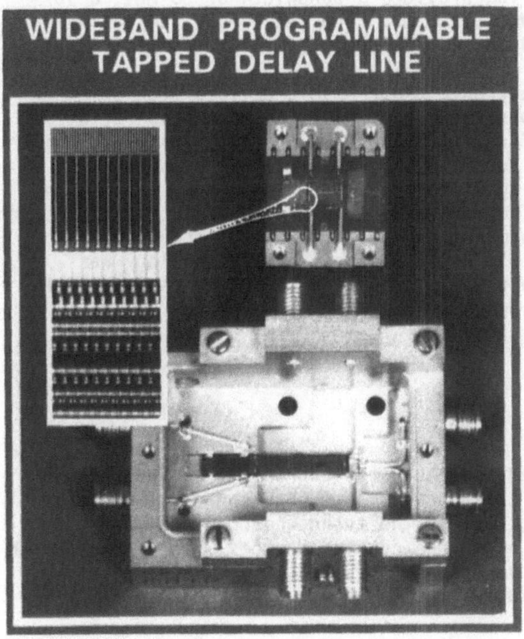

Fig. 17 Photograph of SAW/FET, from OATES, et.al.[24]

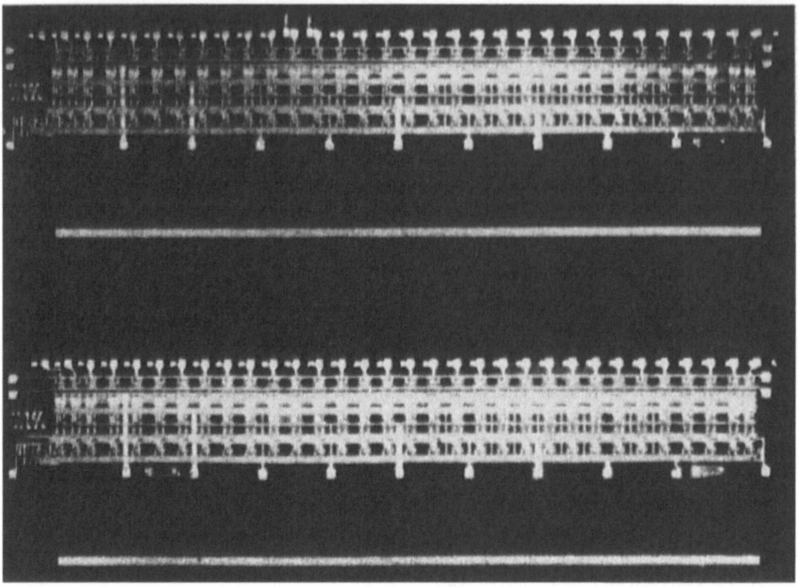

Fig. 18 Monolithic AlN an Si on sapphire programmable transversal filter, from HAGON, et al. [25]

The technology is fully compatible with NMOS logic circuits, and a large variety of integrated acoustoelectronic Rayleigh wave devices could be implemented. The particular device provided a bandwidth of 50 MHz, a TB product of 75, and a dynamic range of 60 dB. The technology is readily expandable to a bandwidth of 100 MHz and TB products in excess of 500.

Similar integrated acoustoelectronic programmable devices with somewhat different technologies were demonstrated as well. The micrograph of a programmable line[25] in Fig. 18 uses adjacent AlN and Si films on sapphire substrates. This technology, discontinued in the mid-1970's, showed considerable promise for achieving wideband fully integrated acoustoelectronic devices. Silicon exhibits a weak piezoresistive effect and adequate acoustoelectronic performance can be achieved by establishing a large current density through piezoresistive taps. The technology was perfected at Motorola[26], but a power dissipation of 100 mW per tap limits it to devices of moderate TB products.

IV. Image Processing Devices

A. Purpose

In the mid-1970's it was realized that acoustoelectronic techniques could potentially be used to read out LWIR focal planes, and several successful experiments demonstrating the utility of the technique occurred. This approach was set aside in favor of charge-coupled devices. The CCD approach has proved to be a rather difficult technology, and it is therefore appropriate to reexamine the relative merits of acoustoelectronic techniques.

Acoustoelectronics could in addition provide the function of spectral analysis of images on the focal plane, and devices could be realized for performing two-dimensional correlation of images. Indeed, the need for such devices is likely to increase in the coming decade in order to meet the demands for the processing of images in autonomous robots. Such acoustoelectronic devices could exceed the performance of conventional techniques by providing processing bandwidths in excess of 50 MHz.

B. Device Concepts

1. Imagers

Acoustoelectronic imaging devices are based on interactions between carriers and Rayleigh waves similar to those in convolvers and memory correlators. For example, at Thompson-CSF[27] a memory correlator, modified with transparent electrodes, was used for this purpose, and is shown schematically in Fig. 19. Here diodes are biased to deep depletion, and illumination from an image creates photoelectron-hole pairs in the depletion region. The appropriate carriers are swept into the diode contacts, resulting in a local decrease of the depletion depth, and the spatial distribution of the depletion depth corresponds inversely to the distribution of light intensity. A long CW signal is entered from the right, a short impulse from the left, and where they overlap a convolution signal appears at the output at twice the frequency of the input. The strength of the convolution signal is inversely proportional to depletion depth, and the amplitude of the output therefore corresponds to the intensity distribution of the image. Here the output signal is generated only in the overlap region of impulse and CW signal, and the remainder of the interactive region acts as a capacitive shunt to ground. This filling-factor loss is unacceptable in high resolution, high pixel-count imagers, and several techniques were developed for reducing this effect, including the example[28] given in Fig. 20. Here a chirped waveform with a TB product somewhat greater than the pixel count is stored in the diode array causing a depletion pattern that corresponds to the chirped waveform. Light from the scene falls upon the diode array, thereby modulating the depth of the depletion pattern. The depletion pattern is read out with an identical chirped waveform, and the output signal corresponds to the spatial Fourier transform of the stored image. This information may be used directly, or transformed once again with

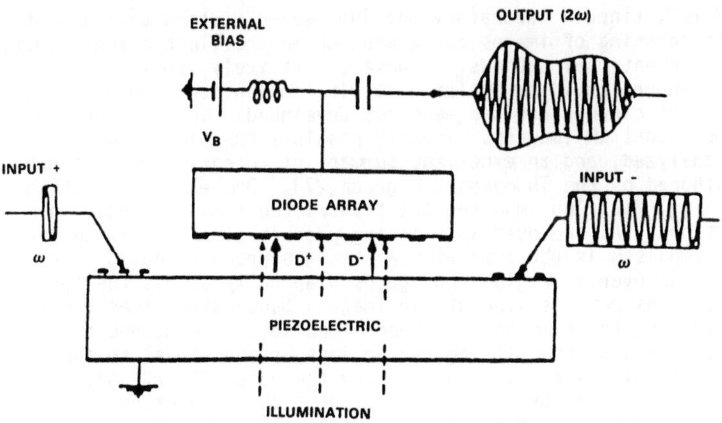

Fig. 19 Acoustoelectric optical imager, from GAUTIER, et.al.[27]

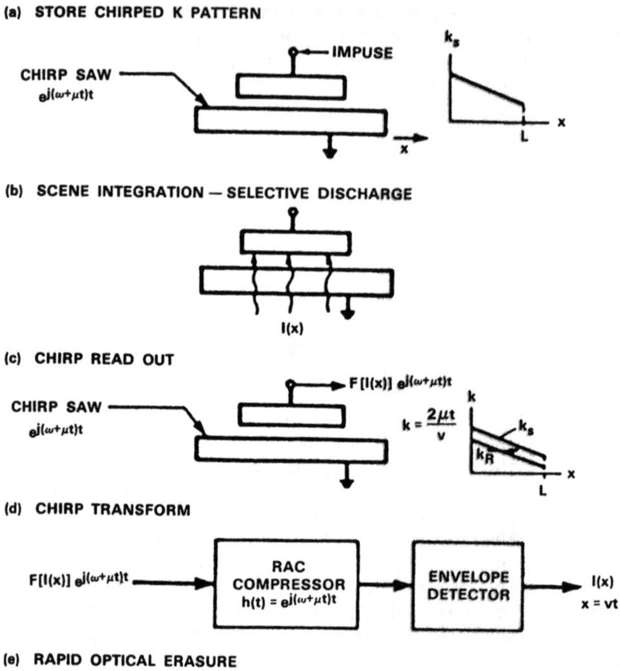

Fig. 20 Method for reducing filling-factor loss, from LEONBERGER, et.al.[28]

a suitable external dispersive delay line in order to recover the image intensity distribution. In this instance, the read-out of each spatial spectral component involves the full interaction region, and filling-factor loss is thereby avoided.

The work at M.I.T. Lincoln Laboratory and Thompson-CSF demonstrated that acoustoelectronic scanning of images could produce in the limit a sensitivity equal to that of conventional methods. However, relatively simple one-dimensional scanners could be implemented with conventional methods, and as a result acoustoelectronic devices were not developed. This is not the case for two-dimensional IR imagers. Several possible two-dimensional approaches were analyzed, and an excellent summary of potential methods are in an article authored by the Thompson-CSF group[27]. One example is shown schematically in Fig. 21. Two impulses are transmitted simultaneously in orthogonal directions. At the overlap a mixing-product signal at twice the frequency of the impulses is obtained with a strength proportional to the image intensity. The overlap region propagates diagonally across the focal plane and thereby scans out one line of the image. Successive lines can be scanned by incrementing the time delay between impulses. The method is however impractical because the filling-factor loss, proportional to the pixel count, would result in a signal well below the noise floor. Although this signal could be improved by the TB-product of a chirp waveform as in Fig. 20, it would be inadequate for scanning out focal planes with pixel counts in excess of 30 x 30.

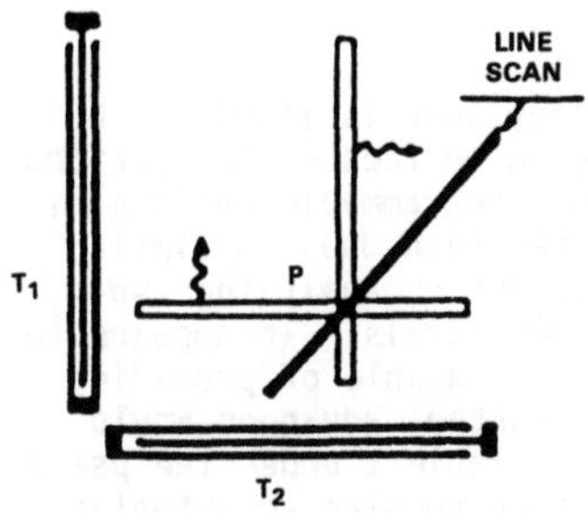

Fig. 21 Two-Dimensional scan method, from Gautier, et.al.[27]

Clearly what is needed is a technique which gives direct access to an individual pixel at the output detector. Perhaps the acoustic charge transport technique[29] shown in Fig. 22, can provide the necessary means for efficient read-out. The transport method relies on the creation of depletion wells by the piezoelectric potential of Rayleigh waves in the surface of semiconductors. In p-type semiconductors, depletion wells are produced by positive potentials, and photoelectrons that fall into a well would be carried along by the Rayleigh wave. Of course, the piezoelectric field intensity must be sufficient to hold the maximum anticipated photoelectrons. If it is not, excess electrons would spill into the trailing well as shown. This technique could be used for a two-dimensional scan of an image. Suppose intense impulses are entered into the two tranducers in Fig. 21. The two bars in the figure could represent a potential well on the surface of a featureless photodetector, and the intersection would represent a well with twice the depth. Photoelectrons created in this well would be transported diagonally to the end of the line scan, where a suitable detector is positioned. Now imagine that two orthogonal CW signals are entered into the transducer. This provides a two-dimensional field of depletion wells, all moving in step in the diagonal direction towards detectors that could be positioned at the edge of the focal plane. Here each pixel would be individually accessed, and filling factor loss would thereby be eliminated. However, other problems remain, such as the bandwidth of the output signals, the immense data rate at the output, the lifetime of the carriers, and the problem of providing adequate image integration time.

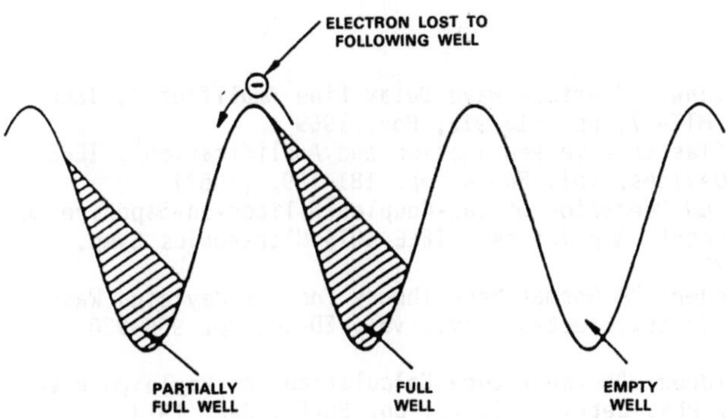

Fig. 22 Acoustic charge transport, from SCHWARTZ, et.al.[29]

Conclusions

Acoustoelectronic devices represent a niche technology that is likely to be used wherever digital circuits or conventional imaging devices cannot perform adequately. A niche exists in signal processing for programmable devices with bandwidths in excess of 50 MHz and TB products greater than 300. A similar niche exists for imaging devices capable of scanning out or analyzing two-dimensional images with a field greater than 300 x 300 pixels. It appears that current acoustoelectronic techniques and materials are capable of providing bandwidths up to 100 MHz and pixel counts of 1000. Further advances would require a new material whose properties are listed in Table 1 under the pseudonym of nonobtanium. The principal properties of other popular acoustoelectronic materials are listed as well. No single material meets the combined requirements of nonobtanium, although the combination of $LiNbO_3$ or BGO with either Si or GaAs comes close. The likelihood of the discovery of a new piezoelectric semiconductor with the properties of nonobtainium is small. Therefore efforts in the coming decade should be directed towards the further development of techniques that combine commercially available materials. The combining of materials is made possible by the unique property of Rayleigh waves, where the piezoelectric fields can be accessed everywhere for transfer to the adjacent semiconductor.

TABLE 1

ACOUSTOELECTRONIC MATERIAL PROPERTIES

	k^2 x10^{-3}	α dB/μs, 1GHz	Mobility cm^2/V^{-s}	ϵ / ϵ_o
Non-Obtanium	> 10.0	< 1.0	> 5000	< 10.0
*$LiNbO_3$	48.0	1.0	—	36.0
BGO	13.6	1.6	—	38.0
*Quartz	1.2	2.6	—	4.5
ZnO (SI)	10.0	(3.0)	200	9.0
CdS (SI)	4.8	8.5	340	9.5
*GaAs (SI)	0.2	4.2	8500	11.0
*Si	0.0	1.6	1500	12.0

*Commercially Available

References

[1] K.M. Lakin and H. J. Shaw: "Surface Wave Delay Line Amplifiers", IEEE MTT Trans., vol. MTT-17, pp. 913-915, Nov. 1969

[2] R.M. White: "Surface Elastic Wave Propagation and Amplification", IEEE Trans. Electron Devices, vol. ED-14, pp. 181-189, (1967)

[3] R.W. Ralston: "Stable CW Operation of Gap-Coupled Silicon-on-Sapphire to $LiNbO_3$ Acoustoelectric Amplifiers", IEEE 1975 Ultrasonics Symp. Proc., pp. 217-221

[4] G.S. Kino and T.M. Reeder: "A Normal Mode Theory for the Rayleigh Wave Amplifier", IEEE Trans. Electron Dev., vol. ED-18, pp. 909-920 Oct. 1971

[5] G.S. Kino and L.A. Coldren: "Noise Figure Calculation for Rayleigh Wave Amplifier", Appl. Phy. Lett., vol. 22, pp. 50-52, Jan. 1973

[6] L.A. Coldren: "Monolithic Acoustic Surface-Wave Amplifier", Appl. Phys. Lett., vol. 18, pp. 317-321, April 1971

[7] L.A. Coldren: "CW Monolithic Acoustic Surface Wave Amplifier Incorporated in a Δ/v Waveguide", Appl. Phys. Lett., vol. 23, pp. 117-118, Aug. 1973
[8] C.F. Quate and R.B. Thompson: "Convolution and Correlation in Real Time with Nonlinear Acoustics", Appl. Phys. Lett., vol. 16, pp. 494-496, 15 June 1970
[9] W.C. Wang: "Surface Wave Convolver via Space Charge Nonlinearity", IEEE 1972 Ultrasonics Symp. Proc., pp. 319-322
[10] M.C. Maerfeld: "Rayleigh Wave Non-Linear Components", (To be published in Springer-Verlag, Berlin, Heidelberg), 1985
[11] I. Yao and S.A. Reible: "Wide Bandwidth Acoustoelectronic Convolvers", IEEE 1979 Ultrasonics Symp. Proc., pp. 703-707
[12] L.R. Adkins: "Strip Coupled AlN and Si on Sapphire Convolvers", IEEE 1973 Ultrasonics Symp. Proc., pp. 150-154
[13] D.P. Morgan: "Signal Processing Using the SAW Diode Convolver", IEEE 1973 Ultrasonics Symp. Proc., pp. 138-142
[14] B.T. Khuri-Yakub and G.S. Kino: "A Monolithic Zinc Oxide on Silicon Convolver", Appl. Phys. Lett., vol. 25, pp. 188-190, July 15, 1974
[15] E. Stern and R.C. Williamson: "A New Adaptive Signal Processing Concept", Electron. Lett., vol. 10, pp. 58-59, Mar. 1974
[16] A.G. Bert, B. Epsztein and G. Kantorowicz: "Charge Storage of Acoustic rf Signals", Appl. Phys. Lett., vol. 21, pp. 50-52, 15 July 1972
[17] K.A. Ingebrigtsen, R.A. Cohen and R.W. Mountain: "A Schottky-Diode Acoustic Memory and Correlator", Appl. Phys. Lett., vol. 26, pp. 597-599, 1 June 1975
[18] D.H. Hurlburt, R.W. Ralston, R.P. Baker and E. Stern: "An Acousto-electric Schottky-Diode Memory Correlator Subsystem", IEEE 1978 Ultrasonics Symp. Proc., pp. 35-38
[19] C. Maerfeld and P.H. Defranould: "A Surface Wave Memory Device Using P-N Diodes", IEEE 1975 Ultrasonics Symp. Proc., pp. 209-213
[20] H.C. Tuan, B.T. Khuri-Yakub and G.S. Kino: "A New Zinc Oxide on Silicon Monolithic Storage Correlator", IEEE 1977 Ultrasonics Symp. Proc., pp. 498-502
[21] R.S. Wagers, B.F. Hall and M.R. Melloch: "Hybrid GaAs/$LiNbO_3$ SAW Memory Correlator", IEEE 1984 Ultrasonics Symp. Proc., pp. 294-298
[22] J.D. Maines: "Surface Acoustic-Wave Devices: Past Successes and Future Prospects", IEEE 1974 Ultrasonics Symp. Proc., pp. 140-148
[23] J.B. Green, G.S. Kino and J.D. Shott: "The SAW/FET: A New Programmable SAW Transversal Filter", IEEE 1982 Ultrasonics Symp. Proc., pp. 437-441
[24] D.E. Oates, et. al.: "Wide-Band SAW/FET Programmable Transversal Filter", IEEE 1984 Ultrasonics Symp. Proc., pp. 314-318
[25] P.J. Hagon, F.B. Micheletti and R.N. Seymour: "Integrated Programmable Analog Matched Filters for Spread Spectrum Applications", IEEE 1973 Ultrasonics Symp. Proc., pp. 334-337
[26] F.S. Hickernell, D.E. Olson and M.D. Adamo: "Monolithic Surface Wave Transversal Filter", IEEE 1977 Ultrasonics Symp. Proc., pp. 615-619
[27] H. Gautier, C. Maerfeld and P. Tournois: "Diode Acoustic Devices Applied to Optical Imaging", IEEE 1978 Ultrasonics Symp. Proc., pp. 231-238
[28] F.J. Leonberger, et. al.: "Acoustoelectrically Scanned Gap-Coupled Si-Diode Array/$LiNbO_3$ Imaging Devices", IEEE 1977 Ultrasonics Symp. Proc., pp. 456-459
[29] R.J. Schwartz, S.D. Gaalema, R.L. Gunshor: "A Surface Wave Interaction Charge Coupled Device", IEEE 1976 Ultrasonics Symp. Proc., pp. 197-199

Systems Impact of Modern Rayleigh Wave Technology

C.S. Hartmann

902 Loganwood Ave., Richardson, TX 75080, USA

When LORD RAYLEIGH [1] mathematically derived what he called surface waves, he was not contemplating use of those waves in radio frequency systems at frequencies between 10 and 1000 megahertz. However, devices using Rayleigh waves at these frequencies are now in everyday use in many households as well as being used as critical elements in a wide variety of military and commercial systems.

This paper will review the systems impact of Rayleigh wave technology from several viewpoints. After a brief introduction, the major systems functions performed by the technology will be reviewed. Next, successful systems applications of the technology will be discussed. In section 4, projected systems applications which appear imminent will be disclosed. Section 5 discusses the architectural impact of the technology on RF systems from a general point of view. Finally, a compilation of future challenges for device technologists will be given including some ideas which have future promise.

This paper will be limited to RF systems applications of Rayleigh wave devices. It will not cover the use of Rayleigh waves in sensors, optical devices, acousto-electric devices, and uses in non-destructive testing. These subjects will be covered by other authors at this conference.

1. Introduction

The acronym "SAW" (Surface Acoustic Wave) will be used freely in the following discussion as an equivalent to "Rayleigh Wave". This acronym is commonly used in literature and was developed to distinguish this type of surface wave from the many other types of surface waves which are know to exist (e.g. surface electromagnetic waves).

In this article, a distinction will be drawn between the "function" of a SAW device which denotes the action performed by the device and the "application" of a SAW device which means a specific system using a SAW device. An IF bandpass filter is an example of a function whereas a television receiver is an example of an application.

The systems impact of Rayleigh wave technology has been the subject of several review papers in the past. WILLIAMSON [2] published a comprehensive review of successful SAW device applications in 1977 in which he enumerated 10 successful applications and an additional 35 device functions which had received at least some development effort.

Other notable reviews include MAINES and PAIGE [3] who discussed signal processing applications of the technology, HAYES and HARTMANN [4] discussed communications applications of SAW technology, COLLINS and GRANT [5] reviewed intercept receivers used in electronic warfare applications, and HARTMANN et. al. [6] published a recent review of all these areas.

The majority of the references to be given here will relate to applications of SAW technology and will largely ignore the voluminous literature on basic SAW device physics. For information of this type, the reader is referred to several special issues of major technical journals which have been published (references [7], [8], and [9]). Another good general reference for this technology are the published proceedings of the annual IEEE Ultrasonics Symposium.

Most SAW device functions have several performance features in common when compared to any competing technology. First, SAW devices generally provide very complex functions on a single chip. For example, bandpass filters with several hundred transfer response zeros (equivalent to several hundred individual inductors and capacitors) are very common. Second, SAW devices generally have excellent reproducibility which results in much more consistent systems performance. Third, SAW devices are an inherently mass-producible technology using standard semiconductor technology which means that the cost of the devices declines dramatically as production volume increases. For example, many high-performance SAW devices used in consumer electronics now sell for less than $1.00. Fourth, the frequency range of contemporary SAW devices spans 10 to 1000 MHz which encompasses the most desirable frequencies for modern RF filtering and signal processing. Finally, the technology generally results in systems implementations which have smaller size, lower weight, lower cost and lower power consumption than previous systems. These features are very important in the ongoing trend toward implementing many systems in portable configurations in consumer, commercial and military uses.

While SAW devices have numerous advantageous features, they also have limitations. First, SAW devices generally have high insertion loss (typically 20 to 30 dB) which has generally limited SAW device utilization to IF signal processing functions. A second limitation is that most SAW devices are fixed tuned at manufacturing, which limits the flexibility of many systems using the devices. Third, the design and tooling of a SAW device are generally costly and time consuming when compared to conventional LC (inductor and capacitor) filter technology. The required tooling time and cost are often major hurdles in persuading a systems designer to use the technology in a new application. Finally, SAW devices only provide bandpass functions; that is - high pass, low pass and all pass functions are not generally available. All of these restrictions (a) have limited the number of functions and applications for which SAW technology are used, and (b) have only been overcome in a limited number of circumstances as will be discussed later, and (c) form the basis for much of the continuing research work in the field.

2. Major Rayleigh Wave Device Functions in Systems

Table 1 provides a compilation of Major Rayleigh wave device functions which will be considered here. While many other device functions have been investigated [2], this list only includes those functions which have enjoyed successful use in RF systems.

The first three categories in Table 1 (transversal bandpass filters, delay lines, and defined impulse response filters) are ones in which each half-cycle in the device impulse response

Table 1. Major Rayleigh wave device functions in systems

Functions	References
Transversal Bandpass Filter Functions	
IF Filters	4-6,12-16,70
Front End Filters	15, 18
Amplitude and/or Delay Compensation Filters	12-14, 17, 70
Filter Banks	5, 6, 22
Clock Recovery Filters	19, 20
Pulse Waveform Spectrum Shaping	21, 50
Antenna Duplexing	18, 61
Transmitter/Signal Generator Output Filters	42, 58, 71
Delay Line Functions	
Analog Signal Memory	6, 23
Comb Filter	23, 25
Frequency Discriminator	24
DPSK Demodulation	5, 24
Multi-Frequency Oscillator	23, 25
Fixed Oscillator	4
Voltage Controlled Oscillator	4-6, 23, 70
Tapped Delay Lines	28, 29, 44
Defined Impulse Response Filter Functions	
Pulse Expansion and Pulse Compression	26, 27, 70
Spread Spectrum Correlators	28
Spread Spectrum Waveform Generation	28, 50
Digital Communications Synchronization	28, 51
Digital Modulated RF Waveform Matched Filters (PSK, QPSK, ASK, FSK, MSK, QAM, etc.)	28
Linear FM and Nonlinear FM Matched Filter	26, 27
Programmable Tapped Delay Line Filters	29
Convolvers	30-32
Resonator Device Functions	
Local Oscillators	11, 33-34, 41
Transmitter Frequency Control	41, 70
Variable Frequency Oscillators	41
High Short-Term-Stablity Oscillators	33-34
Narrowband Filters	35-37, 70-71

can be independently specified. Most of these devices are "First Generation" SAWs in which the impulse response [10] of the device has a simple one-to-one relationship with the electrode pattern comprising the interdigital transducers which are used.

The first category in Table 1, transversal bandpass filters, is currently the largest use of SAW devices. These devices have enjoyed success due to the capability of SAW technology for providing independent control of amplitude and phase response, and doing so in a highly reproducible manner. This listing is surely incomplete because the design and use of first generation bandpass filters is such a routine event that many applications are no longer disclosed in technical literature. The best source of information on new uses is the sales literature of commercial suppliers of SAW devices.

The second category in Table 1, delay lines, makes use of the fact that the time delay of a SAW device is very reproducible and is simply defined by the photomask used for fabricating the device. In many systems applications, the delay line function is combined with either a bandpass filter function or a defined impulse response filter function.

The third category in Table 1, defined impulse response filters, was the first major systems applications of SAW technology. The 20 dB to 30 dB improvement which could be obtained by adding pulse compression to a radar system easily justified the cost of much of the early research work in this technology.

The fourth category in Table 1, resonator devices, make use of the ability to construct high Q resonant cavities for Rayleigh waves by using periodic grating structures as narrowband mirrors for trapping the waves.

The references cited in Table 1 are meant to be a representative rather than an exhaustive list. The references were chosen based on their emphasis on the systems applications of the devices, and where possible because of their tutorial nature.

3. Successful Systems Applications of Rayleigh Wave Technology

Table 2 lists the consumer applications of SAW technology which currently account for the majority of production volume. The IF (intermediate frequency) filter which is designed for TV receivers is also widely used in video cassette recorders and cable television set-top convertors. Filters of this type are produced in Europe, USA, and Japan. The resonator which was developed to stabilize the output modulator of video cassette recorders is now also used widely in CATV set-top convertors, satellite TV receivers, and personal computers. The first high volume use of UHF SAW resonators was for control of the second local oscillator in CATV convertors. Similar resonators are now being used to stabilize low power transmitters widely used for garage door openers, wireless security systems, and medical

Table 2. Consumer applications of Rayleigh wave devices

Application	Functions Performed	References
TV	IF Filter	12-14, 38, 70
VCR	IF Filter & Output Modulator Resonator	12-14, 39, 70
CATV Convertor	IF Filter, 2nd LO & Output Mod. Resonators	12-14, 39, 41, 70
Sat. TV Receiver	IF Filter & Output Mod.	39, 40, 70
Personal Computer	Output Modulator	39
Garage Door Control	Transmit Resonator	41
Wireless Security	Transmit Resonator	41
Medical Alert	Transmit Resonator	41
Cordless Phone	VCXO & Input Filter	70
Pocket Pagers	Front End Filter	70

alert devices. Recent consumer applications include coupled resonator input filters for cordless telephone and pocket pager system, and VCXOs (voltage controlled Xtal oscillators) in cordless phone systems.

Table 3 lists various commercial applications of SAW devices. These applications all require state-of-the-art performance and in many cases further improvement in device performance is still desired. However, in all cases the SAW device is being used because it outperforms competing technology. The first category, CATV/MATV (cable television and master antenna television) headends, use precise filters for generating the VSB (vestigal side-band) modulated TV signals and also use IF filters in the TVRO (TV-receive-only) receivers used to receive signals from satellite transponders.

The next three application listed in Table 3, local area networks, satellite data receivers, and digital microwave systems, all use SAW devices for carefully shaping the spectrum on digital data transmission systems. These filters simultaneously restrict both the adjacent channel interference in the frequency domain and the intersymbol interference between adjacent bits of data in the communications stream. SAW devices are critical elements in allowing data transmission at ever increasing rates while maintaining existing microwave channel spacing. Currently, 16-QAM systems are available which achieve better than 3 bits per second per Hz while meeting FCC adjacent channel interference requirements. Under development are 64-QAM and 256-QAM systems which are projected to achieve 5

Table 3. Commercial applications of Rayleigh wave devices

Application	Functions Performed	References
CATV/MATV Headend	VSB Modulator Filters & TVRO IF Filters	40, 70, 77
Local Area Network	Transmit/Receive Filters	70
Sat. Data Receiver	IF Filter	40, 70
Digital Microwave	QAM Spectral Shaping	21, 76
Fiber Optic Repeater	Clock Recovery Filter	19, 20
TV Test Equipment	VSB & Nyquist Filters	42
RF Synthesizers	Oscillators & Filters	34
Spectrum Analyzers	Oscillators & Filters	34

and 7 bits per second per Hz respectively. Success of these developments depends on developing SAW filters with amplitude and delay distortion levels that are currently unavailable.

The final three applications in Table 3 are all electronic test equipment. It is generally desired that test equipment have performance capabilities which exceeds that of the equipment being tested, and thus there is demand for the best available SAW components. The desire for performance improvement of SAW devices for test equipment use will continue for many years.

Table 4 lists successful military applications of Rayleigh wave devices. Radar is the first category shown and includes

Table 4. Military applications of Rayleigh wave devices

Application	Functions Performed	References
Radar		
Pulse Compression	Pulse Expansion and Compression Filters	2, 3, 6, 26-27, 70
Monopulse	Matched IF Filters & Single Channel Monopulse Tapped Delay Line	6, 44
Doppler	Stable Local Oscillators & Clutter Reference Delay Line Processing	6
ESM Receivers		
Channelized	Filter Bank	2, 5-6,22,45
Queued Analysis	Wideband Delay Line	5-6
Microscan	Linear FM Filters	5-6, 45-49
IFM	Delay Lines	5, 45
ECM Jammers	Pulse Memory Delay Line	2, 5, 23
ECCM		
Direct Sequence Spread Spectrum	Transmit Pulse Shaping Filters, Fixed Matched Filters, Programmable Tapped Delay Lines, Convolvers	2, 28-32, 50
Fast Freq Hopping	Fast Hop Synthesizer	53, 54
Navigation	Front End & IF Filters, Stable Oscillators	2, 4, 6, 15-16
Ranging - DME	Pulse Expansion & Compression Filters	2, 6, 26
Sonobouy	Voltage Controlled Osc.	2
Weather Balloons	Stable Oscillators	72
IFF	Stable Oscillators	
Voice and Data Communications	IF Filters, Stable Oscillators, & Digital Demodulators	2, 4, 6, 16, 24
Missile Seekers	Monopulse IF Filters	
Bomb Proximity Fuses	Tapped Delay Line	2, 55

three types of radar systems. The first, pulse compression radar, was made a practical reality by high performance SAW pulse expansion and compression filters. Processing gains up to 40 dB increased the range ten fold without increasing the transmitter power or trading range resolution. The second type, monopulse radar, significantly improves angular tracking accuracy based on using multi-channel receivers with identical phase and amplitude response characteristics. SAW devices provide the reproducible filter components which are needed. The third type, doppler radar, allows determination of the radial velocity of a target which can be used to distinguish it from surrounding clutter. SAW devices provide the stable reference oscillators needed to process doppler information, or in a simpler system, the radar returns from surrounding clutter are delayed in a SAW delay line and used for a doppler phase reference. In many modern radar systems, pulse compression, monopulse beam-sharpening and doppler detection are all used simultaneously.

The next three categories in Table 4 are all related to EW (electronic warfare) systems. The first of these categories ESM (electronic support measures) is concerned with reception and analysis of electromagnetic emissions of enemy radar, communications and other hardware. Typically one must monitor frequencies from 2 MHz to 18 GHz for both CW and pulsed emitters with greater than 80 dB dynamic range determining frequency, amplitude, time of arrival, angle of arrival, and other signal parameters with 100% probability of intercept and with 1000 to 10,000 emitters [5]. This requires multiple receivers covering multiple bands with many simultaneous frequency channels per band and with parallel channels for determining angle of arrival. The channelized receiver has often been viewed as the ideal approach but it has only become practical with the advent of SAW devices. The queued analysis receiver stores the incoming signal in a wideband SAW delay line while a coarse frequency analysis of an incoming signal is performed. A fine analysis receiver is then tuned to the correct frequency range to analyze the signal when it emerges from the delay line. This technique saves a great amount of hardware but causes extra delay in signal analysis. The microscan receiver uses SAW dispersive filters (chirp filters) to perform real time spectrum analysis. Microscan receivers with analysis bandwidths up to 500 MHz have been implemented, but use of these systems has been hampered by the inability of digital processing equipment to capture data at the 500 MHz output rate. Finally, delay lines and mixers are used to implement IFM (instantaneous frequency measuring) receivers. These receivers are very simple but generally lack the sensitivity and frequency accuracy which is desired. Many ESM receiver systems use combinations of these techniques to achieve the best possible tradeoff between performance and the size, weight, cost and power consumption of the system.

The third major application listed in Table 4 is ECM (electronic counter measures) or jammers. The major application of SAW technology is as short term analog memory, either as a pulse memory for deception type jamming or for a frequency memory for set-on jamming.

The fourth category in Table 4, ECCM (electronic counter counter measures), deals with defeating the enemy's jamming techniques. The favorite ECCM technique involves use of spread spectrum modulation, because the processing gain inherent in the matched filter receiver used for spread spectrum has inherent suppression of jamming energy. Two techniques are commonly used, direct sequence which involves digital modulation of transmitted waveforms at high rates, and frequency hopping in which the transmit frequency is continually changing during transmission. The availability of SAW convolvers and programmable tapped delay lines for implementing programmable matched filters is expected to speed acceptance of spread spectrum for military communications.

Most of the remaining items on Table 4 are self explanatory. The ranging - DME (distance measuring equipment) and missile seekers use technology similar to radar systems. The bomb proximity fuses use a tapped delay line as a simple range timing device for determining distance to the desired target.

While Table 4 refers to successful military applications of SAW technology, use of SAW technology in commercial versions of many of these systems is also occurring. For example, navigation, DME, weather balloons, IFF, and possibly spread spectrum are all systems which have commercial counterparts.

4. Projected Applications of Rayleigh Wave Technology

Predicting the future is always a dangerous occupation but several trends are emerging which point toward several major new applications for SAW technology. Some of these will be discussed here.

Television will continue to evolve toward higher performance. First, it is likely that the high IF TV tuning architecture proposed by ASH [38] will be implemented in the next 5 years but with a change in the first IF frequency from 450 MHz to somewhere above 900 MHz. This double conversion receiver architecture results in significantly simpler front-end tuner circuits and in greatly improved interfering signal rejection. The UHF SAW IF filter and a SAW resonator for the second local oscillator will be key components in this new architecture.

High resolution television is now practical based on the large-area high-definition display tubes developed in Japan [74]. However, new standards must be developed for the higher transmission bandwidth which will be needed [75]. This higher bandwidth is feasible in the 12 GHz band which is designated for direct broadcast satellite use. It appears that high resolution TV will be the vehicle that makes DBS a commercial success. Obviously, SAW devices of a higher performance level will be required for this new service.

Interest continues in providing specialized television programming on a pay-per-view basis. CATV systems in the USA currently provide limited service of this type but these systems suffer from problems of high equipment cost, poor picture quality, and theft of service. The flexibility inherent in SAW technology is capable of providing the needed scrambling and descrambling at reasonable cost [56].

Cellular Telephone is a rapidly growing communications medium in Europe, North America and Japan. Current generations of cellular mobile telephones do not use SAW devices, but recent advances [58] in SAW technology will result in new radio designs which are smaller, lighter and lower in cost.

Digital Communications provides communications links between computers. This is an exploding field due to the recent proliferation of minicomputers, personal computers, word processors, computer aided design systems, computer aided manufacturing systems, and the many other new uses for digital technology. Broadband (meaning RF modulated) local area networks are rapidly dominating personal computer communications and SAW devices are already being used. While local area networks use coaxial cable, wireless interconnection of digital equipment is projected using the spread spectrum techniques originally developed for military communications. The signal security, immunity to interference, immunity to multipath distortion, and the possibility of shared spectrum use are all features that make this an attractive possibility. SAW convolvers would likely play an important role in such systems. In the USA, the FCC has investigated use of spread spectrum [73] and has recently enacted rules for development of unlicensed low-power systems.

Electronic Labels use phase or amplitude encoded SAW devices connected to an antenna as the passive label [59]. The label is interrogated using RF pulses. Such labels are very attractive for the many applications in which computer-read optical labels have failed because line-of-sight access to the label could not be guaranteed. The trend toward automation and computerized handling of a wide range of commercial items is giving major impetus to this technique.

5. Systems Architecture Impact of SAW Device Technology

SAW devices have become widely accepted in RF systems and are now impacting the architecture of these systems. For example, many systems are now implemented with a "building block" approach using gain blocks, filtering blocks, demodulator blocks, etc.. While this trend is partially the result of advances in semiconductor integrated circuit technology, SAW technology has been an important factor in this trend.

Another impact of SAW technology is that modern systems now require levels of performance for phase & amplitude and/or impulse response characteristics which were unrealistic 5 to 10 years ago. This is evident in consumer products such as television, in commercial products such as digital communications equipment, and military products such as EW receivers. Systems designers now routinely specify phase & amplitude performance which would have been considered impossible in the past.

A third impact of SAW technology is that sophisticated signal processing functions are now being used routinely. This trend initially started in military equipment with such items as monopulse and doppler processing for radar. In commercial equipment, the adaptive equalizers which are routinely used in telecommunications modems and digital microwave radio equipment are recent examples of this trend. Similar techniques are planned for consumer products such as TV and satellite receivers. In many cases, the SAW device only performs a part of the signal processing with the remainder being done using digital techniques. However, even in systems which rely on digital signal processing, SAW filters with highly reproducible amplitude and phase response are needed to ensure proper operation of the digital processor.

A fourth impact on new systems architecture is the trend toward multi-channel systems. This is evident in consumer products such as television where the "parallel sound" systems are planned for improved sound fidelity. Examples abound in military hardware such as monopulse radar, EW channelized receivers, and missile seekers. This trend is made possible by the low cost, small size, and highly reproducible performance which SAW devices provide. This is an area of flexibility which has not been fully exploited in both commercial and consumer products.

Another impact of SAW technology is the move to "high IF" receiver architectures. By using higher IF frequencies, one can substantially eliminate image response and other mixer spurious response problems which plague traditional IF frequency choices. By removing these spurious responses, one can either partially or totally eliminate varactor-tuned front-end filters with their attendant alignment and frequency tracking problems. The result is better systems performance, improved reproducibility and cost reduction.

A sixth impact of SAW technology is significant growth in RF hybrid modules which constitute complete RF systems or subsystems. For example, complete receivers in hybrid circuit form have been rare because one could not implement a high performance receiver without the use of multiple coils which are too large for hybrid circuit use. Now however, most of these critical functions can be performed by SAW devices. Initial SAW-based RF hybrids are simple oscillators and transmitters which use the SAW device for frequency control [60].

Another impact of SAW technology is an accelerated trend toward portable equipment. This is true of telephones, satellite receivers, military radar and many other items. The desire for portability has existed for a long time and now many complex RF systems are being implemented in portable form because of SAW technology.

Many of these "architecture impacts" of SAW technology are also the result of advances in semiconductor technology. Some SAW device technologists fear that semiconductors may ultimately replace SAW devices but this is generally an unfounded fear. The analog filtering and frequency control functions of SAW technology are simply not available in semiconductor form. The two technologies are complementary and not competitive in most areas.

6. Future Challenges for SAW Device Technologists

Table 5 gives a list of areas of desired technology improvement which would result in significantly expanded uses for SAW technology. While progress has been made in all the areas shown, major breakthroughs are desired.

The first item, improved low-loss filters, is motivated by one of the oldest problems in the technology. While much progress has been made using three phase transducers [4], [6], [15], [40], and other techniques [18], [58], further improvements are desired. Desired improvements include (a) very low loss (1 dB or less), (b) high power capability (1 to 10 watts), (c) elimination of 3 phase matching, and (d) achieving a tuneable filter. One promising technique for low loss filters is the single phase unidirectional transducer [64].

The second item, improved packaging, is important from several viewpoints. Achieving a lower cost package is important in consumer applications [63] because packaging is the dominant cost in consumer SAW devices. Also, there is a widespread desire to achieve a chip-mount package for compatibility with other components which are now available in this form for automated assembly. Third, many current packages degrade performance either due to parasitic capacitance and resistance or due to lack of hermiticity and degraded cleanliness [62]. These problems will all become worse as the trend toward UHF and microwave SAW devices continues.

Table 5. Future challenges for SAW device technologists

Functions	References
Improved Low-Loss Filters	4, 18, 58, 61, 64
Improved Packaging	55, 62, 63
Microwave Frequency Devices	15, 58, 67, 68
Programmable Filters	29-32, 65, 69
More Accurate Filter Design Techniques	10, 12-14, 21, 42
Improved Materials & Processing	12-14, 55, 66-68

The third item in Table 5, microwave frequency devices, reflects the trend in modern RF systems toward ever higher frequencies and ever widening signal bandwidths. Devices are now available up to 1 GHz from several manufacturers. This progress has primarily resulted from E-Beam photomask improvements. While further improvement is expected, the technology is approaching some limits in terms of propagation loss, surface defect density, and alignment tolerance for multi-level devices. One promising approach is to use other surface modes such as the Sezawa wave [66] or surface transverse shear waves [67], [68] which have higher inherent propagation velocity.

The fourth item on Table 5, programmable filters, continues to be an area of major research effort. The most successful programmable devices to date are the elastic SAW convolvers [30], [31], [32], but many drawbacks remain. An interesting possibility in this area is the GaAs acoustic charge transport devices which use SAW for a low-power high-speed clock [65]. Also, the recently disclosed holographic grating devices [69] may result in the ability to reconfigure a filter as desired.

The fifth and sixth items in Table 5 need little explanation. In many of the applications discussed above, there is a continuing desire for more accurate SAW filters, or more stable filters or even more reproducible filters. These can only result from improvements in design techniques and from better materials and from improved device fabrication techniques.

7. Summary

The impact of Rayleigh wave technology on RF systems has already been substantial. This paper has discussed 9 major consumer applications, 9 major commercial applications, and 18 major military applications of the technology. This has resulted in a world wide market for SAW devices which is estimated to be nearly $100 million. Several large new applications are projected which will continue to fuel a rapid market growth.

As discussed above, SAW devices have impacted the architecture of RF systems. The result has been reduced systems costs, size and weight while major new features and performance enhancements have been achieved.

SAW technology is now in its rapid growth phase which is characteristic of any successful new technology. This is based largely on the success of first generation SAW devices. Second generation devices (resonators, low-loss filters, convolvers, etc.) are now entering a phase of large scale use. This, coupled with the advent of even newer product generations will continue to fuel future growth.

REFERENCES

1. Lord Rayleigh, London Math. Soc. Proc., 17 (1885)
2. R. Williamson: "Case Studies of Successful Surface-Acoustic-Wave Devices", 1977 IEEE Ultrasonics Symp. Proceedings, pp. 460-468
3. J. Maines and E. G. S. Paige: "Surface-Acoustic-Wave Devices for Signal Processing Applications", Proc. IEEE 64, pp. 639-652 (1976)
4. R. Hayes and C. Hartmann: "Surface-Acoustic-Wave Devices for Communications", Proc. IEEE 64, pp. 652-671 (1976)
5. J. Collins and P. Grant: "Current and Future Sub-Systems For Intercept Receivers", 1980 IEEE Ultrasonics Symp. Proceedings, pp. 204-215
6. C. Hartmann, R. Kansy, W. Daniels, and B. Potter: "SAW Devices for Military, Communications, Radar and EW Systems", Microwave Journal 25, No. 7 pp. 73 (1982)

7. Proc. IEEE 64, pp. 577-832 (May 1976)
8. IEEE Trans. on MTT MTT-21, pp. 159-314 (April 1973) Joint Issue with IEEE Trans on SU SU-20 (Apr 1973)
9. IEEE Trans. on MTT MTT-17, pp. 797-1052 (Nov 1979) Joint Issue with IEEE Trans on SU SU-16 (Oct 1979)
10. C. Hartmann, D. Bell, Jr., and R. Rosenfeld: "Impulse Model Design of Acoustic Surface-Wave Filters", IEEE Trans. MTT MTT-21, pp. 162-175, (1973)
11. D. Bell, Jr. and R. Li: "Surface-Acoustic-Wave Resonators", Proc. IEEE 64, pp. 711-721, (1976)
12. A. DeVries and R. Adler: "Case History of a Surface-Wave TV IF Filter for Color Television Receivers", Proc. IEEE 64, pp. 671-676 (1976)
13. K. Hazama, J. Yamada, M. Ishigaki, and T. Toyama: "Design and Mass Productive Fabrication Techniques of High Performance SAW TV IF Filter", 1978 IEEE Ultrasonics Symp. Proceedings, pp. 504-508
14. G. Tobolka, W. Faber, G. Albrecht, and D. Pilz: "High Volume TV-IF Filter Design, Fabrication, and Applications", 1984 IEEE Ultrasonics Symp. Proceedings, pp. 1-12
15. B. Potter: "L-Band Low-Loss SAW Filters", 1979 IEEE Ultrasonics Symp. Proceedings, pp. 533-536
16. B. Potter and D. MacDonald: "SAW Filter Technology and Applications", Proc. 35th Freq. Control Symp., pp. 352-357 (1981)
17. T. Chiba: "SAW Filter as Applied to the Group-Delay Equalizer in Television Rebroadcast Transmitters", 1979 IEEE Ultrasonics Symp. Proceedings, pp. 545-549
18. M. Hikita, T. Tabuchi, H. Kojima, A. Nakagoshi, and Y. Kinoshita: "Low Loss SAW Filter for Antenna Duplexer", 1983 IEEE Ultrasonics Symp. Proceedings, pp. 77-82
19. R. Rosenberg, D. Ross, P. Trischitta, D. Fishman, and C. Armitage: "Optical Fiber Repeatered Transmission Systems Utilizing SAW Filters", 1982 IEEE Ultrasonics Symp. Proceedings, pp. 238-246
20. J. Minowa, K. Nakagawa, K. Okuno, Y. Kobayashi, and M. Morimoto: "400MHz SAW Timing Filter for Optical Fiber Transmission Systems", 1978 IEEE Ultrasonics Symp. Proceedings, pp. 490-493
21. E. Ehrmann-Falkenau, H. Stocker, C. Ruppel, and W. Mader: "SAW-Filters for Spectral Shaping in a 140 MBit/s Digital Radio System Using 16 QAM", 1983 IEEE Ultrasonics Symp. Proceedings, pp. 17-22
22. W. Porter and B. Smilowitz: " Design and Performance of a Channelizing Filter Bank in a Pulsed Signal Environment", 1983 IEEE Ultrasonics Symp. Proceedings, pp. 154-159
23. C. Boyne and F. Lewis: "A Wideband Frequency Memory Using an Overmoded SAW Oscillator", 1981 IEEE Ultrasonics Symp. Proceedings, pp. 139-143
24. P. Brossard, J. Henaff, and D. Lombard: "N Phase - PSK Differential Demodulators Using ASW Delay - Lines", 1977 IEEE Ultrasonics Symp. Proceedings, pp. 532-536
25. K. Kishimoto, M. Ishigaki, K. Hazama, and S. Matsuura: "SAW Comb Filter for TV Frequency Synthesizing Tuning System", 1980 IEEE Ultrasonics Symp. Proceedings, pp. 377-381

26. H. Gerard, W.Smith, W. Jones, and J. Harrington: "The Design and Application of Highly Dispersive Acoustic Surface-Wave Filters", IEEE Trans MTT MTT-21, pp. 176-186 (1973)

27. R. Williamson and H. Smith: "The Use of Surface-Elastic-Wave Reflection Gratings in Large Time-Bandwidth Pulse-Compression Filters", IEEE Trans. MTT MTT-21, pp. 195-205 (1973)

28. D. Bell and L. Claiborne: "Phase Code Generators and Correlators", Chapter 7, Surface Wave Filters, pp. 307-346, Wiley, 1977

29. C. Panasik and J. Toplicar: "Adaptive Interference Suppression Using SAW Hybrid Programmable Transversal Filters", 1983 IEEE Ultrasonics Symp. Proceedings, pp. 170-174

30. D. Defranould and C. Maerfeld: "A SAW Planar Piezoelectric Convolver", Proc. IEEE, 64, pp. 748-751 (1976)

31. I. Yao: "High Performance Elastic Convolver with Extended Time-Bandwidth Product", 1981 IEEE Ultrasonics Symp. Proceedings, pp. 181-185

32. H. Gautier and C. Maerfeld: "Wideband Elastic Convolvers", 1980 IEEE Ultrasonics Symp. Proceedings, pp. 30-36

33. T. Parker: "Random and Systematic Contributions to Long Term Frequency Stability in SAW Oscillators", 1983 IEEE Ultrasonics Symp. Proceedings, pp. 257-262

34. W. Shreve, R. Bray, S. Elliott, and Y. Chu: "Power Dependence of Aging in SAW Resonators", 1981 IEEE Ultrasonics Symp. Proceedings, pp. 94-99

35. W. Tanski: "Multipole SAW Resonator Filters: Elements of Design and Fabrication", 1981 IEEE Ultrasonics Symp. Proceedings, pp. 100-104

36. R. Rosenberg and L. Coldren: "Fast Synthesis of Finite-Loss SAW Resonator Filters", 1977 IEEE Ultrasonics Symp. Proceedings, pp. 882-887

37. L. Coldren and R. Rosenberg: "Surface-Acoustic-Wave Resonator Filters", Proc. IEEE 67, pp. 147-158 (1979)

38. D. Ash: "An Improved High-Performance TV Receiver", Proc. IEEE 70, No. 11, pp. 1345-1357 (1982)

39. Y. Ebata, K. Sato, and S. Morishita: "A $LiTaO_3$ SAW Resonator and Its Application to Video Cassette Recorder", 1981 IEEE Ultrasonics Symp. Proceedings, pp. 111-116

40. T. Hinkle: "Satellite TVRO Receiver Design", Coop's Satellite Digest, Aug & Sept, 1984

41. RF Monolithics, Inc., Dallas, Tx., SAW Resonator Application Handbook (1984)

42. W. Drummond and S. Roth: "Application of High Performance SAW Transversal Filters in a Precision Measurement Instrument", 1978 IEEE Ultrasonics Symp. Proceedings, pp. 494-499

43. S. Elliott, M. Mierzwinski, and P. Planting: "The Production of Surface Acoustic Wave Resonators", 1981 IEEE Ultrasonics Symp. Proceedings, pp. 89-93

44. F. Nyffeler, J. Ebneter, and R. Baechtiger: "Application of SAW Devices for Combined Filtering/Multiplexing and Ranging in a Novel CIWS Radar IF Signal Processor", 1983 IEEE Ultrasonics Symp. Proceedings, pp. 165-169

45. J. Tsui and W. Brumfield: "EW Receivers using Acousto-Optic and Ultrasonic Components", 1984 IEEE Ultrasonics Symp. Proceedings, pp. 345-350
46. P. Grant, A. Gentles, and J. Collins: "High Accuracy Surface Acoustic Wave Spectrum Analyser", 1984 IEEE Ultrasonics Symp. Proceedings, pp. 108-111
47. M. Jack, P. Grant, and J. Collins: "Theory, Design and Application of SAW Fourier-Transform Processors", Proceedings IEEE 68, pp. 450-468 (1980)
48. D. Arsenault and V. Dolat: "Compact Multiple-Channel SAW Sliding-Window Spectrum Analyzer", 1981 IEEE Ultrasonics Symp. Proceedings, pp. 220-225
49. G. Moule, T. Browning, and R. Bale: "A 1GHz Bandwidth SAW Compressive Receiver", 1980 IEEE Ultrasonics Symp. Proceedings, pp. 216-219
50. W. Smith: "SAW Filters for CPSM Spread Spectrum Communication", 1977 IEEE Ultrasonics Symp. Proceedings, pp. 524-528
51. J. Cafarella: "Application of SAW Convolvers to Spread-Spectrum Communications", 1984 IEEE Ultrasonics Symp. Proceedings, pp. 121-126
52. M. Kowatsch: "Design of a Convolver-Based Packet Voice Spread Spectrum System", 1984 IEEE Ultrasonics Symp. Proceedings, pp. 127-131
53. M. Zaken, N. Silbermann, E. Kogan, P. Romik, and E. Sofer: "A UHF Fast Frequency Synthesizer Using SAW Filters", 1980 IEEE Ultrasonics Symp. Proceedings, pp. 230-234
54. B. Darby, R. Bale, C. Eustace, J. Hannah, and E. Patterson: "Performance of Frequency Hop Synthesisers Based on Chirp Mixing", ", 1977 IEEE Ultrasonics Symp. Proceedings, pp. 514-523
55. A. Janus and F. Cho: "Manufacturing Status of SAW Devices", 1977 IEEE Ultrasonics Symp. Proceedings, pp. 553-562
56. L. Ragan and C. Hartmann: "Television Scrambling and Unscrambling Method and Apparatus", U. S. Patent 4,403,252 (1983)
58. M. Hikita, T. Tabuchi, H. Kojima, A. Sumioka, A. Nakagoshi, and Y. Kinoshita: "High Performance SAW Filters with Several New Technologies for Cellular Radio", 1984 IEEE Ultrasonics Symp. Proc. pp. 82-92
59. P. Cole and R. Vaughn: "Electronic Surveillance System", U.S. Patent 3,706,094 (1972)
60. RF Monolithics, Inc., Dallas, Tx., Brochure on SAW Resonator Hybrid Oscillator Modules (1985)
61. Y. Kinoshita, A. Nakagoshi, H. Kojima, and M. Hikita: "High Power SAW Filter for Antenna Duplexer", 1983 IEEE Ultrasonics Symp. Proceedings, pp. 83-86
62. T. Meeker and W. Grise: "Packaging and Reliability of SAW Filters", 1983 IEEE Ultrasonics Symp. Proceedings, pp. 117-124
63. C. Erikson, Jr. and D. LeSiege: "Vertical Inline SAW Package (VSIP)", 125-128
64. C. Hartmann, P. Wright, R. Kansy, and E. Garber: "An Analysis of SAW Interdigital Transducers With Internal Reflections and the Application to the Design of Single-Phase Unidirectional Transducers", 1982 IEEE Ultrasonics Symp. Proceedings, pp. 40-45

65. M. Hoskins and B. Hunsinger: "Monolithic GaAs Acoustic Charge Transport Devices", 1982 IEEE Ultrasonics Symp. Proceedings, pp. 456-460
66. T. Shiosaki, T. Yamanoto, and A. Kawabata: "Higher Order Mode Rayleigh Waves Propagating on ZnO/Substrate Structures", 1977 IEEE Ultrasonics Symp. Proceedings, pp. 814-818
67. K. Yen, K. Lau, and R. Kagiwada: "Recent Advances in Shallow Bulk Acoustic Wave Devices", 1979 IEEE Ultrasonics Symp. Proceedings, pp. 776-785
68. B. Auld and D. Thompson: "Temperature Compensation of Surface Transverse Waves for Stable Oscillator Applications Using Shallow Gratings", 1984 IEEE Ultrasonics Symp. Proceedings, pp. 213-217
69. D. Oates, P. Gottschalk, and P. Wright: "Holographic-Grating Acoustic Devices", 1984 IEEE Ultrasonics Symp. Proceedings, pp. 97-102
70. Toshiba Surface Acoustic Wave Devices Catalog No. C84A-'84
71. W. Tanski, R. St.Cyr, A. Vulcan, and C. Gloeckl: "The Results of a Two-Year Space-Qualification Study on a 500 MHz SAW Resonator Filter", 1984 IEEE Ultrasonics Symp. Proceedings, pp. 197-202
72. K. Lau: "MM&T Program for Precision, Low Cost SAW Delay Lines for UHF Applications", Final Report on U.S. Army Contract DAAK20-82-C-0294 (1985)
73. W. Scales: "Potential Use of Spread Spectrum Techniques in Non-Government Applications", MITRE Corporation Report No. MTR-80W335, Contract No. FCC-0320, (1980)
74. K. Tokuji: "Development of High-Definition TV Displays", IEEE Trans. on Broadcasting, BC-28, pp. 51-64 (1982)
75. T. Fujio: "High Definition Television Systems: Desirable Standards, Signal Forms, and Transmission Systems", IEEE Trans. on Communications, COM-29, pp.1882-1891 (1981)
76. W. Grafinger, J. Nossek, and G. Sebald: "Design and Realization of a High Speed Multilevel QAM Digital Radio Modem with Time-Domain Equalization", IEEE International Conf. on Communications, Paper 31.5, June 1985
77. See for example Data Sheet for P/N 850969, SAWTEK, Inc. Orlando, Fla.

The Fabrication Technology of Rayleigh Wave Devices

R. Veith

Siemens AG, Balanstraße 73, D-8000 München 80,
Fed. Rep. of Germany

1. Introduction

The **need grows** for communication forces telecommunication engineers to transmit more information over a limited number of channels in cable, fiberoptic, terrestrial radio and satellite networks. With a given bandwidth per channel the only way to increase information capacity is to improve the dynamic range or signal to noise ratio. This requires stable frequency sources and filters precisely matched to a signal spectrum in amplitude and phase.

Rayleigh wave devices offer solutions to these problems which cannot be obtained with conventional techniques /1,2/ and therefore have become key components in modern telecommunication systems /3/. Rayleigh waves are electrically excited and detected on the surface of piezoelectric crystals with interdigital electrode structures (Fig.1). They propagate at about 3.300 m/s and achieve a delay of 30 µs on a path 10 cm long. At a frequency of 100 MHz this delay corresponds to an absolute phase rotation of 10^6 degrees. By varying the position of the transducer in steps of 0.1 µm using modern micro-

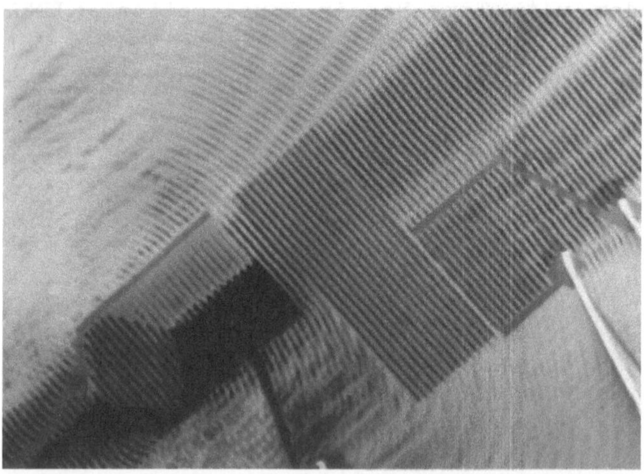

Figure 1: Visualization of Rayleigh waves in an operating TV IF filter using the scanning electron beam microscope in the stroboscopic voltage contrast mode /4/.

lithographic processes this phase rotation can be set with a resolution of 1 degree. The same relative resolution of 10^{-6} can be achieved for setting the amplitude of the transmitted Rayleigh wave by varying the overlap of interdigitated electrodes.

Making use of the interference of waves from locally distributed sources, Rayleigh waves offer the potential to construct transfer functions with a precision unrivalled by any other technology. The fabrication technology of Rayleigh wave devices has the objective to exploit this unique potential as far as possible.

2. The Role of Evolution in Fabrication Technology

Before looking into the details of Rayleigh wave device fabrication it is useful to discuss the evolution of a fabrication process in general. A fabrication process must continuously improve its products both economically and in performance in order to remain competitive. Fabrication technology thus involves a permanent learning process that can be described in a simple evolution model (Fig. 2). This model assumes a fabrication process in the state N and a variation to be introduced. If this variation leads to an improvement it is selected to change the process to the state (N+1), else it is disregarded.

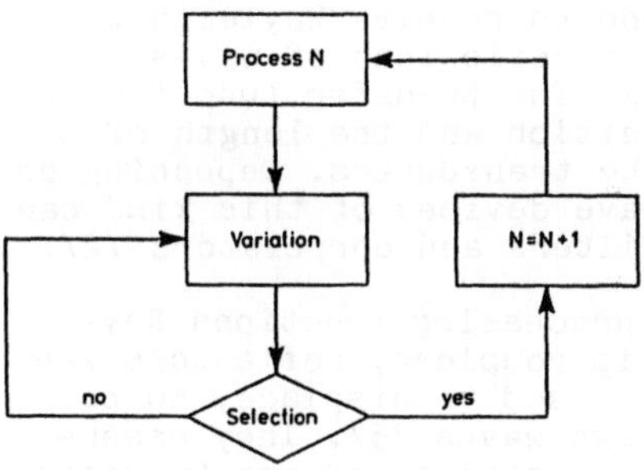

Figure 2:
Simple model for the evolution of a fabrication process

According to this model a learning curve can be defined as

$$L = N(t) \quad (1)$$

The slope of this learning curve determines the success of a fabrication technology. It is controlled by three parameters:

$$dL/dt = S = (1/T)\ J\ C \quad (2)$$

The first parameter 1/T is the rate at which variations are introduced. To allow an **unambiguous** evaluation only one variation should be introduced per process cycle. The slope of the learning curve then becomes inversely proportional to the process cycle time T.

The second factor I is the probability that a variation improves the process. In a complex and sophisticated process the probability of **improvements** by accidental variations will be very low. Computer aided simulations can however try and sort variations before they are introduced and increase the probability of an improvement to near unity. Exact simulation therefore is a powerful method to accelerate the evolution of a fabrication process.

The third factor C measures the probability that a process variation is correctly evaluated. It is a function of process noise. With increasing production scatter an actual **improvement** is more likely to be misinterpreted and dismissed. A fabrication scatter **equal to or larger than the effect of a** reasonable variation will thus turn the slope of the learning curve to zero. Reproducibility therefore is a prerequisite for a positive slope of the learning curve. It is important to note that process noise increases with the number of variables. The fabrication process with the minimum number of steps will therefore achieve the best reproducibility and the fastest evolution.

These parameters can now be applied to select fabrication techniques most appropriate for Rayleigh wave devices.

3. Rayleigh Wave Device Fabrication

A Rayleigh wave device basically consists of a piezoelectric substrate and a single layer metallisation for the interdigital transducers used to transmit and to receive Rayleigh waves (Fig. 3). Acoustic absorbers prevent reflexions of waves emitted towards the substrate edges. The transfer function of the device is determined by the position and the length of interdigitated electrodes within the transducers. Depending on the transducer lay-out, Rayleigh wave devices of this kind can be used as delay lines, bandpass filters and correlators /2/.

To perform more complex signal processing functions Rayleigh wave devices employ multistrip couplers, reflectors and waveguides. Multistrip couplers are used to displace, to compress or to expand a beam of Rayleigh waves /5/. They essentially consist of parallel conductive strips and are therefore fabricated together with interdigital transducers in the same metallisation process.

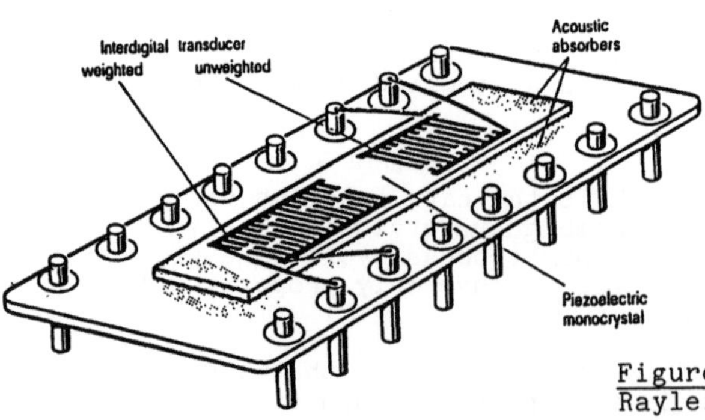

Figure 3:
Rayleigh wave device

Reflectors are used in resonators and reflective array compressors for pulse-compression systems. Ion etched grooves have been used extensively as reflectors because the reflexion coefficient can be weighted by controlling the groove-depth /6/ and because grooves can be used to recess transducer electrodes in order to reduce unwanted mechanical reflexions /7/. Ion etching however increases the complexity of the fabrication process and therefore the alternative use of metal strip reflectors has been investigated.

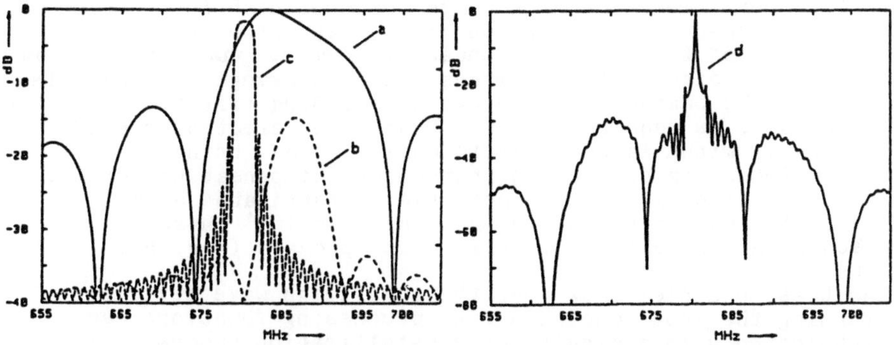

Figure 4: Rayleigh wave resonator. Computed frequency responses of transducer (a), transducer reflexions (b), metal strip reflectors (c) and resonator (d).

In resonators reflexions from transducers lead to spurious resonances. These reflexions can be reduced by shifting the resonance frequency with respect to the center frequency of the transducers /8/. Because reflexions have only half the transducer bandwidth, resonators can be operated at a frequency with zero reflexions (Fig. 4).

Reflective array compressors achieve long dispersive delays needed in pulse-compression systems. In order to obtain low

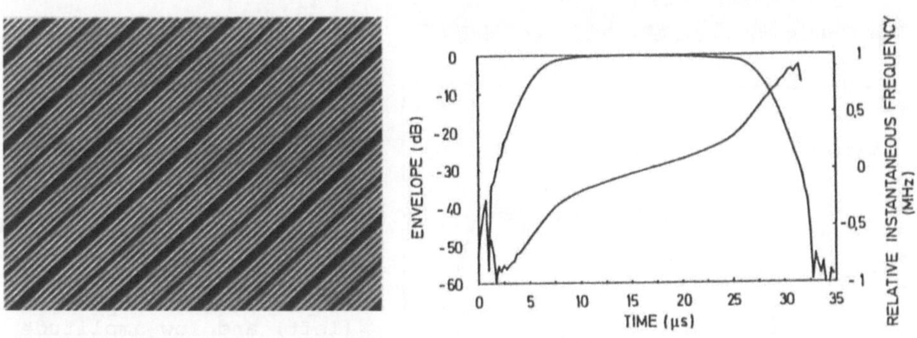

Figure 5: Reflective array compressor with position weighted metal strip reflector (left) and measured phase and amplitude weighted impulse response (right).

sidelobes of the compressed pulse, the reflector array must be amplitude weighted. Instead of groove-depth weighting, reflective metal dot arrays /9/ or position weighted reflectors employing interference from adjacent metal reflector strips can be used /10/. Figure 5 shows a reflective array compressor with 20 µs dispersive delay which is part of a pulse compression system with 40 dB sidelobe suppression.

Waveguides consisting of narrow metal strips on piezoelectric substrates are used in convolvers to achieve high acoustic power densities and to increase the nonlinear interaction of two counter-propagating waves /11/. Dispersion due to the finite width of the waveguide **causes quadratic phase distortions** in the output signal of convolvers. In thin waveguides this dispersion can be compensated for by mass loading effects, but a two layer process is required because transducer electrodes need a thicker aluminum layer to avoid ohmic losses. Using small aperture focussing dispersive transducers compensating for waveguide dispersion /12/ it was possible to fabricate a concolver with 100 MHz bandwidth and less than 10 ° phase distortion in a single layer metallisation process (Fig. 6).

As these examples show even those Rayleigh wave devices performing the most complex signal processing functions can be fabricated with a single layer metallisation process. This process is therefore considered to be adequate for the

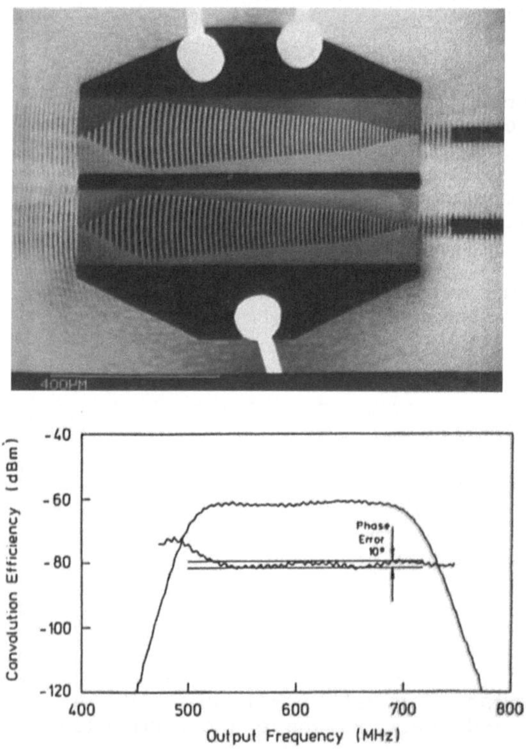

Figure 6: Rayleigh wave convolver with small aperture focussing dispersive transducer compensating for waveguide dispersion (left) and low amplitude and phase ripple convolution efficiency (right).

fabrication of Rayleigh wave devices. The fabrication then consists of 4 steps: the choice of a substrate material, the computer aided design of the electrode patterns, the metallisation and patterning of substrates and the packaging of devices.

3.1 Substrate Materials

Apart from electrode structures the transfer function of Rayleigh wave devices depends only on the acoustic propagation velocity. The substrate material therefore is primarily significant for the reproducibility of Rayleigh wave devices. Piezoelectric single crystals obviously guarantee the best reproducibility. Unfortunately there is no piezoelectric single crystal providing both temperature-stability and strong electromechanical coupling. However quartz, $LiTaO_3$ and $LiNbO_3$ /13/ cover a broad and useful range of these parameters (Table 1). Quartz is used as a temperature stable substrate for devices with a relative bandwidth up to a few percent. $LiNbO_3$ for widebandwidth applications and $LiTaO_3$ as a material inbetween. Though being single crystals Rayleigh wave substrates were not perfect from the beginning. X-ray topography proved to be a very efficient method to reveal crystal imperfections /14/.

Table 1: Substrate materials for Rayleigh wave devices

Material	Orientation	Velocity /m/s/	Piezoelectric coupling, k^2	Temperature coefficient
Quartz	ST,X	3158	0.0016	- 0.03 ppm/K^2
$LiTaO_3$	112°Y,Z	3290	0.0023	-18 ppm/K
$LiNbO_3$	Y,Z	3488	0.048	-94 ppm/K
$LiNbO_3$	128°Y,X	3980	0.056	-75 ppm/K

$LiNbO_3$ is a ferroelectric material and has to be poled to align the polar axis. In the presence of small angle grain boundaries residual ferroelectric 180 ° domains remain in the crystal and cause strong reflexions of Rayleigh waves . With carefully selected defect-free seed crystals, optimized growth parameters and computer controlled Czochralski pulling stations the crystal growth process was improved and almost perfect $LiNbO_3$ crystals can be grown now (Fig. 7).

Synthetic quartz stones for 3" wafers were first grown from small seed crystals and therefore consisted of different growth regions with varying dislocation densities. Rayleigh wave propagation velocity variations in the different growth regions were the cause of phase distortions in pulse-compression filters extending across the whole wafer and consequently showed up in spurious sidelobes of the compressed pulse. Improved sidelobes could be achieved with extended seeds treated carefully to provide no surface defects as sources for dislocations (Fig. 8).

Today 3" single crystal substrates with satisfactory lattice perfection are available. High frequency applications and sub-

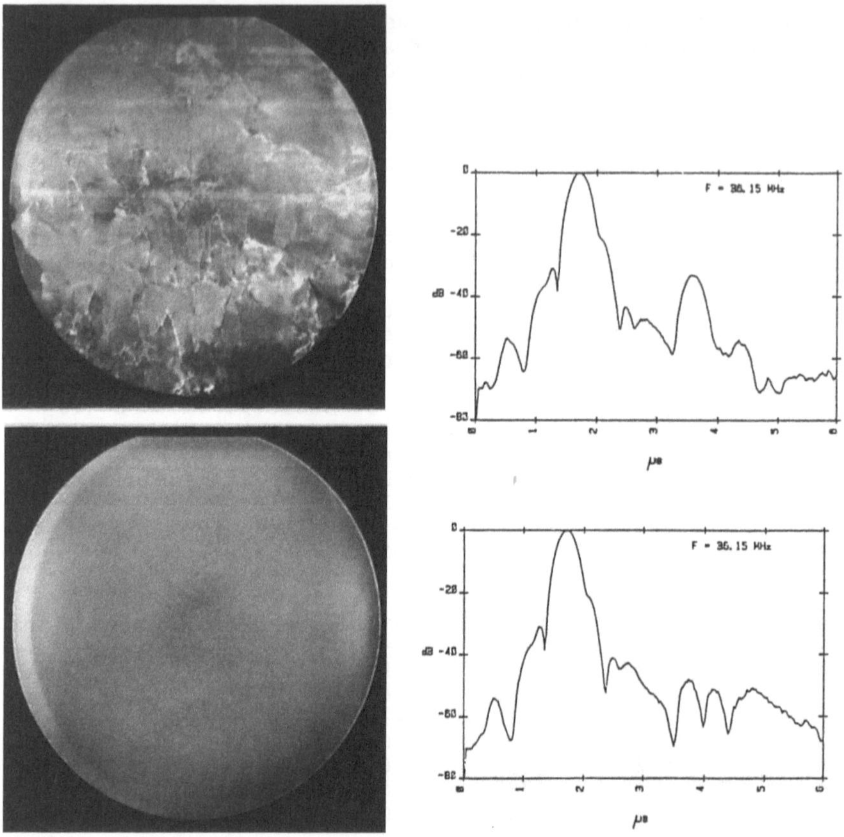

Figure 7: Effect of lattice imperfections in $LiNbO_3$ on the performance of Rayleigh wave TV IF filters. X-ray topography of 3" Y,Z wafer with small angle grain boundaries (upper left), impulse response with echo due to reflexion from ferroelectric domain, (upper right), perfect wafer (lower left) and corresponding impulse response (lower right)

micron lithography however require improved polishing techniques yielding damage free surfaces with surface flatness less than 0.5 µm/cm. Provided these requirements can be met larger wafer diameters would certainly improve the economics of Rayleigh wave device fabrication.

3.2 Design

Computer aided design tools do not only define the performance of Rayleigh wave devices. Simulations of expected results, being part of the design, are also necessary to improve the fabrication process. As an example of the state of the art the design of a 64 QAM filter is shown in Fig. 9. This filter was designed for a digital transmission system using 64 different signal states to achieve high spectral efficiency. In order to

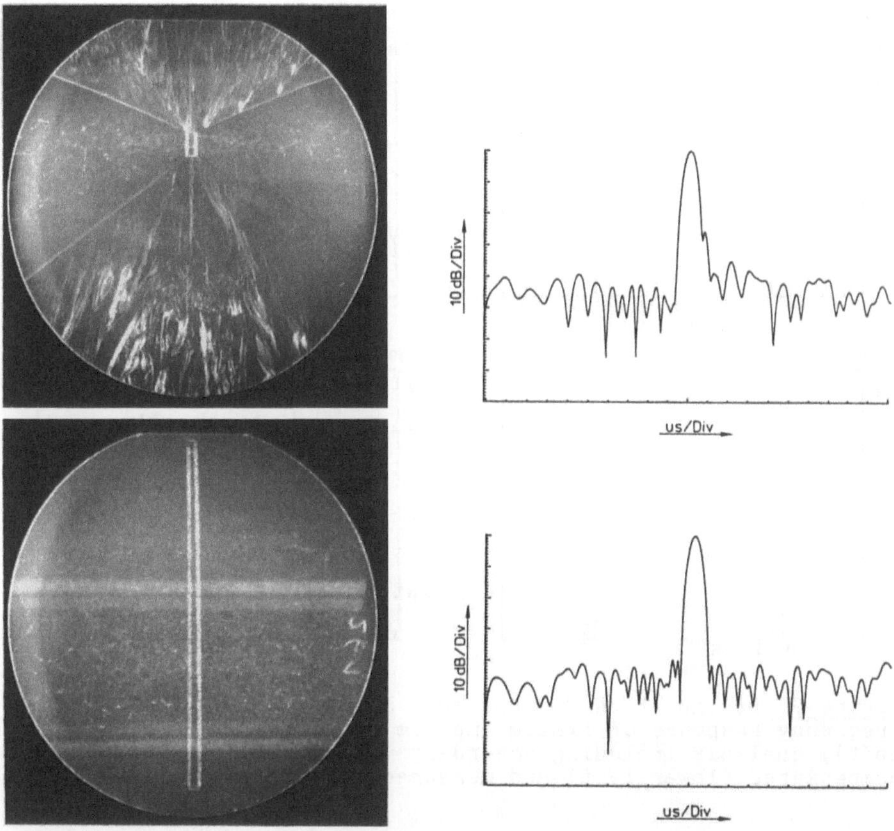

Figure 8: Effect of lattice imperfections in quartz on the performance of Rayleigh wave pulse-compression filters. X-ray topography of 3" ST,X wafer with small seed and different growth regions (upper left), compressed pulse with sidelobe due to non-uniform Rayleigh wave velocity, uniform wafer with extended seed (lower left) and corresponding compressed pulse (lower right)

prevent intersymbol interference a 6 dB bandwidth of the signal spectrum equal to twice the symbol rate and a cosine roll-off at the band edges are required. Spectrum shaping is performed at the intermediate frequency stage with a Rayleigh wave filter /15/. This application required a pass band tolerance of ± 0.1 dB and adjacent channel rejection of 40 dB.

The design is executed in 3 steps /16/. In the first step an ideal filter is designed using a simple delta function model. Second order effects are not considered in this step. A simplex algorithm is used to achieve the required fixed points in the pass band and the optimum stop band rejection The second step is an analysis including all relevant second order effects such as electrical loading, charge distribution in the transducer electrodes, reflexions and diffraction. A least squares algo-

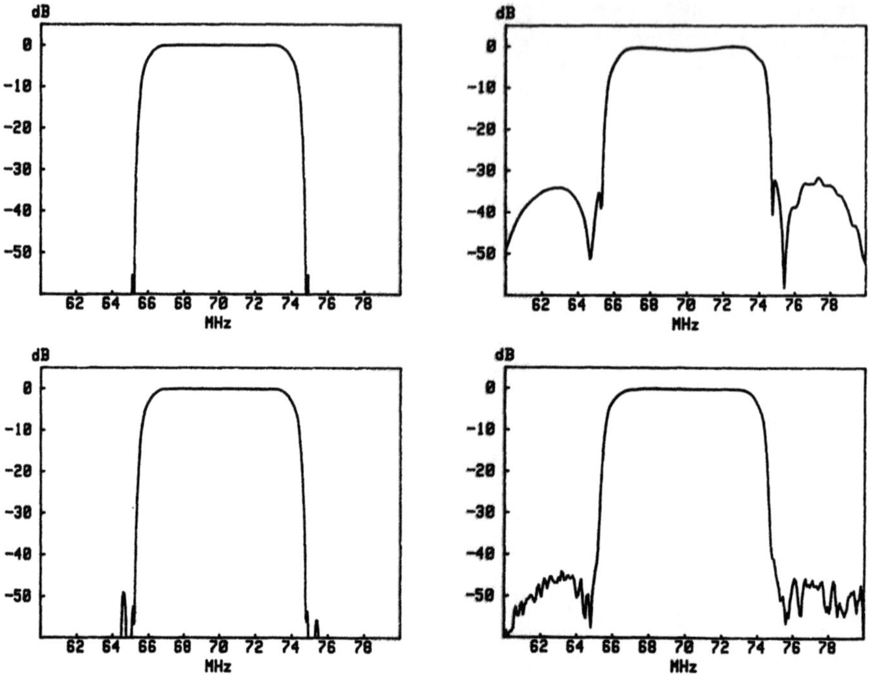

Figure 9: Design of 64 QAM spectral shaping filter. Computed frequency response of finite impulse response filter (upper left), analysis including secondary effects (upper right), compensated (lower left) and measured filter (lower right)

rithm is then used to calculate small variations of the transducer electrodes compensating for the calculated second order effects. The measured Rayleigh wave filter is in very good agreement with the simulation (Fig. 9).

The same design tools were applied to reduce the substrate size of large volume Rayleigh wave devices /17/. TV IF filters originally used multistrip couplers to suppress bulk wave transmission in Y,Z $LiNbO_3$ (Fig. 10 a). A detailed analysis however revealed that the multistrip coupler mainly reduced diffraction. Using diffraction compensation, 128 °**rotated** Y,X $LiNbO_3$ and omitting the multistrip coupler it was possible to reduce the substrate size by a factor 2 and to increase stop band rejection at the same time (Fig. 10 b).

A further reduction of the substrate size yields slightly inferior stop band rejection and therefore has to be improved (Fig. 10 c). Deviations between calculated and measured results are due to second order efects not included in the analysis models. Whereas systematic deviations are to be reduced in the design, scattering can only be removed with improved fabrication techniques.

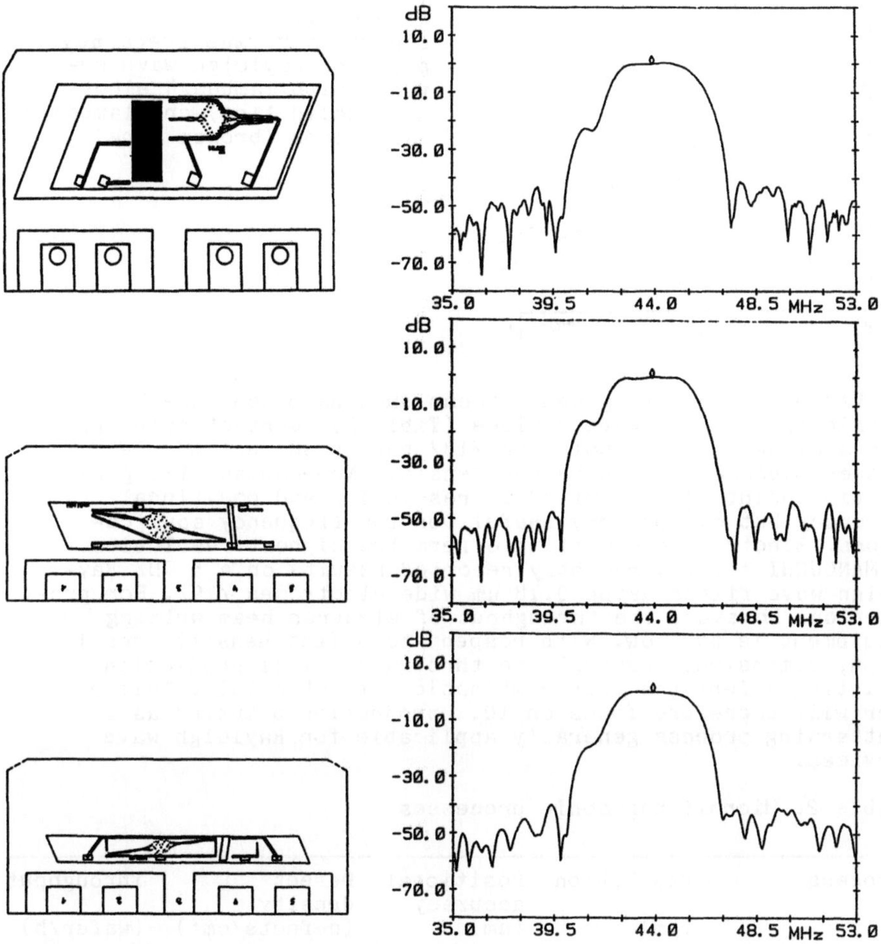

Figure 10: 3 generations of Rayleigh wave TV IF filters. Chip-size and package lay out (left), measured frequency response of NTSC standard filters (right).

3.3 Patterning

Patterning is the most challenging step in the fabrication of Rayleigh wave devices. Rayleigh wave devices may have pattern areas of several cm² and therefore require extremely low defect densities. Submicron line-widths are necessary to achieve GHz frequencies (Fig. 11). Line-widths and positions have to be controlled within less than 0.1 µm. Though exceeding the limits of present integrated circuit fabrication technology, these requirements can be met because Rayleigh wave devices use a single layer metallisation on optically flat substrates with typically 50 to 500 nm thick aluminum electrodes.

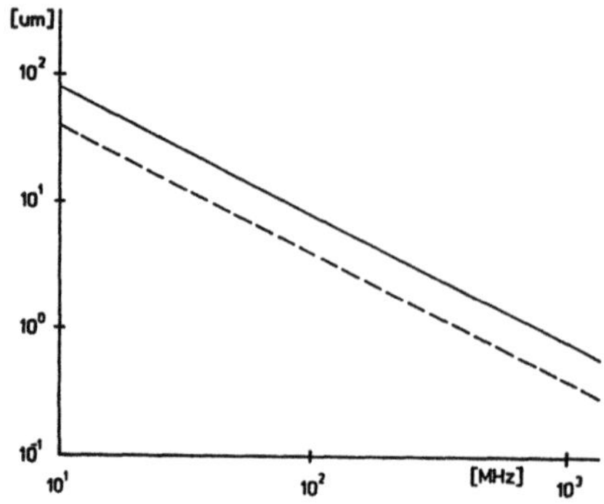

Figure 11:
Linewidth **versus frequency** plot for Rayleigh wave devices with lambda/4 electrodes (solid line) and lambda/8 electrodes (broken line)

Various microlithographic techniques have been used to fabricate Rayleigh wave devices (Table 2). Contact printing can achieve 0.5 μm resolution /18/ but is subject to yield losses because of high defect density. Non-contact 1:1 projection printing is limited in resolution and positional accuracy but can be very useful for low frequency applications. Electron beam writing offers the highest resolution. YAMANOUCHI et. al. recently reported results on a 5 GHz Rayleigh wave filter using 0.18 μm wide electrodes /19/. For production, however, the throughput of electron beam writing equipment is too low. With respect to defect density, resolution, dimensional control and throughput, 10:1 projection printing offers the best combination of potentials. This paper will therefore focus on 10:1 projection printing as a patterning process generally applicable for Rayleigh wave devices.

Table 2: Microlithographic processes

Process	Resolution (μm)	Positional accuracy (μm)	Defect density (defects/cm²)	Throughput (wafer/h)
Contact printing	0.5	0.2	10	100
Optical 1:1 projection	1.5	0.5	0.1	100
Optical 10:1 projection	0.8	0.1	0.1	20
E-beam writing	0.2	0.2	0.1	1

3.3.1 Resolution

The resolution of optical 10:1 projection printing lenses is diffraction limited. In order to resolve a line grating in photoresist the modulation transfer function should exceed a value of 0.75 /20/. The Zeiss 10:1 lens 10-77-82 with numerical aperture 0.28 and exposure wavelength 436 nm gives a resolution limit of 0.8 μm (Fig. 12).

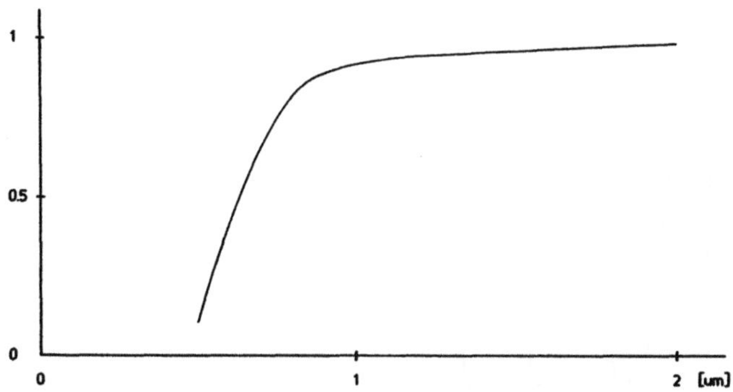

Figure 12: Modulation transfer function of optical 10:1 projection lens Zeiss 10-77-82 with 0.28 numerical aperture and 436 nm exposure wavelength

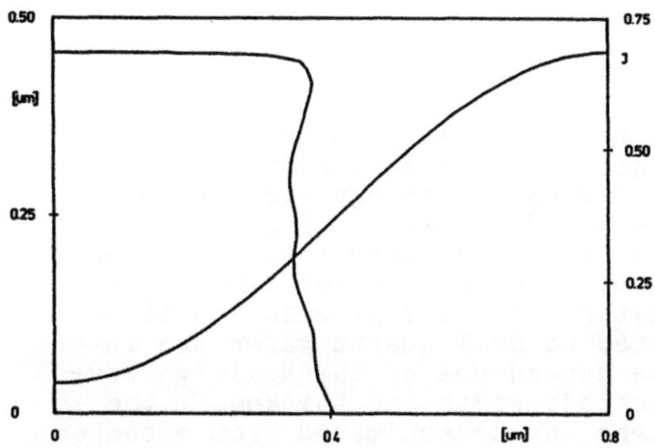

Figure 13: Simulated light intensity and photoresist profile for 0.8 µm line grating exposed with 10:1 projection lens.

For an exposure with this lens, Fig. 13 shows simulations of the light intensity profile and the corresponding photoresist profile for 0.8 µm lines. Although the intensity profile **is sinusoidal** at the resolution limit, steep photoresist walls can be obtained because of the strong nonlinearity of the photoresist. Close linewidth control, however, will only be achieved if photoresist thickness, exposure time and development parameters are kept reproducible. Photoresist processing is therefore executed in a computer controlled system with hands-off operation. In order to maintain the accuracy of the resist pattern, the lift-off technique is used to fabricate the electrode structures.

Using this process a 960 MHz two pole resonator filter on ST-quartz with 0.82 µm lines was fabricated /21/ (Fig. 14). In the future new lenses with 0.42 numerical aperture, 365 nm exposure wavelength and 0.4 µm resolution limit will be available. This will enable the fabrication of Rayleigh wave devices up to 2 GHz and, employing 1.6 times faster surface skimming bulk waves, devices up to 3 GHz.

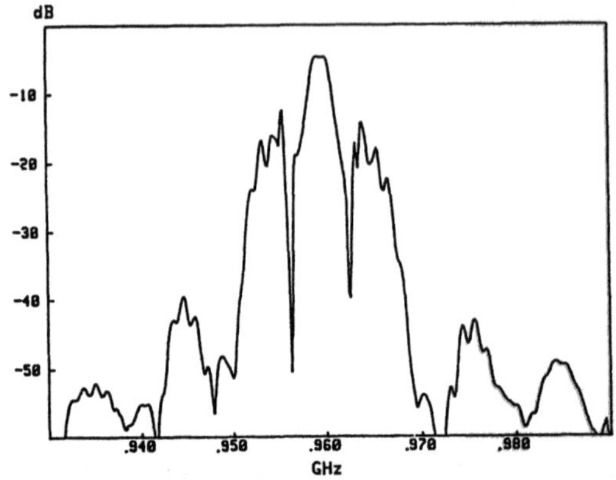

Figure 14: Measured frequency response of 960 MHz two pole resonator filter with 0.82 μm electrodes (matched to 50 Ohm source and load impedance)

3.3.2 Dimensional Control

The propagation velocity of Rayleigh waves in an array of aluminum electrodes is changed due to mass loading, short circuiting of piezoelectric fields and energy storage in the electrode walls. Linewidth and thickness of aluminum electrodes therefore must be carefully controlled. In order to test the reproducibility of these dimensions a batch of 461 MHz Rayleigh wave resonators with 1.7 μm wide and 68 nm thick electrodes on a selected natural quartz wafer was investigated /22/. Knowing the dependence of the Rayleigh wave propagation velocity on electrode width and thickness, the scattering of these parameters can be estimated from a center frequency measurement. The standard deviation of the center frequency was ± 15 ppm (Fig. 15) corresponding to a standard deviation of ± 0.03 μm for the linewidth and ± 0.4 nm for the aluminum thickness. Allowing for a larger deviation and additional variations of the Rayleigh wave velocity from wafer to wafer, a center frequency accuracy of ± 150 ppm can be achieved for quartz resonators at this frequency.

The image field of a 10:1 projection lens is typically 10x10 mm^2. Larger electrode patterns have to be composed using a laser interferometer controlled x-y stage that can be positioned to ± 0.1 μm. Composing from different reticles however leads to alignment errors of 0.3 μm. In pulse-compression filters extending over a length of several cm this may result in phase distortions and increased sidelobes of the compressed pulse (Fig. 16 a). Using a direct pattern generation on the wafer process by composing larger transducers or reflectors from a few apertures arranged on a single reticle /23/, long range positional errors can be reduced to 0.1 μm and sidelobes of the compressed pulse can be improved by 10 dB (Fig. 16 b). Besides improving reproducibility this process shortens the cycle time for the implementation of a new design to a few days.

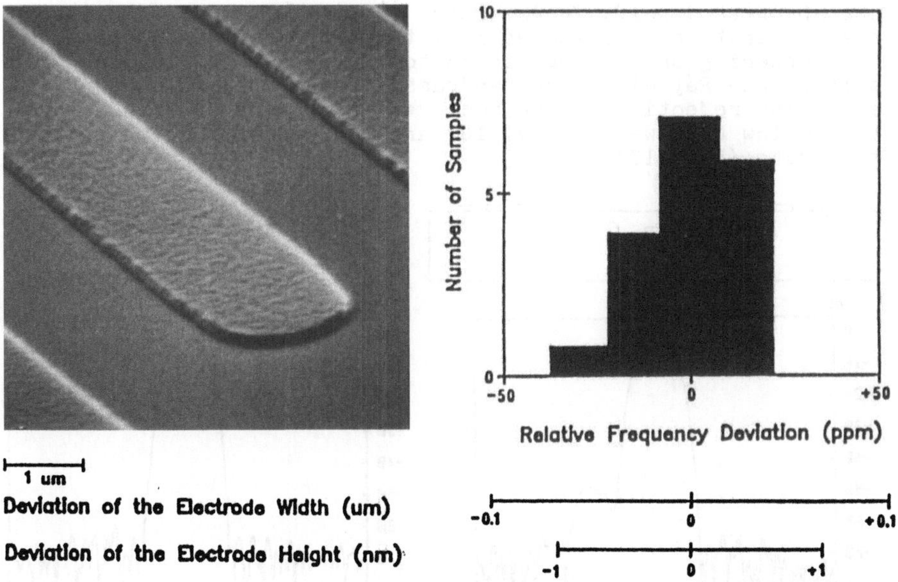

Figure 15: Center frequency distribution of a batch of 461 MHz Rayleigh wave resonators on quartz with 1.7 µm wide and 68 nm thick aluminum electrodes with corresponding scales for linewidth and thickness distribution

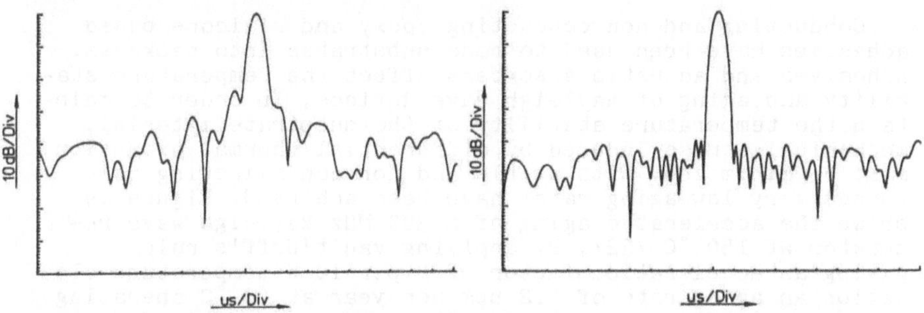

Figure 16: Effect of positional accuracy on the performance of pulse-compression filters. Compressed pulse from filters composed from 3 reticles with 0.3 µm positional errors (left), same design fabricated by direct pattern generation on the wafer with 0.1 µm positional error (right)

3.4 Packaging

Packaging is equally important for the performance and the cost of Rayleigh wave devices. In order to reduce reflexions of waves from substrate edges, organic compounds are usually applied as acoustic absorbers to parts of the substrate surface. Careful design and control of the acoustic impedance as well as the shape of the absorber is necessary to trade off extraction of acoustic energy from the substrate against reflexions from the absorber itself. Transmission of bulk waves emitted into the substrate and reflected from the bottom can be suppressed by roughening or canting the bottom surface /24/. In high performance Rayleigh wave bandpass filters more than 60 dB stop band rejection can be achieved if spurious responses due to shallow bulk waves (Fig. 18) are reduced by a notch in the substrate (Fig. 17).

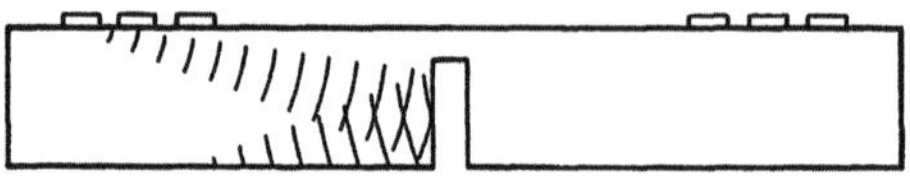

Figure 17: Suppression of shallow bulk wave transmission with a notched substrate

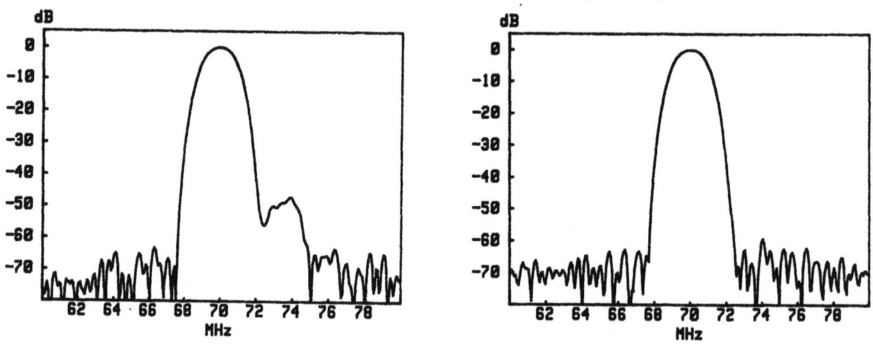

Figure 18: Effect of shallow bulk waves on the performance of Rayleigh wave filters. Substrate with (right) and without notch (left)

Conducting and non conducting epoxy and silicone based adhesives have been used to bond substrates into packages. Adhesives and acoustic absorbers affect the temperature stability and aging of Rayleigh wave devices. In order to maintain the temperature stability of the substrate material, mechanical stress induced by differential thermal expansion must be minimized. With stable and non contaminating compounds very low aging rates have been achieved. Figure 19 shows the accelerated aging of a 300 MHz Rayleigh wave resonator at 150 °C /22/. By applying van t'Hoff's rule, giving an acceleration factor of 2 per 10 K temperature elevation, an aging rate of 0.2 ppm per year at 60 °C operating temperature after preaging for 40 days can be derived from this curve.

A variety of metal glass feedthrough packages is available for all kinds of Rayleigh wave devices providing hermetic sealing in a proven technology (Fig. 20). In very high volume production, where packaging becomes the dominant cost factor, the handling of single packages is too costly. Integrated circuits therefore use lead frames which combine typically 10 plastic molded packages. **Similar** packages have been designed for Rayleigh wave TV IF filters /25/. Instead of molding devices are encapsulated into plastic lids (Fig. 21) and sealed using an epoxy resin thus providing a free substrate surface for the propagation of Rayleigh waves. A 5-pin single-in-line package is used for standard TV IF filters. A 10-pin dual-in-

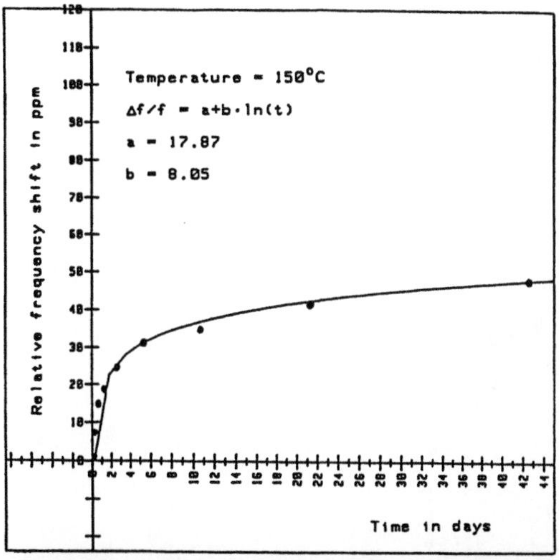

Figure 19:
Accelerated center frequency aging of a 300 MHz Rayleigh wave resonator

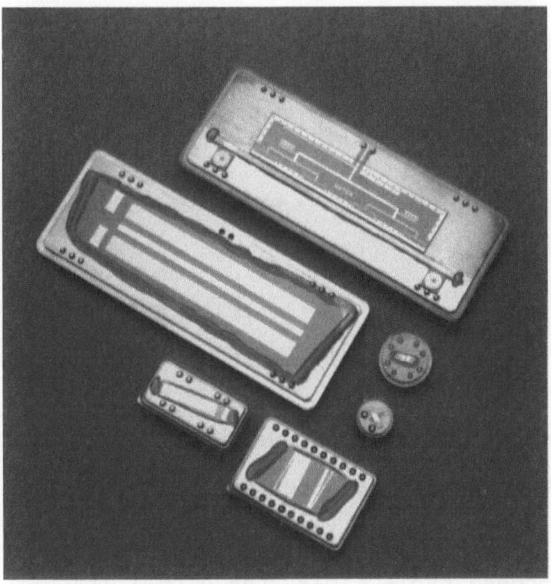

Figure 20:
Hybrid metal package styles for Rayleigh wave resonators, delay lines, bandpass filters, reflective array compressors and convolvers

line version provides additional pins for a quasi-parallel sound filter with two channels for video and sound signals. This version was also especially designed for improved electromagnetic isolation and is therefore used for direct broadcasting satellite receiver filters up to 600 MHz.

Lead frame packages are automatically assembled with computer controlled positioning of adhesives, substrates, wire bonds and absorbers. Apart from **being cost effective lead** frame packages thus provide the high degree of reproducibility adequate for Rayleigh wave devices.

Figure 21: Plastic leadframe packages for Rayleigh wave devices

4. Conclusion

Rayleigh wave devices have become key components in modern telecommunication systems because of their unrivalled capability to precisely match a signal spectrum in amplitude and phase.

Due to the extensive use of computer integrated manufacturing concepts, short process cycle times, the capability to accurately simulate each fabrication step and its inherent high reproducibility the fabrication of Rayleigh wave devices holds an outstanding rank among contemporary high technologies.

Today Rayleigh wave devices are mainly used in the intermediate frequency stage where the typical insertion loss of 15 to 30 dB can be compensated for by amplification. In the future Rayleigh wave devices will expand to the front end of systems. Requirements for high frequencies above 1 GHz, and low insertion loss below 6 dB as well as the **necessity of retaining reproducible fabrication techniques will open a** challenging new field for research in new materials, new types of waves, new structures and new patterning technologies.

Acknowledgement

The author recognizes that G. Eberharter. W. E. Bulst, H. Stocker, D. Pilz and G. Tobolka did not only contribute in innumerable discussions to the basic ideas outlined in this paper, but also took the most active part in implementing them into a state of the art fabrication technology for Rayleigh wave devices. He is also indebted to G. Riha, H. P. Grassl, E. Ehrmann-Falkenau, W. Till, P. Zibis, E. Willibald, G. Nösekabel, U. Knauer and W. Faber for contributing unpublished results. Thanks are due to P. Gasser for the preparation of drawings and to S. Neumayer for typing the manuscript.

References

1. H. Matthews: Surface Wave Filters-Design, Construction and Use (John Wiley and Sons, New York 1977)
2. D.P. Morgan: Surface Wave Devices for Signal Processing (Elsevier Science Publishers Amsterdam 1985)
3. C.S. Hartmann: "Systems Impact of Modern Rayleigh Wave Technology", this issue
4. G. Eberharter, H.P. Feurbaum: "Scanning Electron Microscope Observations of Propagating Acoustic Waves in Surface Acoustic Wave Devices", Appl. Phys. Lett. 37, pp. 698-699, 1980
5. F.G. Marshall, C.O. Newton, E.G.S. Paige, "Surface Acoustic Wave Multistrip Components and their Applications", IEEE Trans. MTT, MTT-21, pp. 216-224, 1973
6. R. Williamson, H. Smith: "The Use of Surface-Elastic-Wave Reflection Gratings in Large Time-Bandwidth Pulse-Compression Filters", IEEE Trans. MTT, MTT-21, pp. 195-205, 1973
7. W. Tanski: "Multipole SAW Resonator Filters: Elements of Design and Fabrication", IEEE Ultrasonics Symp. Proc., pp. 100-104, 1981
8. W.E. Bulst, H. Eschler, "Akustisches Wellenfilter", German patent DE 2909705,C2, 1979
9. L.P. Solie: "Surface Acoustic Wave Reflective Dot Array", Appl. Phys. Lett. 28, p. 420, 1976
10. G. Riha, H.R. Stocker, R. Veith, W.E. Bulst: "RAC-Filters with Position Weighted Metallic Strip Arrays", IEEE Ultrasonics Symp. Proc., pp. 83-87, 1983
11. C. Maerfeld: "Rayleigh Wave Non-Linear Components", this issue
12. H.P. Grassl, H. Engan: "Small Aperture Focussing Chirp Transducers vs. Diffraction Compensated Beam Compressors in Elastic SAW Convolvers", accepted for publication in IEEE Trans. SU, 1985
13. A.J. Slobodnik, "Surface Acoustic Waves and SAW Materials", Proceedings IEEE, 64, pp. 581-595, 1976
14. E. Willibald: "Auswirkung von Kristalldefekten in den Piezoelektrika Lithiumniobat und Quarz auf die Ausbreitung Akustischer Oberflächenwellen", Thesis, Technical University Munich, 1983

15. E. Ehrmann-Falkenau, H.R. Stocker, C. Ruppel, W.R. Mader, "SAW-Filters for Spectral Shaping in a 140 MBit/s Digital Radio System Using 16 QAM", IEEE Ultrasonics Symp. Proc., pp. 17-22, 1983
16. W. Till, E. Ehrmann-Falkenau: private communication
17. G. Tobolka, W. Faber, E. Ehrmann-Falkenau: private communication
18. H. Smith, "Surface Wave Device Fabrication", in ref. 1., pp. 165-217
19. K. Yamanouchi, T. Meguro, Y. Wagatsuma: "Characteristics of 5 GHz SAW Filter Fabricated by Electron Beam Exposure", Japan. Journ. Acoustical Soc., pp. 643-644, March 1985
20. W.G. Oldham, A.R. Neureuther: "Projection Lithography with High Numerical Aperture Optics", Solid State Technology, pp. 106-11, 140, May 1981
21. P. Zibis, G. Nösekabel, U. Knauer: private communication
22. W. E. Bulst, E. Willibald: "Ultrareproducible SAW Resonator Production", 36th Annual Frequency **Control Symp.** Proc., pp. 442-447, 1982
23. H.R. Stocker, R. Veith, E. Willibald, G. Riha: "Surface Wave Pulse Compression Filters with Long Chirp Time", IEEE Ultrasonics Symp. Proc., pp. 78-82, 1981
24. A.J. Slobodnik, T.L. Szabo, K.R. Laker: "Miniature Surface Acoustic Wave Filters", Proceedings IEEE, 67, pp. 129-146, 1979
25. G. Tobolka, W. Faber, G. Albrecht, D. Pilz: "High Volume TV IF Filter Design, Fabrication and Applications", IEEE Ultrasonics Symp. Proc., pp. 1-12, 1984

Part V

Rayleigh Waves in Ultrasonic Microscopy

Rayleigh Waves in Scanning Acoustic Microscopy

H.L. Bertoni

Department of Electrical Engineering and Computer Science,
Polytechnic Institute of New York, Brooklyn, NY 11201, USA

I. Introduction

The role played by the Rayleigh wave in the acoustic microscope came as a fortuitous surprise when the microscope was used in the reflection mode to examine integrated circuits and other materials having Rayleigh wave velocity greater than the velocity in water. The primary contrast mechanism for such objects is due to the presence of the Rayleigh wave, and it has made possible the use of the microscope for quantitative materials characterization. The scanning acoustic microscope (SAM) was conceived by QUATE, who together with co-workers did much to develop it. Reference [1] gives an excellent discussion of the microscope and their early work.

A schematic diagram showing the acoustic microscope operating in the reflection mode is shown in Fig. 1. A piezoelectric transducer at one end of the lens rod is excited by a short pulse of RF voltage, which launches elastic waves propagating towards the concave spherical surface at the other end of the rod. The waves are refracted into the coupling fluid, which is usually water, and are focused on the object. The waves reflected from the surface propagate back towards the lens where they are again refracted back to the transducer. At the transducer, the pulse of elastic waves is converted to a voltage output V, the transducer terminals having been switched to a detector during the travel time of the pulse. The detector output can be recorded for later processing and/or used to modulate the intensity of the electron beam in a CRT. An image of the object is formed by mechanically scanning the lens rod over the surface of the object, and in synchronism, scanning the electron beam across the face of the CRT.

Because typical materials for the lens rod have a longitudinal wave velocity that is several times the wave velocity V_w in the coupling fluid, the lens has small spherical aberration even when the half-angle θ_m of the lens aperature is greater than 45°. The combination of large θ_m and small aberration implies that the acoustic spot-size at the focus will be close to the diffraction limit $\lambda_w/2$, where $\lambda_w = 2\pi V_w/\omega$ is the wave-length in the coupling fluid [2]. Thus at a frequency of 1GHz, and using water couplant,

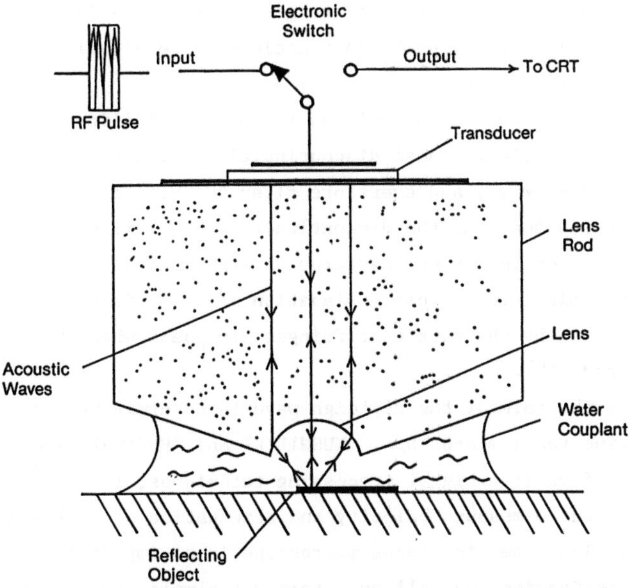

Fig. 1 Scanning acoustic microscope operating in the reflection mode

the spot-size, and hence the resolution, are about 0.7µm, which is comparable to an optical microscope [3].

When the microscope was used in the reflection mode to examine integrated circuits [1,3-5], and grain structure in metals [6,7], contrast was obtained in the images that could not be explained in terms of variations over the object of the magnitude of the reflection coefficient at the water-object interface. Instead it was shown that contrast resulted from the dependence of the output voltage V on defocus distance z, which has been called the V(z) dependence, and variations of this dependence over the object surface [1,8]. Moreover, it was found by WEGLEIN [9] that when the microscope was defocused towards the object through large distances, V(z) exhibited a series of minima whose spacing Δz is characteristic of the Rayleigh wave or SAW velocity of the object surface. This series of minima was called the acoustic material signature (AMS), and was the first use of the microscope for quantitative materials characterization.

Explanation for the V(z) variation was found in terms of Rayleigh critical angle phenomena [10] which occur for angles of incidence θ near $\theta_R = \sin^{-1}(V_w/V_R)$, where V_R is the Rayleigh wave velocity. The Fourier optics explanation [11] views the variation as resulting from the rapid 2π phase variation of the reflection coefficient $R(\theta)$ near θ_R. An equivalent ray-optical explanation [12] views the variation as resulting from the interference between fields of near axial rays that are geometrically reflected from the object, and the fields of ray paths that include the excitation and re-radiation of the Rayleigh wave.

The presence of the Rayleigh wave also explains the high visibility of cracks and grain boundaries [13,14]. These discontinuities serve to reflect and scatter the Rayleigh wave, and therefore alter the V(z) dependence as the microscope is scanned over the discontinuity. The reflected Rayleigh wave also accounts for the fringe patterns of light and dark lines observed parallel to discontinuities. This explanation of the fringes is most graphically demonstrated by the fact that fringes disappear when the Rayleigh wave is eliminated [15].

In this paper we study the role of the Rayleigh wave in determining the output voltage for the line focus microscope. KUSHIBIKI and CHUBACHI have shown that with the help of various signal processing techniques, the line focus microscope can be used to measure velocity and attenuation of Rayleigh and other SAW waves [7]. Thus the line focus microscope is an important tool for material characterization, as well as offering a simple, two-dimensional structure in which to describe the role of the Rayleigh wave. A ray optical description is given here that is similar to that previously employed for the point focus lens [16]. This approach has the advantage of being graphical, and clearly distinguishes the effects of the Rayleigh wave.

II. Separation of Geometrically Reflected and Leaky Wave Fields.

The choice of a wide lens angle θ_m to improve resolution implies that the reflection from the object surface is some weighted average of $R(\theta)$. Of interest here are inorganic solids, which typically have Rayleigh wave velocity V_R substantially greater than V_w. For such solids, the reflection coefficient has dependence similar to that shown in Fig. 2 for YIG. The magnitude of R is less than unity for normal incidence (θ=0) with a sharp rise to unity at the longitudinal critical angle $\sin^{-1}(V_w/V_\ell)$, where V_ℓ is the long wave velocity of the solid. In the absence of dissipation in the solid, $|R|$ again reaches unity for angles at and above the shear critical angle $\sin^{-1}(V_w/V_s)$ where V_s is the shear wave velocity. The phase of R is nearly constant except for a rapid change of almost 2π centered at the Rayleigh critical angle θ_R. As discussed below, the phase variation for θ near θ_R results from the presence of a leaky Rayleigh wave propagating along the object surface.

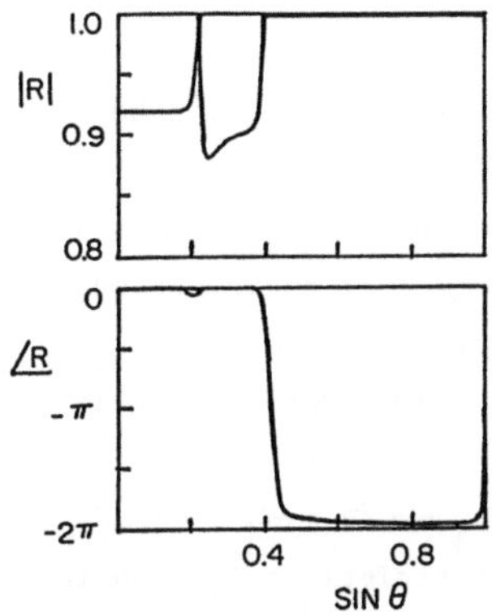

Fig. 2 Magnitude and phase of R(θ) for a water - YIG interface - after Quate, et. al. [1]

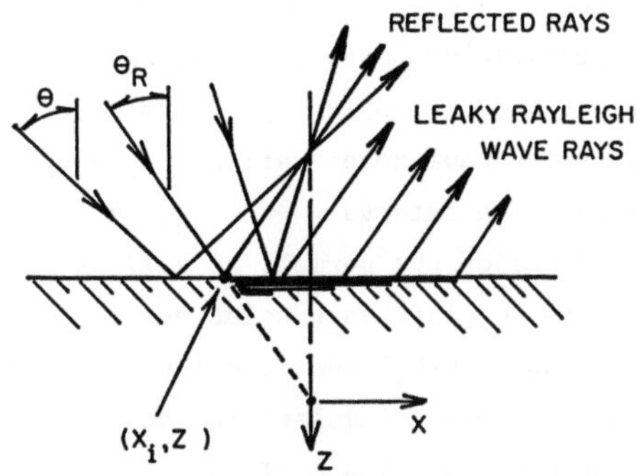

Fig. 3 Reflected field and the leaky Rayleigh wave excited by a convergent beam

A. Leaky Rayleigh Wave

When the object surface is loaded by a fluid, a Rayleigh wave propagating with phase constant k_R will excite a plane wave in the fluid having the same phase constant along the surface. If $V_w < V_R$, then the plane wave will propagate in the fluid at the angle θ_R. Since the plane wave carries energy away from the surface, the Rayleigh wave must attenuate exponentially along the surface. This behavior is suggested in Fig. 3, where the Rayleigh wave is excited by a convergent beam incident from the fluid. The attenuation constant α_ℓ due to leakage of energy into the fluid is approximately given by

$$\alpha_\ell = \frac{k_R}{2\pi} \frac{\rho_w}{\rho_s} \sin \theta_R \tag{1}$$

where ρ_s/ρ_w is the specific gravity of the object relative to the coupling fluid [17]. For $\rho_s/\rho_w > 1$, α_ℓ is small compared to k_R.

Viscous damping and scattering by irregularities in the solid will also attenuate the Rayleigh wave. The attenuation constant α_d due to damping and scattering is approximately the same as for a free-surface Rayleigh wave, and can be added to α_ℓ to give the total attenuation constant α [10,18]. Neglecting the second order change in the phase constant resulting from fluid loading, the complex wave number of the leaky Rayleigh wave is given by

$$k_p = k_R + i\alpha = k_R + i(\alpha_d + \alpha_\ell) \tag{2}$$

for time dependence $\exp(-i\omega t)$.

B. Relation of the Leaky Wave to $R(\theta)$

The analytic properties of the reflection coefficient giving rise to the variation depicted in Fig. 2 are most easily discussed in terms of the transverse wavenumber variable $k_x = k_w \cdot \sin\theta$. As used here k_x ranges over the infinite interval $-\infty < k_x < \infty$, which includes angles of incidence on both sides of the normal, as well as complex angles of incidence corresponding to plane wave solutions in the water that are evanescent along z. Variation of $R(k_x)$ near the longitudinal and shear critical angles is due to branch point singularities at $\pm k_\ell$ and $\pm k_s$, where $k_\ell = \omega/V_\ell$ and $k_s = \omega/V_s$.

Because the wavenumbers $\pm k_p$ of (2) represent characteristic solutions for water-solid interface, they must be poles of $R(k_x)$. In the absence of dissipation in the solid ($\alpha_d = 0$), $|R|$ must be unity for $k_x > k_R$, since the plane waves excited in the solid are evanesent and cannot carry power away from the interface. This property of $|R|$ together with the 2π phase change near k_R in Fig. 2 are obtained if $R(k_x)$ has zeros $\pm k_o$ located at the conjugates of the poles $\pm k_p$, as shown in Fig. 4. When dissipation is present, the zero k_o is found to move nearly vertically upwards in the k_x plane by the amount α_d [10,18]. Thus the zero of $R(k_x)$ is approximately given by

$$k_o = k_R + i(\alpha_d - \alpha_\ell), \qquad (3)$$

which is the conjugate of k_p for $\alpha_d = 0$.

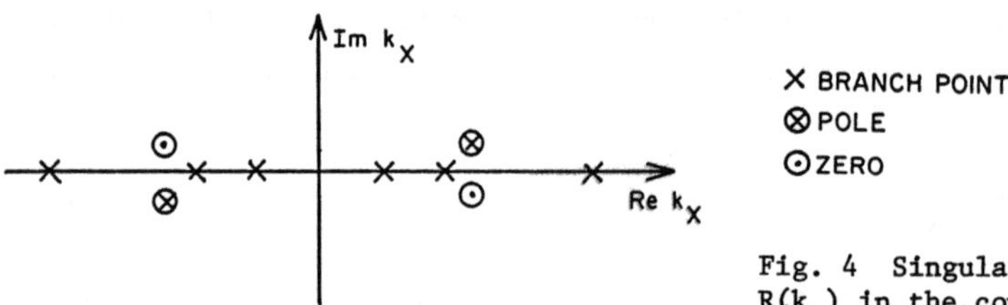

Fig. 4 Singularities of $R(k_x)$ in the complex k_x plane

In addition to the singularities shown in Fig. 4, $R(k_x)$ has branch points $\pm k_w$ and other poles that are far removed from the real k_x axis. When the object is anisotropic, the reflection coefficient will depend on the orientation of the plane of incidence with respect to crystallographic directions in the surface. For some material cuts and directions of propagation in the surface, the free surface Rayleigh wave does not exist, so that the pole is far from the real axis. For other crystal cuts and directions of propagation, additional poles may exist close to the real axis [7,19]. In this paper we restrict the discussion to material cuts for which there exists a free surface Rayleigh wave and a single pole-zero pair near k_R.

C. Approximation for $R(k_x)$

The pole-zero pairs can be separated out from the remaining variation of $R(k_x)$ using the expression

$$R(k_x) = \tilde{R}(k_x)\,\frac{k_x^2 - k_o^2}{k_x^2 - k_p^2} = \tilde{R}(k_x)\left[1 + \frac{k_p^2 - k_o^2}{k_x^2 - k_p^2}\right] \tag{4}$$

where the branch point and other singularities are contained in the function $\tilde{R}(k_x)$. In the vicinity of k_R, where $(k_p^2 - k_o^2)/(k_x^2 - k_p^2)$ is of order unity, $\tilde{R}(k_x)$ is very close to one. For k_x away from k_R, $|k_p^2 - k_o^2|/|k_x^2 - k_p^2|$ is small compared to unity. We may therefore approximate (4) by

$$R(k_x) \approx \tilde{R}(k_x) + \frac{k_p^2 - k_o^2}{k_x^2 - k_p^2} \quad . \tag{5}$$

The separation of $R(k_x)$ given in (5) will be shown to lead to a separation of the leaky Rayleigh wave from the total reflected field.

III. Leaky Wave and Geometrically Reflected Fields

In the case of a line-focus lens, the incident wave in the fluid couplant is in the form of a converging cylindrical wave whose pressure is given by

$$P_{in}(x,z) = \Psi(\theta)\,\frac{\exp(-ik_w\sqrt{x^2 + z^2})}{k_w^{1/2}\,(x^2 + z^2)^{1/4}} \tag{6}$$

for points (x,z) away from the focal point. As indicated in Fig. 3, x and z are the horizontal and vertical displacements from the focal point and $\Psi(\theta)$ gives the angular dependence of the amplitude.

Let $\bar{P}_{in}(k_x,z_i)$ be the Fourier transform of (6) taken in the plane of the object surface. The pressure $P_r(x,z)$ of the reflected wave in this plane is the inverse Fourier transform of the product of $\bar{P}_{in}(k_x,z_i)$ and $R(k_x)$. Using (5) we may separate the reflected pressure into a geometric component $P_G(x,z_i)$ and two leaky wave components $P_L^+(x,z_i)$ and $P_L^-(x,z_i)$ propagating in opposite directions along x. Thus

$$P(x,z_i) = P_G(x,z_i) + P_L^+(x,z_i) + P_L^-(x,z_i) \quad . \tag{7}$$

The geometric contribution to (7) is the inverse transform of the product of $P_{in}(k_x,z_i)$ and $\tilde{R}(k_x)$. This contribution is composed of the geometric optics reflection and the lateral wave fields associated with the branch points $\pm k_\ell$ and $\pm k_s$. By comparison to the leaky Rayleigh wave, the lateral waves have only a small influence on the reflected field, and are neglected for simplicity. The geometrical optics term is nearly the product of the incident pressure and the reflection coefficient for the particular angle of incidence, i.e.,

$$P_G(x,z_i) = \tilde{R}(k_w \sin\theta)\, P_{in}(x,z_i) \quad . \tag{8}$$

Using the ordinary rules of geometric optics one can find the pressure away from the surface.

Figure 3 shows the leaky wave component propagating in the positive x direction, which is excited by the field incident at the angle θ_R near the point $-x_i$, where $x_i = |z_i|\tan\theta_R$. The leaky wave propating to the left is excited by the field incident near x_i. Except near their respective excitation points the fields of these leaky waves are given by [16]

$$P_L^{\pm}(x,z_i) = -2\alpha_\ell \sqrt{\frac{2\pi|z_i|}{k_w(\cos\theta_R)^3}}\; e^{-i\pi/4} e^{ik_p(\pm x + x_i)} U(\pm x + x_i)\; P_{in}(\mp x_i, z_i) \tag{9}$$

where $U(\pm x + x_i)$ is the unit step function. The foregoing spearation of the reflected field into geometric optics and leaky wave fields allows the output voltage due to these wave types to be determined separately.

IV. Microscope Output Voltage

A cross-sectional view of the line focus microscope lens and transducer is shown in Fig. 5. The halfwidth of the transducer is R_T, while that of the

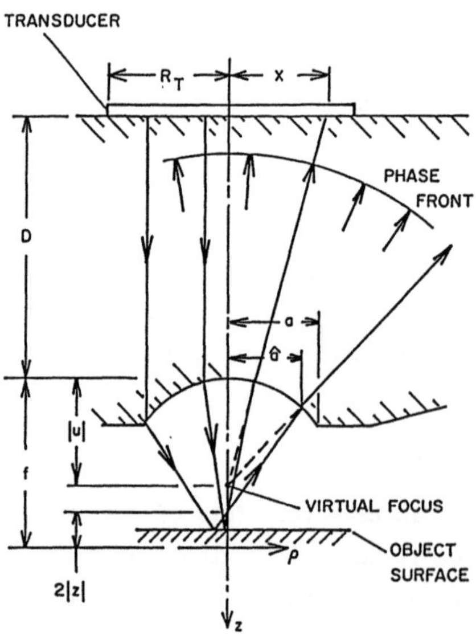

Fig. 5 Ray paths fot the geometrically reflected field

lens aperture is a, which is assumed to be less than or equal to R_T. We have also shown in Fig. 5 ray paths of the geometrically reflected field for the case when the object is closer to the lens than the focal point ($z < 0$).

A. Voltage Due to Geometrically Reflected Fields

The geometrically reflected rays for $0 > z > -f/2$ cross through a line focus at a distance $|z|$ above the object and are subsequently refracted by the lens. In the lens, the refracted rays appear to come from a virtual focus at a distance $|u|$ below the lens surface, where from the simple lens formula

$$u = f(f + 2z)/2nz \tag{10}$$

and n is the ratio of the longitudinal velocity in the lens to V_W. When $z < -f/2$, these rays have a single real focus in the lens rod at a distance $u > 0$ above the lens surface.

The field returning to the transducer for $z \neq 0$ will differ from those for $z = 0$ by the exponential factor $\exp[(ik_W - \alpha_W)2z]$ due to the change in the fluid path length, where α_W is the attenuation constant in the fluid. They will also have the quadratic phase variation $\exp[ik_W x^2/2n(D-u)]$ due to cylindrical divergence of the rays, where x is the displacement from the center of the transducer. Here D is length of the lens rod. In order to conserve energy in a wedge-shaped tube of rays, the field amplitude must vary along the rays. The field returning to the lens surface in the water path is changed by the factor $\sqrt{f/|f + 2z|}$, while propagation from the lens surface to the transducer introduces the factor $\sqrt{|u|/|D - u|}$. Finally, the field is reduced by the reflection coefficient $\tilde{R}$ of the object. Except at focus ($z = 0$), the rays illuminating the transducer come from a narrow wedge of rays in the water, so that we may use the value $\tilde{R}(0) \approx R(0)$ for all rays.

With the foregoing considerations, and using (10) the field illuminating the transducer is given by

$$A\,R(0)e^{(ik_W - \alpha_W)2z} \sqrt{\frac{z_o}{z_o - z}} \exp\left[\frac{ik_W z_o z}{f^2(z - z_o)}\, x^2\right] \tag{11}$$

where A is the field reaching the transducer for a perfectly reflecting object at focus. The term z_o is given by

$$z_o = \frac{f^2}{2n(D - f/n)} \quad . \tag{12}$$

While (11) was derived for $z < 0$, it can also be shown to hold for $z > 0$.

The singular condition at $z = z_o$ is a result of the geometrical optics approximations used to derive (11), and arises when the rays come to a focus at the transducer itself, as discussed for the spherical lens in Ref. [16].

The output voltage due to the field in (11) is obtained by integrating over the illuminated area of the transducer. When the transducer width $2R_T$ is greater than the lens aperture 2a, only the portion of the transducer of width 2a is illuminated for an object at focus (z=0). As the object is moved towards the lens ($z < 0$), the returning ray bundle spreads until the entire transducer is illuminated, which occurs when $z = z_o[(R_T/A)-1]$. When the object is moved away from the lens ($z > 0$), the rays initially converge on the transducer, until at $z = z_o$ they are focused on the transducer. For $z > z_o$ the rays are again divergent at the transducer, and finally illuminate the entire transducer when $z = z_o[(R_T/a) + 1]$, as discussed in Ref. [16]. Integrating with respect to x over the illuminated portion of the transducer leads to the following expression for the output voltage

$$V_{GO} = V_o R(0)\, e^{(ik_w - \alpha_w)2z} \sqrt{\frac{F}{2}}\, \frac{G(X)}{\sqrt{|z/z_o|}} \tag{13}$$

where V_o is a proportionality constant. The parameter F is defined by

$$F \equiv n\lambda_w(D - f/n)/R_T^2 = f^2\lambda_w/2R_T^2 z_o \quad . \tag{14}$$

In (13)

$$G(X) = \begin{cases} C(X) + iS(X) & \text{for } z \leq 0 \text{ or } z \geq z_o \\ -iC(X) - S(X) & \text{for } 0 < z < z_o \end{cases} \tag{15}$$

where C (X) and S (X) are the Fresnel integrals, which arise from integration of the quadratic phase variation in (12). The variable X is given by

$$X = \begin{cases} \left(\frac{2}{F}\, \frac{|z|}{|z - z_o|}\right)^{\frac{1}{2}} & \text{for } z < -z_o[(R_T/a)-1] \text{ or } z \geq z_o[(R_T/a) + 1] \\ \frac{a}{R_T}\left(\frac{2}{F}\left|\frac{z}{z_o}\left(\frac{z}{z_o} - 1\right)\right|\right)^{\frac{1}{2}} & \text{for } -z_o[(R_T/a)-1] < z < z_o[(R_T/a) + 1] \end{cases} \tag{16}$$

and accounts for the partial illumination of the transducer.

The definition (14) gives F as the ratio of the spacing between the transducer and the back focal plane of the lens, (D-f/n), to the Fresnel distance

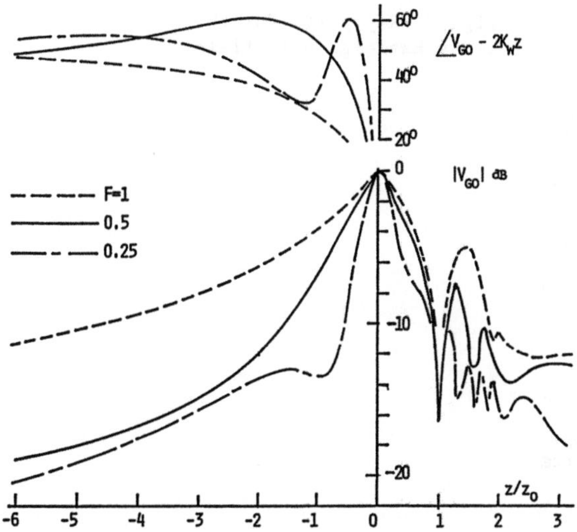

Fig. 6 Magnitude and phase of the output voltage due to the geometrically reflected field for $R_T = a$, $\alpha_w = 0$ and several values of F

of the transducer $R_T^2/n\lambda_w$. The dependence of V_{GO} on lens dimensions is seen to be contained within the parameters z_o, F and R_T/a. Lenses are frequently designed with values of F that are close to unity.

The variation of $|V_{GO}|$ with z is shown in Fig. 6 for $R_T = a$, $\alpha_w = 0$ and several values of F. The phase of V_{GO}, with the linear dependence $k_w 2z$ subtracted out, is also shown for $z < 0$. Least variation in the AMS region $z < 0$ is obtained for F near unity. Small values of F result in ripple of amplitude of V_{GO} in the region $z < 0$. Since the leaky wave does not contribute for $z > 0$, the variation of $|V_{GO}|$ in this region can be compared directly to the measured variation of $|V|$ for line focus lenses [20,21]. As compared to the point focus lens, the amplitude variations are less pronounced in the line focus lens.

B. Voltage Due to Leaky Rayleigh Waves

Fields radiated into the coupling fluid by the leaky wave travel along parallel rays at the Rayleigh angle. The field along the rays is given by its value (9) at the surface together with the phase change and attenuation due to the path length in the water. The rays are refracted at the lens surface towards one of two foci lying a distance $f(\cos\theta_R)^2/n$ above the lens, and at distances $\pm x_R$ from the lens axis, where

$$x_R = f \sin\theta_R \, . \tag{17}$$

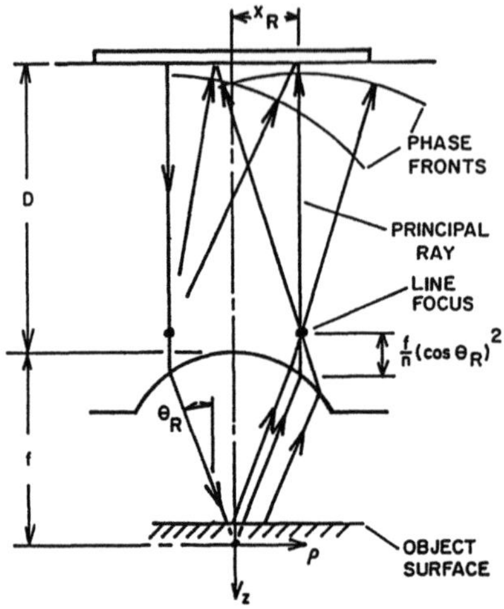

Fig. 7 Ray paths for the leaky Rayleigh wave field

The rays passing through the right-hand focus are shown in Fig. 7. Past the foci, the ray bundles diverge and finally illuminate the transducer. For z more negative than about $-z_o$, each ray bundle fully illuminates the transducer. For the usual case with D>>f/n, the rays that illuminate the transducer are closely grouped about the principle ray that appears to come from the focus and is refracted in the lens rod parallel to the axis. When z becomes positive, the rays miss the transducer entirely, so that the leaky waves do not contribute to the output voltage, as been demonstrated using Fourier optics for a line focus lens [13].

Assuming $D \gg f(\cos\theta_R)^2/n$, ray passage through the focus changes the field amplitude by the factor $\sqrt{f/nD}\cos\theta_R$. Presence of the focus introduces the additional phase change of $-\pi/2$ and a phase variation across the transducer that is quadratic in $(x \pm x_R)$. The leaky wave fields reaching the transducer are thus given by

$$A(2\alpha_\ell f\sqrt{\lambda_w/nD})\, e^{i3\pi/4}\, e^{2(\alpha_w - ik_w)|z|\sec\theta_R}$$

$$\cdot e^{i2k_p|z|\tan\theta_R}\left[e^{ik_w(x - x_R)^2/2nD} + e^{ik_w(x + x_R)^2/2nD}\right] \quad . \qquad (18)$$

The first of the two exponentials containing $|z|$ represents the reduction in path length in water of the principle ray, as compared to case when the object is at the focus. The second exponential represents the propagation factor of a leaky Rayleigh wave over the path length $2|z|\tan\theta_R$ on the object surface. Finally, the exponentials in the brackets give the phase deviation across the transducer due to the cylindrical curvature of the phase fronts for the two ray bundles.

Integrating (18) over the transducer gives the output voltage V_{LW} due to the leaky wave field. After some manipulation one obtains

$$V_{LW} = V_o K(f\alpha_\ell \lambda_w/R_T)\, e^{i3\pi/4}\, e^{i2(k_R \tan\theta_R - k_w \sec\theta_R)|z|}$$
$$\cdot e^{2(\alpha_w \sec\theta_R - \alpha \tan\theta_R)|z|} \quad (19)$$

where V_o is the same constant appearing in (13) for V_{GO}. Here K is given in terms of the Fresnel integrals as

$$K = C\left[\sqrt{\tfrac{2}{F}}\,(1- X_R/R_T)\right] + C\left[\sqrt{\tfrac{2}{F}}\,(1+ X_R/R_T)\right]$$
$$+iS\left[\sqrt{\tfrac{2}{F}}\,(1- X_R/R_T)\right] + iS\left[\sqrt{\tfrac{2}{F}}\,(1+ X_R/R_T)\right] . \quad (20)$$

In (19) the z dependence is entirely contained within the two exponential factors. The first factor gives the phase change with z, while the second gives the amplitude dependence associated with attenuation along the water path, and attenuation of the leaky wave along the surface. Leaky wave attenuation depends on the total attenuation constant α, while the amplitude term $(f\alpha_\ell \lambda_w/R_T)$ in (19) contains only the portion α_ℓ that involves coupling of the Rayleigh wave to the fluid. KUSHIBIKI and CHUBACHI [7] have used the simple exponential dependence of $|V_{LW}|$ to measure α for various materials.

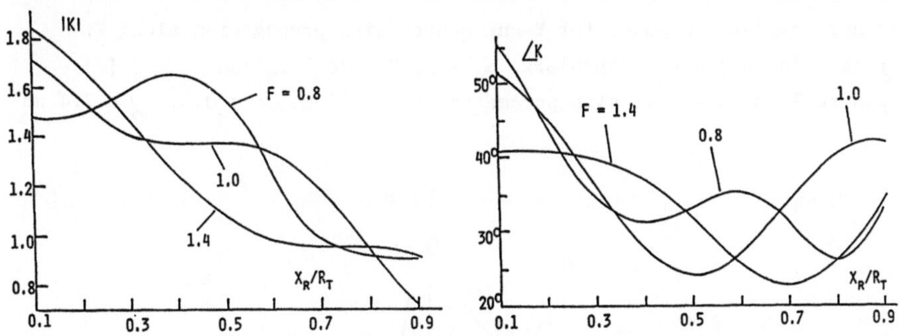

Fig. 8 Dependence of the magnitude and phase of the factor K on the parameter x_R/R_T

The factor K depends on the lens parameters, as well as the Rayleigh angle through x_R. Amplitude and Phase of this factor are shown as a function of x_R/R_T in Fig. 8 for several values of F. When using the microscope for materials characterization, it may be desirable to choose F and R_T, such that K is nearly constant over the expected range of x_R.

C. Total Output Voltage

The magnitude of V_{LW} for $z = 0$ is approximately half the value given by (19), and it decreases rapidly as z increases above zero [13]. Thus the total output voltage V will be approximately equal to V_{GO} of (13) for $z > 0$. For z more negative than about $-z_o$, V is the sum of V_{GO} and V_{LW} given by (19). As z varies in the range $z < 0$, V_{GO} and V_{LW} go in and out of phase so that $|V|$ goes through relative maxima and minima. The minima of $|V|$ occur when the difference in phase between V_{LW} and V_{GO} is an odd multiple of π, which gives

$$2[k_w(1-\sec\theta_R) + k_R\tan\theta_R]|z| + \frac{3}{4}\pi + \angle K - \angle G = m\pi \tag{21}$$

where $m = 1,3,5...$

For z less than about $-z_o$, the variation of $\angle G$ is seen from Fig. 6 to be small, and to approach a constant for $|z| >> |z_o|$. Treating $\angle G$ as constant, leads to the well known expression for the separation Δz between successive minima

$$\Delta z = \frac{\pi}{k_w(1-\sec\theta_R) + k_R\tan\theta_R} = \frac{\lambda w/2}{1-\cos\theta_R} \; . \tag{22}$$

Measuring Δz, many authors have used (22) to find the Rayleigh angle θ_R and from this the Rayleigh velocity [7,9].

As an example of the interaction between the geometrical optics and leaky Rayleigh wave components, we have calculated the individual components and complete response for Y-cut quartz with propagation along Z. For this orientation $V_R = 3826$ m/s, $\alpha_\ell = 3.47 \times 10^{-2}\, k_R$ and $\alpha_d = 0$ [7]. A sapphire lens (n = 7.5) with parameters f = 1.14 mm, F = 0.8, $z_o = 7.4$ µm

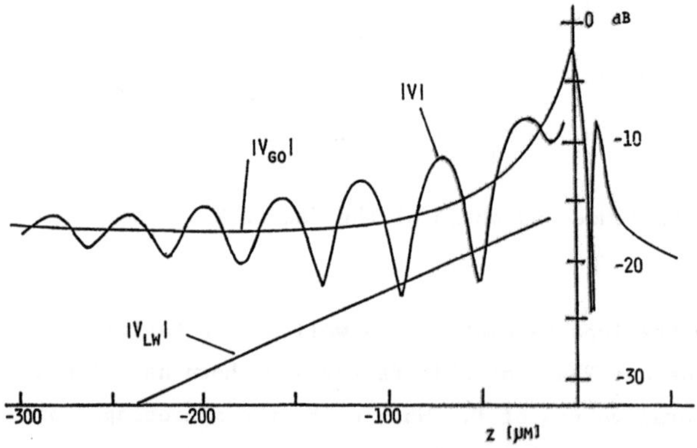

Fig. 9 Variation with z of the output voltage of a 225 MHz line focus lens ($F = 0.8$, $f = 1.14$ mm, $z_o = 7.4$µm and $R_T = a$) looking at Y-cut quartz

and $R_T/a = 1$ is assumed with operating frequency of 225 MHz. The water couplant has $V_w = 1483$ m/s and $\alpha_w = 1.36 \times 10^{-3}\ k_w$, so that $\theta_R = 22.8^o$. The variations of $|V_{GO}|$, $|V_{LW}|$ and $|V|$ are shown in Fig. 9.

It is seen that $|V_{GO}|$ and $|V_{LW}|$ are close to each other only over a limited range of z, so that only a few minima of $|V|$ are deep. Overall, the curve for $|V|$ is very similar to the experimentally observed variation for this object [7]. A sophisticated processing scheme is described in Ref. [7] for accurately determing Δz and α even when most minima are not deep, as in Fig. 9. For denser material having smaller α_ℓ, the slope of the $|V_{LW}|$ curve is less, but its value for $z \to 0$ is smaller, so that depth of minima is not greatly improved. Using a lens with smaller value of F reduces $|V_{GO}|$ and can be expected to give deeper minima.

Image contrast can be understood in terms of the V(z) curve in Fig. 9, and its dependence on V_R, α_ℓ and α_d. For example, neighbouring grains in a polycrystalline material will have different crystallographic orientations, and hence different values of V_R on the surface. If the microscope is defocused to a minimum of V(z) for one grain, that grain will appear dark. Because the neighbouring grain has a different V_R, the defocus distance will not correspond to a minimum and the grain will appear brighter. In effect, the microscope is imaging variations in V_R from grain to grain. Similarly, micrographs of integrated circuits are images of the variation of the velocity of the Rayleigh wave resulting from the pattern of surface layers used in their fabrication. Variations of α_ℓ, α_d or surface height also result in variations of V(z) for a given z, and hence produce contrast in the images.

IV. Rayleigh Wave Reflection From Discontinuities

When the Rayleigh Wave is incident on a lateral discontinuity, such as the edge of a surface plating, a crack, or a boundary between grains, it is partially reflected and partially transmitted. Reflections have been observed as a fringe pattern of parallel bands of light and dark in micrographs [5,14, 15]. Reflection of focused Rayleigh beams has itself been used as a basis for imaging [10]. Because the Rayleigh wave reflection and transmission coefficients Γ_R and T_R contain information about the discontinuity, it is important for materials characterization to be able to measure them.

We assume the discontinuity to lie in a plane perpendicular to the direction of Rayleigh wave propagation, and to be located a distance $x > 0$ from the lens axis. The discontinuity may be located outside the illuminated spot of width $2x_i$ where $x_i = |z| \tan \theta_R$, or within it, as shown in Figs.

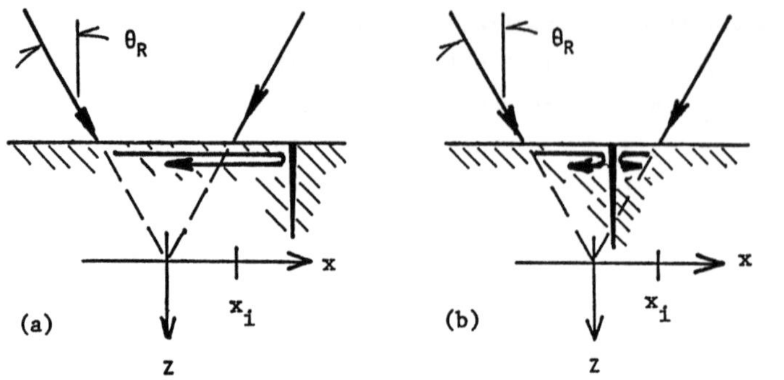

Fig. 10 Scattering of Rayleigh waves by a crack: a) $x > x_i$; b) $0 < x < x_i$

10(a) and (b), respectively. Analysis of the output voltage in the presence of a discontinuity has been carried out using a Fourier optics approach [13,22]. A simpler ray-optical analysis is given here for the case when $|x-x_i|$ is greater than about λ_R, and the lens is defocused to $z < 0$ by at least several times λ_w.

When the discontinuity is located at a point $x > 0$ outside the illuminated spot, the reflected Rayleigh wave contributes to the fields of the rays originated near the point $-x_i$. Because of additional path length, the reflected Rayleigh wave fields at $-x_i$ differ from those due to excitation at $+x_i$ by the factor $\Gamma_R \exp(i2k_p x)$. Since this Rayleigh wave contribution affects only the field of the rays passing through the left hand focus, the leaky wave output voltage V_{LW} will be modified by the multiplicative factor

$$H = 1 + \frac{1}{2} \Gamma_R \exp[(ik_R - \alpha)2x] \tag{23}$$

applicable for $x > x_i$.

When the discontinuity lies within the illuminated spot, as in Fig. 10(b), the fields of the rays passing through both the right and left hand foci are modified. For simplicity, we assume that k_R and α are the same on both sides of the discontinuity, as would occur for a crack in a homogeneous medium. Now the total Rayleigh wave field reaching the point $\mp x_i$ is modified by the factor

$$T_R + \Gamma_R \exp(\pm i2k_p x) \tag{24}$$

For this case the leaky wave output voltage V_{LW} will be modified by the multiplicative factor

$$H = T_R + \frac{1}{2} \Gamma_R \{\exp[(ik_R - \alpha)2x] + \exp[(-ik_R + \alpha)2x]\} \tag{25}$$

applicable for $0 < x < x_i$.

The total output voltage is a function of both the defocus distance z and the lateral displacement x of the discontinuity from the lens axis. For $z < 0$ the functional dependence is of the form

$$V(x,z) = V_{GO}(z) + V_{LW}(z)\, H(x) \tag{26}$$

In order to demonstrate the nature of the x dependence, we assume that the lens is defocused to a point z where $x_i = 2\lambda_R$, and that $V_{GO} = V_{LW}$. Taking a stainless steal object with a crack depth of $\frac{1}{2}\lambda_R$, then $\alpha\lambda_R = 0.069$ and from ANGEL and ACHENBACH [23], $|T_R| \simeq 0.23$ and $|\Gamma_R| \approx 0.42$. Neglecting the phase of T_R and Γ_R, the x variation of $|V|$ is shown in Fig. 11.

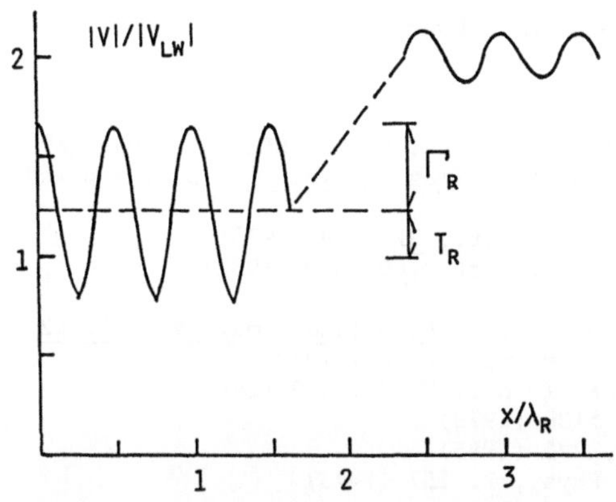

Fig. 11 Variation of output voltage with lateral displacement x from a crack

The dashed portion of the curve in Fig. 11 indicates the transition region about x_i where the simple ray expressions do not hold. For the simple case chosen a substantial variation of $|V|$ is obtained for $x < x_i$, and a smaller variation about a large average for $x > x_i$. This behavior is also seen in the Fourier optics solution [22]. From Fig. 11 it is seen that the values of Γ_R and T_R can be obtained from the variation of $|V|$ for $x < x_i$. It is further seen from Fig. 11 that as the microscope is scanned to obtain an image, the crack will produce an image of width $2x_i$. Thus the reflection and transmission of Rayleigh waves explains why cracks appear so large in micrographs [13,14].

V. Conclusion

When viewing solid objects having $V_R > V_W$ using a wide-angle acoustic lens, image contrast results from the presence of the leaky Rayleigh wave. One contrast mechanism is due to the V(z) dependence, which is the result of interference between the leaky Rayleigh wave fields and the geometrically reflected fields. Another mechanism involves scattering of the Rayleigh wave by cracks and other discontinuities. In the case of cracks, visibility is determined by crack depth rather than crack opening. Thought of another way, the microscope lens is a versatile method for exciting and detecting Rayleigh waves, so that it can be used for quantitative materials characterization. While its use for measuring V_R and α is well documented, it should also be possible to measure Γ_R, T_R and related Rayleigh wave properties using the microscope.

Acknowledgement

This work was supported in part by the National Science Foundation under Grant No. ECS-8414081. The author thanks Dimitri Chizhik for his help in carrying out the numerical calculations reported here.

References

1. C.F. Quate, A. Atalar and H.K. Wickramasinghe, Proc. IEEE, 67, 1092 (1979)
2. M. Nikoonahad and E.A. Ash, in Acoustical Imaging, Vol. 12, E.A. Ash and C.R. Hill, Eds., 47 (Plenum, New York 1982)
3. V. Jipson and C.F. Quate, Appl. Phys. Lett., 32, 789 (1978)
4. R. Hammer and R.L. Hollis, Appl. Phys. Lett., 40, 679 (1982)
5. A.J. Miller IEEE Trans., SU-32, 320 (1985)
6. G.A.D. Briggs, C. Ilett and M.G. Somekh, in Acoustical Imaging, Vol. 12, E.A. Ash and C.R. Hill, Eds., 89 (Plenum, New York 1982)
7. J-I Kushibiki and N. Chubachi, IEEE Trans., SU-32, 189 (1985)
8. A. Atalar, Jnl. Appl. Phys., 49, 5130 (1978)
9. R.D. Weglein, IEEE Trans., SU-32, 225 (1985)
10. H.L. Bertoni and T. Tamir, Appl. Phys., 2, 157 (1973)
11. A. Atalar, Jnl. Appl. Phys., 50, 8237 (1979)
12. W. Parmon and H.L. Bertoni, Elect. Lett., 11, 684 (1979).
13. J.M.R. Weaver, M.G. Somekh, G.A.D. Briggs, S. Peak and C. Ilett, IEEE Trans., SU-32, 302 (1985)
14. K. Yamanaka and Y. Enomoto, Jnl. Appl. Phys., 53, 846 (1982)
15. P. Sivaprakasapillai, M. Nikoonahad and E.A. Ash, Proc. of the Ultrasonics Symposium, 632 (IEEE 1983)
16. H.L. Bertoni, IEEE Trans., SU-31, 105 (1984)
17. K. Dransfeld and E. Salymann, in Physical Acoustics, Vol. VIII, W.P. Mason and R.N. Thurston, Eds., 260 (Academic, 1970)
18. T.D.K. Ngoc and W.G. Mayer, Jnl. Acoust. Soc. Amer., 67, 1149 (1980)
19. G.W. Farnell, in Physical Acoustics, Vol VI, W.P. Mason and R.N. Thurston, Eds., 109 (Academic, 1970)
20. J. Kushibiki, A. Ohkubo and N. Chubachi, Elect. Lett., 18, 6 (1982)
21. Proc. of the Ultrasonics Symposium, 623 (IEEE 1982)
22. M.G. Somekh, H.L. Bertoni, G.A.D. Briggs and N.J. Burton, Proc. Royal Soc., to be published
23. Y.C. Angel and J.D. Achenbach, Jnl. Acoust. Soc. Amer., 75, 313 (1984)

Ultrasonic Micro-Spectroscopy via Rayleigh Waves

N. Chubachi

Department of Electrical Engineering, Faculty of Engineering, Tohoku University, Sendai 980, Japan

1. Introduction

Ultrasonic spectroscopy is a method to reveal the nature of materials by the analysis of the ultrasonic spectrum. It appeared in the literature [1] , that Nernst and Keutel attempted already in 1910 to measure the dissociation rate of NO_2 gas by the experiment of ultrasonic dispersion. Ultrasonic spectroscopy, therefore, has a long history, so that many ingenious techniques have been developed so far to study the properties of liquids, biological tissues and solid materials as a function of frequency [2,3,4,5].

In the last decade, the developments of high frequency ultrasonic techniques have made it possible to measure the ultrasonic velocity and attenuation in the frequency range above 100 MHz. In liquids, an optical method, so-called high-resolution Bragg reflection method have been established to make precise measurements of both ultrasonic velocity and absorption in the frequency range from 60 MHz to 1.7 GHz [6]. Recently, a new reflection correlation method has been proposed to measure ultrasonic velocity and attenuation for a small quantity of liquids (less than 0.3 cm^3) in the frequency range above 100 MHz [7]. In solid materials, on the other hand, the acoustic microscopy has been developed to measure the acoustic properties of materials on a microscopic scale. Interference fringe patterns obtained by acoustic microscopes provide useful information on velocity distributions in the samples [8, 9]. By means of the nonscanning reflection acoustic microscope, acoustic properties of various solid surfaces can be determined quantitatively by analyzing the V(z) curve which is the record of the output signals received by the piezoelectric transducer with the variation of distance z between the acoustic lens and the sample [10]. Recently, a line-focus-beam acoustic microscope system has been established for the quantitative measurement of velocities and attenuations of acoustic surface waves with an accuracy of better than 0.2 % absolutely in the velocity (relatively 0.025 %) and 4 % in the attenuation measurements [11].

In this paper, first, the general concept of "ultrasonic micro-spectroscopy (UMS)" is outlined. Secondly, a realizable and practical construction of the UMS system which is applicable to the investigation of Rayleigh waves and other surface wave modes is discussed on the basis of the experiments performed in our laboratory. Finally, a new directional point-focus-beam acoustic microscope is proposed for the UMS system with which quantitative measurements of Rayleigh-wave velocity and attenuation can be made including the angular dependence on a small area of the samples. Some preliminary results are demonstrated to show the realization of the system.

2. Ultrasonic Micro-Spectroscopy - UMS

The propagation of an ultrasonic wave traveling in an elastic medium can be expressed in a form of the following equation,

$$\boldsymbol{v} = A \cdot \exp j(\omega t - K \cdot r), \tag{1}$$

where $\boldsymbol{v}$ is the particle velocity, A is the amplitude, ω is the angular frequency, t is the time, K is the wave vector for the particular mode and $\mathbf{r}$ is the positional vector.

Let us consider that the wave vector K is expressed as a complex value which includes the attenuation coefficient α_i in a form as follows,

$$K_i = 2\pi/\Lambda_i + j\alpha_i \qquad (\, i = 1,\ 2,\ \text{or}\ 3\,) \tag{2}$$

and the velocity c_i has a relation of $\omega/c_i = 2\pi/\Lambda_i$. Then, we may define ultrasonic spectroscopy as "a method to investigate the structure of materials by analyzing the frequency dependence of K measured under the various conditions of temperatures, pressures and others". In the anisotropic materials, it may be naturally to investigate the angular dependence of K, because K has different values even for the same mode (longitudinal or transversal) depending on the propagation direction of the waves.

In the ultrasonic spectroscopy, the researches which have been performed so far are mostly macroscopic. Recently, it has been desired to study the ultrasonic properties of the materials with a small quantity or a small part of the samples to develop the new material physics and science. This means that microscopic researches should be introduced into ultrasonic spectroscopy. In this trend, the developments of acoustic microscopy can be expected to cultivate a new method to be called as " ultrasonic micro-spectroscopy (UMS) ", because we have already established to measure the ultrasonic properties in a small quantity of liquids as well as spatial distributions of the acoustic properties in solid materials including biological tissues with acoustic microscope systems. Thus, we may define the UMS as "a new field in the ultrasonic spectroscopy where ultrasonic properties of a small quantity or small part of the sample are measured quantitatively, so that the spatial spectrum is further obtained in addition to the frequency spectrum of ultrasonic properties". This may be said in other words that "in the UMS the $\mathbf{r}$-dependence of K is investigated quantitatively in addition to the ω-dependence of K in the Eq. (1)".

3. Construction of Reflection UMS System for Rayleigh Waves

Let us discuss the realizability of the system which can obtain both phases, that is, spatial spectrum and frequency spectrum for material characterization by UMS.

The line-focus-beam acoustic microscope has been already developed to measure the frequency dependence of ultrasonic properties in the wide frequency range from 50 MHz to 400 MHz. Three different lenses with curvatures of 0.75, 1.25, and 1.5 mm are employed in the present set-up to cover the frequency range. An application of this system to material characterization is described here to show that this kind of the system can be used to compose the UMS system.

Structural variations in the materials such as microscopic defects of grains, pores, and roughness near the surface will affect the frequency dependence of the ultrasonic properties in the samples. Figure 1 shows the

(a) (b)

Fig. 1. Acoustic images for two samples of Mn-Zn ferrite wafers observed by a point-focus-beam acoustic microscope at 750 MHz. (a) grain'L' sample, defocus: z = -12 μm, (b) grain'S' sample, defocus: z = -7 μm.

acoustic images for two samples of Mn-Zn ferrite wafers, Grain'L' sample and Grain'S' sample, observed by a point-focus-beam scanning acoustic microscope in the reflection mode at a frequency of 750 MHz. From these photographs, we can determine the size of grains appearing on the surface. The average grain sizes for the Grain'L" and Grain'S' samples are measured to be 50 μm and 8 μm, respectively.

The frequency dependence of the attenuation for leaky Rayleigh waves measured on these samples by the line-focus-beam microscope are shown in Fig. 2 in the frequency range from 50 to 300 MHz. The attenuation increases strongly with the frequency as seen in the figure. Considering the variation of wavelength for the leaky Rayleigh waves with frequency, it can be said that according to the scattering criteria [12] for the Grain'S' sample, the scattering phenomena below the frequency lower than 100 MHz is in the Rayleigh scattering region ($\Lambda > 2\pi D$ where the D is the average grain sizes and Λ is the wavelength). On the other hand, for the Grain'L' sample, the phenomena are considered to be almost entirely in the intermediate region ($\Lambda < 2\pi D$).

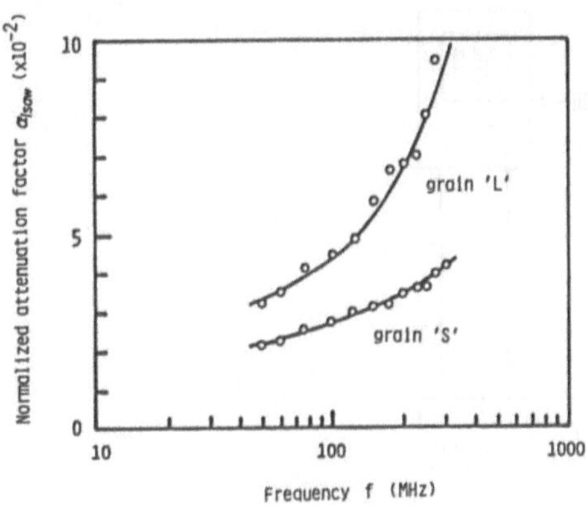

Fig. 2. Frequency dependence of the attenuation for leaky Rayleigh waves for the Mn-Zn ferrites, grain'L' and grain'S' samples.

As seen in this case, it is necessary to use a point-focus-beam acoustic microscope to observe the spatial distribution of grain size for analyzing the attenuation properties as a function of frequency. This is because the line-focus-beam is directional and has high resolution basically only along the direction perpendicular to the line axis. At present, therefore, two kinds of acoustic microscopes developed for different purposes, namely, the quantitative measurement system (the line-focus-beam acoustic microscope) and the high resolution imaging system (the conventional point-focus-beam acoustic microscope), are combined for convenience to realize a practical UMS system as shown in Fig. 3. In this way, both the frequency dependence as well as spatial dependence (including the angular dependence) of ultrasonic properties on the surface of materials can be investigated with the combined system.

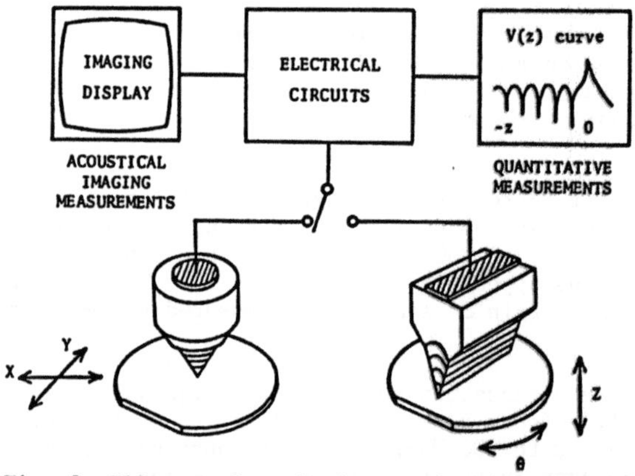

Fig. 3. Illustration of ultrasonic micro-spectroscopy (UMS) system.

4. Directional Point-Focus-Beam Microscope for new UMS System

A pair of point-focusing transducers for generating and receiving leaky SAWs on a sample have a possibility to develop a new UMS system. One of the configuration, which we propose here, is shown in Fig. 4, where a pair of

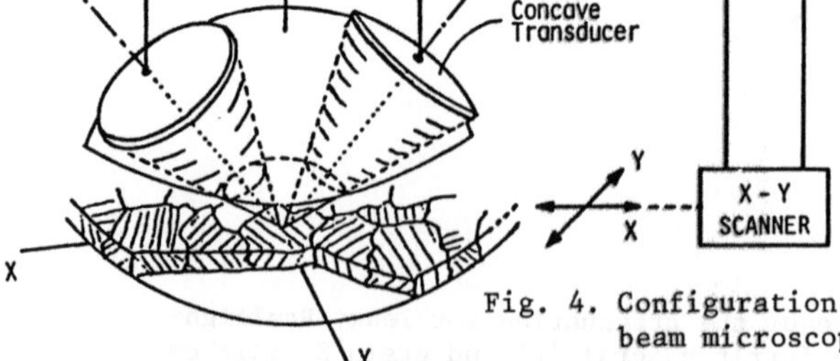

Fig. 4. Configuration of directional point-focus-beam microscope proposed for new UMS system

concave transducers are employed for generating and receiving point-focusing beams. These concave transducers are fabricated on a spherical glass shell with a concave radiation surface as shown in Fig. 5. The system is basically a directional acoustic microscope for reflection mode, and also a kind of mechanically scanned acoustic interference microscope (SAIM) because the received signal via leaky SAWs is mixed with the electrical reference signal to obtain the interference signal as seen in Fig. 4. With this system both the two-dimensional acoustic images and the V(z) curves at a small area on a sample can be obtained.

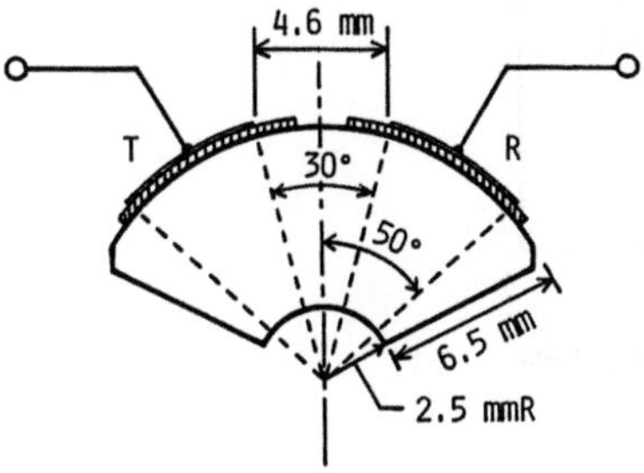

Fig. 5. Design of point focusing transducers. A pair of ZnO concave transducers deposited on a spherical surface of glass shell with concave radiation surface for focusing beams.

Figure 6 shows the acoustic images of the surface of Mn-Zn ferrite plate observed at a frequency of 80 MHz. These images were taken at the same defocus length of -80 μm and only the sample was rotated by 90° so that directional images can be obtained. As expected, different contrasts are imaged in the same area on the surface due to the anisotropy of the grain faces, although the resolution is insufficient.

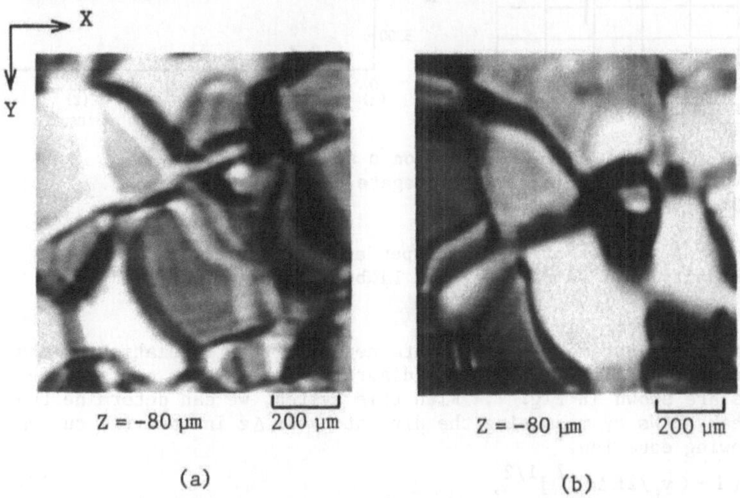

Fig. 6. Acoustic images of the surface of Mn-Zn ferrite plate observed at 80 MHz. The sample was rotated by 90° at the same defocus length of -80 μm. (a) Rayleigh waves propagating along X-direction. (b) Rayleigh waves propagating along Y-direction.

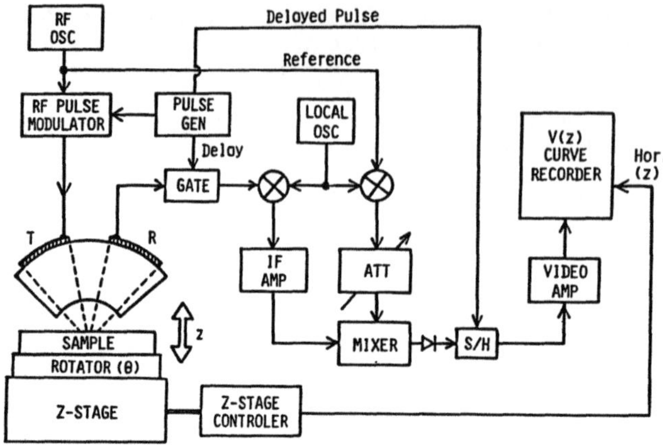

Fig. 7. Block diagram of the directional SAIM system for V(z) curve measurements.

Fig. 8

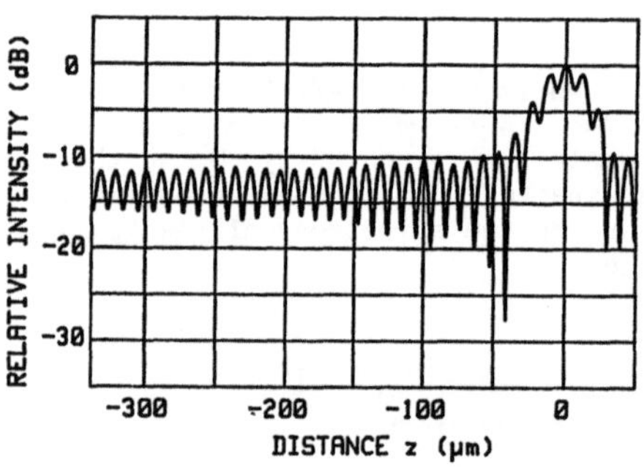

Fig. 9

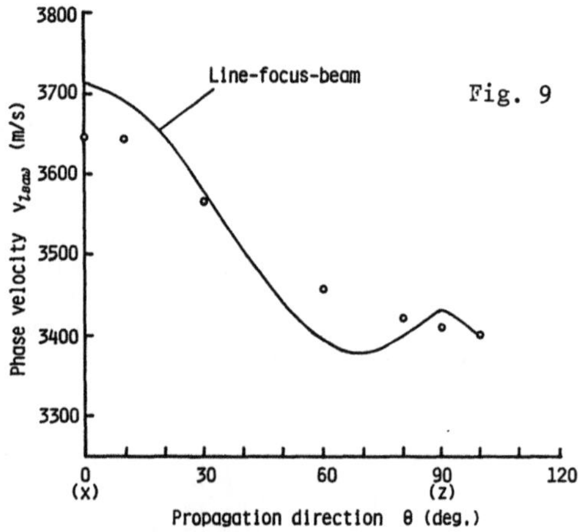

Fig. 8. An example of V(z) curve obtained on a sample of Y-cut $LiNbO_3$ plate. Rayleigh waves propagate along Z axis ($\theta = 90°$ in Fig. 9.).

Fig. 9. Experimental results of angular dependence of Rayleigh-wave velocity on a sample of Y-cut $LiNbO_3$ plate.

The V(z) curve measurements can be obtained with the translation of the sample surface along z-axis. The block diagram of the system employed for the experiments are shown in Fig. 7. With this system, we can determine the velocity of leaky SAWs by measuring the dip intervals Δz in the v(z) curves using the following equation.

$$v_{Lsaw} = v_L / [1 - (v_L/2f \Delta z)^2]^{1/2}, \quad (3)$$

where v_L is the velocity of water coupler and f is the ultrasonic frequency.

Figure 8 shows a V(z) curve, as an example , obtained for a sample of Y-cut $LiNbO_3$ (90° prop.). High peaks and deep troughs by interference can be clearly obtained in the V(z) curve pattern by adjusting the reference signal level with the variable attenuator. The SAW velocity has been calculated from the dip intervals by the FFT analysis developed at our laboratory to be 3410 m/sec for this sample.

Preliminary experiments were performed at a frequency of 80 MHz to investigate the dependence of leaky SAW velocity with respect to the propagation direction. Angular dependence of the ultrasonic velocities measured on the sample of Y-cut $LiNbO_3$ with rotating the mount-table is shown in Fig. 9. Although there are some discrepancies between the values obtained with the present system (circle) and those obtained with the line-focus-beam acoustic microscope (line) [13], we consider that the results are satisfactory at the present stage.

5. Concluding Remarks

The UMS system developed at Tohoku University has been described for material characterization. In this paper, UMS has been studied and discussed mainly for the applications to SAW materials, although the UMS itself has various fields of applications. A new directional point-focus-beam acoustic microscope proposed here is considered to be promising for the development of the UMS system in the fields of material science and SAW technology. The UMS system will prevail in the fields of material science and non-destructive evaluation of electronic materials for industry. It is also expected that UMS will cultivate a new microscopic world in ultrasonic spectroscopy in the near future.

Acknowledgment

The author wishes to express heartily his gratitude to Mr. Toshio Sannomiya for the continuous assistance throughout this work. This work was supported in part by grant-in-aids for scientific research from the Ministry of Education, Science and Culture, Japan.

References

1 D.Tabuchi,J.Chem.Phys.55,2218(1971).
2 D.Tabuchi,N.Inoue,T.Okuwa and K.Ohno, Acoustica, 32,236(1975)
3 F.Dunn,P.D.Edmonds,W.J.Fry,Biological Engineering,ed.by H.P.Shwan, (McGraw-Hill,1969).
4 A.F.Brown,Ultrasonics,204(1973).
5 M.A.Breazeale,J.H.Cantrell,Jr., and J.S.Heyman,Ultrasonics,ed.by P.D. Edmonds,(Academic Press,1981).
6 K.Takagi and K.Negishi,J.Phys.D.Appl.Phys.15,757(1982)
7 N.Chubachi,T.Sannomiya and K.Imano,to be presented at Autumnal meeting of Acoust.Soc.,Japan(Oct.1985).
8 S.A.Goss and W.D.O'Brien,Jr.,J.Acoust.Soc.Am.65(2),507(1979).
9 N.Chubachi,Japan.J.Appl.Phys.21,Suppl.3,7(1982).
10 C.F.Quate,A.Atalar,H.K.Wickramasinghe,IEEE,67,1092(1972).
11 J.Kushibiki and N.Chubachi,IEEE Trans.SU-32,189(1985).
12 E.P.Papadakis,Physical Acoustics,IV-B,ed.by W.P.Mason,(Academic Press, 1968).
13 J.Kushibiki,A.Ohkubo,and N.Chubachi,Electron.Lett.18,6(1982).

Part VI

Rayleigh Waves in Non-Destructive Testing

NDT with Rayleigh and Optical Waves

G.S. Kino

Edward L. Ginzton Laboratory, W.W. Hansen Laboratories of Physics, Stanford University, Stanford, CA 94305, USA

I. Introduction

It is a great pleasure to speak on the applications of Rayleigh waves to nondestructive testing. This conference attests to the breadth and scope of the contributions Lord Rayleigh's discovery of surface waves has made to so many acoustic applications. Yet, the discovery of Rayleigh waves is only one small part of his many theories which have been fundamental to the development of modern acoustical and optical techniques. We can only marvel at his immense output.

LORD RAYLEIGH [1] established the basic properties of acoustic surface wave propagation along the surface of an elastic solid. He showed that the wave is nondispersive, propagates at a velocity slightly less than the shear wave velocity, and that most of its energy is contained within a wavelength of the surface. His mathematical results lead to the following conclusions:

(1) The diffraction of a Rayleigh wave is two dimensional, so that the amplitudes of the fields fall off relatively slowly as $1/r^{1/2}$ rather than with the $1/r$ fall-off rate for bulk waves with distance r from the source;

(2) Since the cross-sectional area of the wave is approximately the product of its width and its wavelength, the amplitudes of the disturbances for a given power in the waves tend to be much stronger than for a bulk wave. A bulk wave will typically be many wavelengths in diameter and so have a much larger cross section and correspondingly lower fields for a given power or energy in the wave;

(3) Since a Rayleigh wave propagates near the surface, it is easily accessible from the surface, and therefore easily excited and measured.

It was realized during the 1950's that Rayleigh waves could be useful for nondestructive testing. Much of the early work in this field was carried out in Russia and Germany. Major contributions on the application of Rayleigh waves to nondestructive testing were made by VIKTOROV [2], who wrote a landmark book, published in Russia in 1966 and in the United States in 1967, summarizing his investigations and those of others. In that work he pointed out that Rayleigh waves in the ultrasonic range could be used to measure the presence of flaws such as cracks and holes, near the surface of a sample. As an example, he gave results, illustrated in Fig. 1, for the reflection of a Rayleigh wave from a crack; these had originally been observed by POHLMAN [3]. The top photograph shows an optical microscope picture of a fatigue crack, and the lower shows a transverse section of the crack. The upper oscilloscope photo shows signals from the sample before

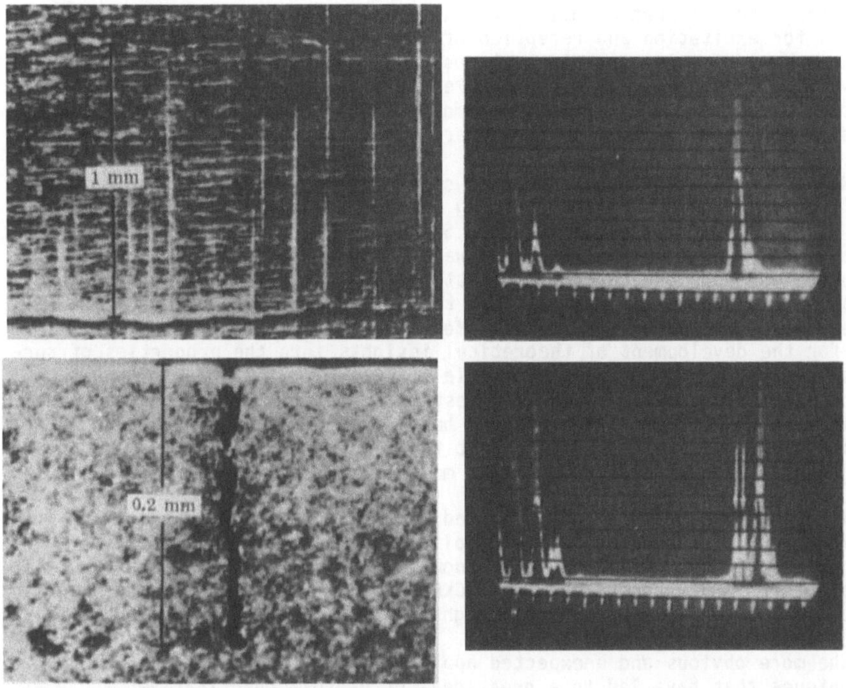

Fig. 1. Top left: optical microscope picture of fatigue crack. Lower left: transverse section of the crack. Upper right: oscilloscope signals without cracks present. Lower right: oscilloscope signals with cracks present.

the fatigue crack was introduced. The initial pulse represents the transmitted pulse; the last pulse represents the reflection from the edge of the plate. The lower oscilloscope picture shows the presence of extra pulses after cracks have been introduced. The presence of these cracks is indicated by arrows. The basic techniques used for measuring cracks have changed in sophistication, but are still based on the same principles today.

VIKTOROV [2] pointed out that the change in attenuation and velocity of the Rayleigh waves could be used as an indication of the material state near the surface of a solid. For example, the measurement of velocity could give an assessment of residual stress near the surface, as well as thermal and mechanical properties of the surface layer of the sample. The basic ideas given in Viktorov's book have been expanded upon, modern computer technology has been brought to bear on the results, and the theory has been developed, as we shall discuss further in this paper.

Viktorov developed a body of theory, some of which has not been repeated to this day. For instance, he and his co-workers developed the theory for the propagation of Rayleigh waves around the circumference of a cylinder (VIKTOROV [2]), which is relevant to the understanding of propagation around rounded corners. Viktorov was also concerned with the propagation of acoustic surface waves at the interface between a solid and a liquid, the change in attenuation due to a change in surface quality, and techniques for excitation and reception of acoustic surface waves. He extrapolated many of these techniques to use with ultrasonic Lamb waves and showed how Lamb waves could be employed for measuring defects in thin plates. Thus, Viktorov established most of the basic ideas that have been used in the field in the eighteen years since the publication of his book.

More recently, the development of surface acoustic wave technology for application to nondestructive testing was accelerated by the development of the surface wave devices (WHITE [4], SMITH ET AL [5]). This led to a body of experience with surface acoustic waves, and established the fact that, even at very high frequencies, the attenuation of surface waves is not severe on high-quality materials. New techniques were developed for the measurement of surface wave amplitude, for the construction of transducers, and for the development of theoretical insights into the properties of surface waves. It was therefore possible to take over this body of knowledge and apply it to the nondestructive testing field. Before that time, theoretical and experimental analyses of bulk waves were, to some extent, understood. It was felt, however, that the theoretical problems associated with surface acoustic waves were far more difficult.

In our own laboratory, where we had a great deal of experience with surface acoustic wave devices for applications to communications and radar, we realized that we could use this knowledge and apply it to nondestructive testing problems. Professor Ash (WICKRAMASINGHE ET AL [6,7]) recognized this application very early and brought the same type of experience in his laboratory to bear on these problems. We shall describe in this paper some of the more obvious and unexpected applications of surface acoustic wave techniques that have led to a great deal of useful, quantitative, and experimental NDT measurement techniques. We shall use examples, mainly drawn from our work at Stanford University, to illustrate the principles, power, and quantitative accuracy of the surface wave techniques that have been developed. Some examples are the use of Rayleigh waves to measure scattering from cracks, their sizes and stress intensity factors, the application of Rayleigh waves to the measurement of stress, the use of the acoustic microscope to measure material properties associated with Rayleigh wave propagation, the development of wedge, edge-bonded, and EMAT transducers for application to NDT, and the imaging of surface defects with Rayleigh waves. These techniques have led to a variety of methods for determining crack depth and width, and are perhaps the most important application of surface acoustic waves.

Starting with Professor Ash's work (WICKRAMASINGHE ET AL [6,7]), optical probes and photoacoustic techniques have been used in combination with surface acoustic waves to measure the properties of cracks and to measure material properties. More recently, the use of the acoustic microscope has proven to be a useful technique for exciting and receiving Rayleigh waves, and has made it possible to measure Rayleigh wave properties in regions of relatively small extent. This has led to very useful imaging techniques, which has made it possible to determine regions where adhesion is poor, to measure stress, and to measure material properties.

II. Excitation to Rayleigh Waves

The standard technique for exciting Rayleigh waves on a piezoelectric material employs interdigital transducers. Most materials that are of interest to nondestructive testing, such as metals, plastics, and ceramics, are not piezoelectric. Therefore, it is not possible to make an interdigital transducer to excite waves directly on these materials. Instead, it is necessary to excite the waves by other means. The simplest technique is to immerse the sample to be tested in water and use a standard transducer to excite bulk waves at the so-called Rayleigh angle θ_R to the normal:

$$\sin \theta_R = \frac{V}{V_R} \tag{1}$$

where V_R is the Rayleigh wave velocity and V is the velocity of acoustic waves in water. This technique excites the Rayleigh waves. As they propagate along the material, however, they re-excite bulk waves in water, and hence become "leaky" Rayleigh waves - waves which are continually reradiating, and are therefore attenuated with distance. The theory of excitation, and of the leak rate for such Rayleigh waves, has been worked out by BERTONI ET AL [8] and FRASER ET AL [9]. Fraser and his co-workers have developed efficient transducers. Relatively inefficient wedge transducers, based on Viktorov's early work, have been available commercially for many years.

An alternative to using water, which is often more practical, is to use a transducer mounted on a solid material, in the form of a wedge. A wedge transducer is illustrated in Fig. 2. Typically, the transducer is made on a solid material, such as RTV rubber, polystyrene, or quartz; the basic condition required is that acoustic wave velocity in the wedge material must be lower than the Rayleigh wave velocity of the substrate. Such techniques have made it possible to work with operating frequencies in the 1-100 MHz range. The length over which optimum excitation takes place can be predicted by the theory. It is typically of the order of ten Rayleigh wavelengths. The contact to the substrate which is being examined can be dry when a material like RTV rubber is used, but it is normally easier to employ grease or water to ensure good contact to the substrate. It is then possible to move the transducer over the surface of the substrate so as to examine the various features of interest.

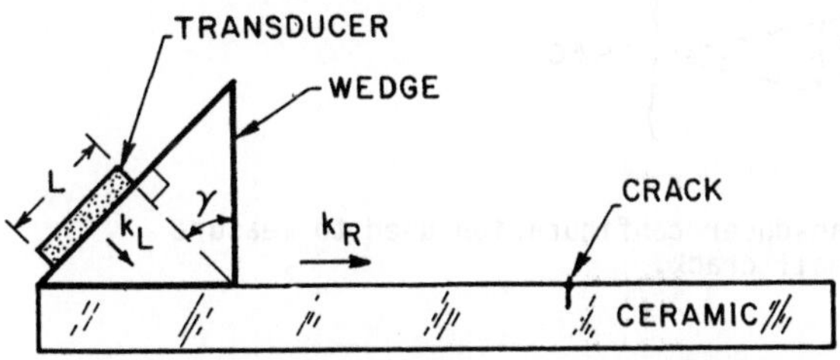

Fig. 2. A schematic of a wedge transducer for exciting Rayleigh waves on a ceramic substrate.

As we describe various applications of Rayleigh waves, we will discuss other excitation techniques, closely related to the wedge transducer, which are more practical to use in, for instance, acoustic surface wave imaging. We shall also describe the EMAT (electromagnetic acoustic transducer) which avoids the necessity of making contact with the substrate.

III. The Measurement of Stress Intensity Factor of a Crack

An important problem of great interest in structural mechanics is the determination of the size of a crack. Surface cracks weaken a material and lead to its eventual fracture. The larger the crack, the lower the breaking stress. Therefore, it is important to be able to measure the size of the crack and predict fracture stress. The relationship between fracture stress and crack size has been well developed for brittle materials, although the theory is more complicated for ductile materials such as metals. If it is assumed that the Rayleigh wavelength is large compared to the size of the crack, then it is possible to develop quasi-static theories for Rayleigh wave scattering from a crack based on the fracture mechanics theories in the literature (SMITH ET AL [10]). These theories predict the Rayleigh scattering relation. The reflection coefficient of a surface wave from a crack is approximately proportional to the cube of its radius. Typically, surface cracks have a half-penny shape. The basic quantitative theories to determine these effects for Rayleigh waves were developed by KINO [11] and AULD [12]. Their theories were, in turn, developed from earlier theories that had been developed for plane wave scattering from cracks. A particularly important theory of this type was given by BUDIANSKY ET AL [13], who showed that it is not necessary to know the geometry of the crack to predict the breaking stress associated with it. The reflection coefficient of a plane acoustic wave from a buried crack is directly related to a geometrical factor, the stress intensity factor, from which the breaking stress can be predicted. For surface waves this basic result is still approximately true.

RESCH ET AL [14] carried out a series of measurements on glass substrates. They induced cracks with a Knoop hardness indentor, and measured the reflection coefficient of a surface acoustic wave in the two-transducer configuration shown in Fig. 3. This avoided the difficulty of having the weak received signal dominated by the ringing of the transmitting transducer.

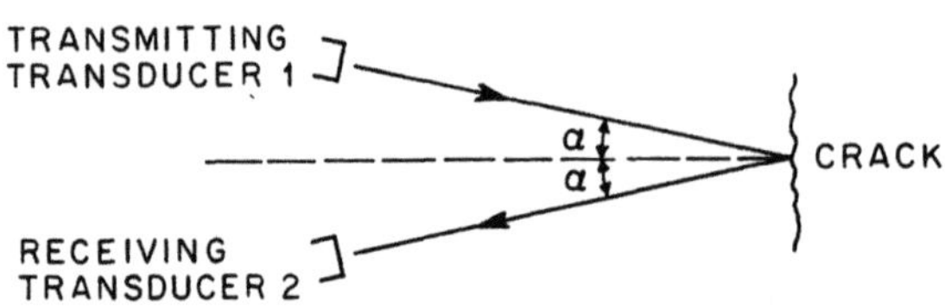

Fig. 3. Experimental two-transducer configuration used to measure scattering from a small crack.

In their initial work with an acoustic surface wave incident at a few degrees to the normal to the crack, they were able to predict the breaking stresses to within ten percent. Later work on ceramics gave far more detailed information on the material properties of the ceramics. TIEN ET AL [15] developed a more complete surface wave scattering theory for this case, using an energy conservation theory from fracture mechanics by BUDIANSKY ET AL [16] and numerical data developed by SMITH ET AL [10]. They predicted that the size of the crack for fracture was approximately 2.5 times the measured size of the crack before the application of stress. A detailed experimental study using surface acoustic waves yielded the result shown in Fig. 4, which shows that the crack increased in size as stress was applied. When the stress was released, there was hysteresis;

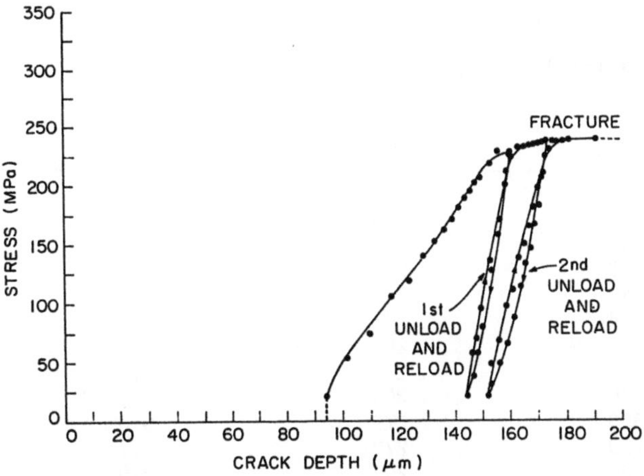

Fig. 4. Stress vs. acoustic estimates for the crack depth for an as-indented sample. Two unload-reload cycles were performed.

eventually the crack fractures at approximately 2.5 times its initial size. These results were in agreement with the theory of MARSHALL ET AL [17,18] and LAWN ET AL [18], who showed that there is residual stress due to the Knoop hardness indentor which was used to induce the crack; this causes the effect shown. On the other hand, if the crack was annealed, it was possible to use the surface acoustic wave technique to determine how the crack broke in this case. As shown in Fig. 5, there was now very little hysteresis and very little residual stress. However, because the crack was partially in contact over its surface, there was no scattering of acoustic waves from this region, so acoustically the crack behaved as if it were still increasing in size with stress.

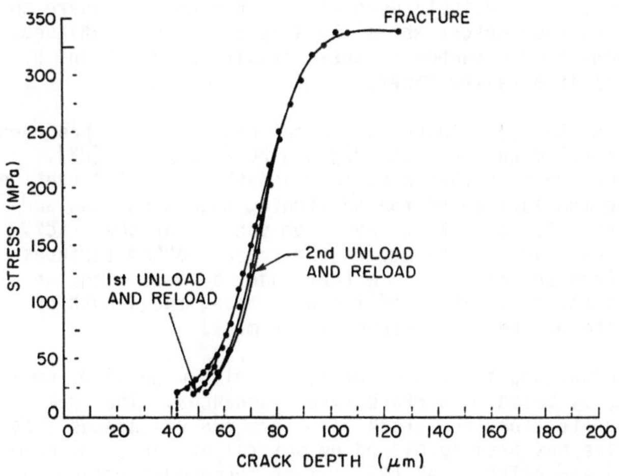

Fig. 5. Stress vs. acoustic estimates of the crack depth of a vacuum annealed sample. Two unload-reload cycles were performed.

Table 1

FRACTURE STRESSES: AS-INDENTED SAMPLES

β	Acoustic σ_F (MPa)	Actual σ_F (MPa)	Deviation
0	246	238	3%
0	250	239	5%
0	247	242	2%
0	230	233	1%
5	242	241	0.4%
10	224	246	9.0%
15	242	259	7.0%
20	261	281	7.0%
25	270	256	5.0%
30	303	279	9.0%
35	280	278	0.7%
40	304	297	2.0%
45	299	294	2.0%

It was possible to show by these techniques, based on the ideas of BUDIANSKY ET AL [16], that even if the surface acoustic wave was propagating at an angle β normal to the crack, providing the applied stress was applied at the same angle to the crack, the predictions would still work. Predictions of this type are shown in Table 1. They show, very clearly, the power and accuracy of the surface acoustic wave technique for nondestructive testing. Later work by MARSHALL ET AL [19] showed that the theory could be extended to deal with the nature of machining damage in brittle materials. In this case, long linear cracks are developed when a grit from the diamond wheel scores the surface. Now the residual stresses develop as before, but the crack grows to approximately four times its initial size. Again, the surface acoustic wave techniques have given evidence that these theories are accurate.

A similar set of results were developed for fatigue cracks in metals, and the technique is now being routinely used at Stanford and elsewhere to determine the growth of fatigue cracks while the sample is being fatigued. KHURI-YAKUB [20] has summarized a number of these results by Resch and by others (RESCH ET AL [21]) in a review paper.

A very different set of NDT techniques for larger cracks, which has been highly successful, was carried out by SILK [22] and HUDGELL ET AL [23]. These authors made use of the fact that a surface acoustic wave incident on a crack propagates along the surface of the substrate, then along the surface of the crack, and radiates as a bulk wave from the tip of the crack. Detection of this bulk wave can be obtained on the top or bottom surfaces of the substrate, away from the crack. In either case, by measuring the extra time delays due to the propagation of the wave along the length of the crack, a good estimate of the crack depth can be made.

Yet another set of techniques for cracks of relatively large size, compared to the wavelength, is based on surface wave resonances. The idea that surface waves propagate along the crack and there are resonances associated with the crack size has been developed by several workers, for example, DOMARKAS ET AL [24] and AYTER ET AL [25]. Such techniques proved to be somewhat sensitive to the surface roughness of the crack and have not been fully developed. However, the techniques using radiation from the crack tip have become very important and extremely useful in practical NDT problems.

IV. Acoustic Surface Wave Imaging

Instead of trying to infer the size of the crack from theory or from reconstruction of scattering data, another possibility is to form a direct image of the crack. To carry out these measurements, a new type of surface acoustic wave edge-bonded transducer array was constructed by KINO ET AL [26]. An illustration of an edge-bonded array with thirty-two transducer elements is shown in Fig. 6. Each element is approximately a Rayleigh wavelength deep, and is mounted on a substrate material of the same kind as that being examined, in this case aluminum. Each element of the array behaves like a line source and excites a spreading acoustic surface wave which travels along the substrate. A thin plastic coupling strip is used to couple these excited surface acoustic waves to the specimen being examined. Just as with the wedge transducer, grease is used to ensure good contact. The surface wave travels along the substrate being examined, is reflected from the crack, and then the reflected signal is received again by the array.

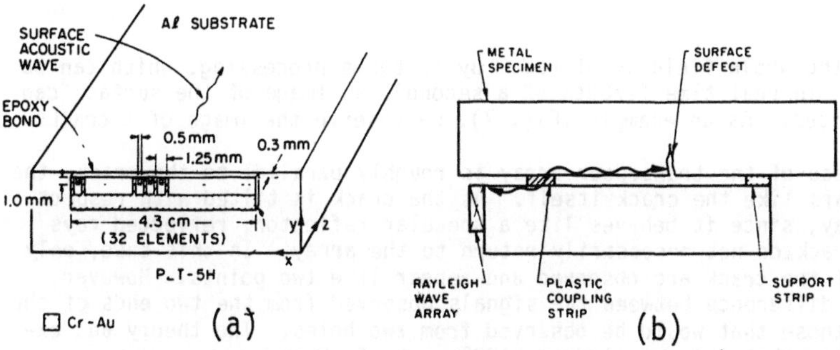

Fig. 6. Edge-bonded transducer (EBT) surface wave array. (a) Schematic of array construction. (b) Diagram of a sample in contact with the array.

The edge-bonded transducer has several major advantages in this and other applications. First, it can be constructed on a wide variety of substrates matched to the application. Second, because of its simple single-layer construction on a high-impedance substrate, it tends to be very broadband and hence can excite short pulses. Third, in imaging applications, the wedge transducer excites waves at an angle to the normal, which in turn, are intended to excite or receive focused Rayleigh waves propagating along the substrate. However, this process of conversion to Rayleigh waves causes severe aberrations of the focused beam. On the other hand, since the velocity of surface waves on the edge-bonded transducer substrate and the substrate being examined are identical, no aberrations occur in the transfer process.

A synthetic aperture imaging system has been employed with the edge-bonded transducer. An individual element is excited, and then all signals received on the same element are stored digitally. Each element is excited in turn, and the signals received by it are stored digitally until informa-

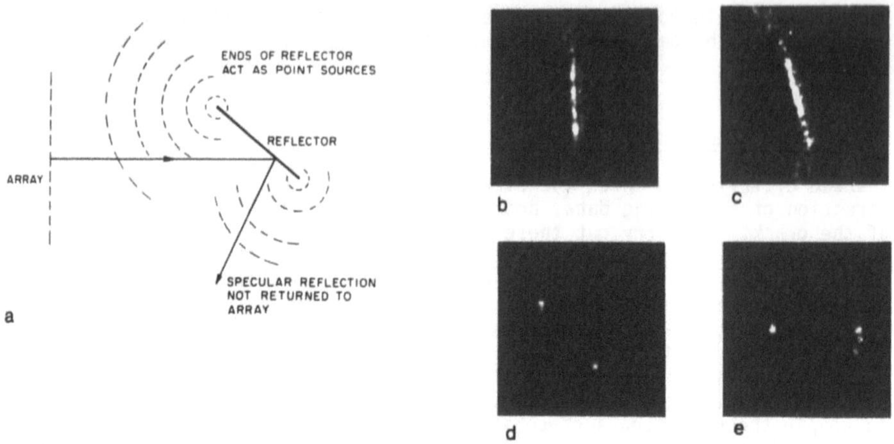

Fig. 7. (a) When imaging specular reflectors at a large angle of incidence, only the two ends are seen. Images of a 11 mm EDM slot at (b) 0° angle of incidence; (c) 15° ; (d) 45° ; and (e) 90° .

tion from the whole field is stored. By suitable processing, which can be carried out in real time (1/60th of a second), an image of the surface can be constructed. As an example (Fig. 7), we observe the image of a crack.

When the face of the transducer array is roughly parallel to the crack, the image appears like the crack itself. As the crack is tilted with respect to the array, since it behaves like a specular reflector, reflected rays from the crack do not necessarily return to the array. In this case, only the ends of the crack are observed and appear like two points. However, there is a difference between the signals observed from the two ends of the crack and those that would be observed from two holes. The theory and experiment show that they are indeed 180° out of phase.

We have already discussed how wave conversion at the crack tip can give information on the depth of the crack. This type of procedure can be carried out in an imaging mode, as shown in Fig. 8. In this case, a longitudinal wave is incident on the crack from the opposite side of the substrate being examined. The surface wave imaging device is used to image the crack; several images are seen. By measuring the time difference between the different images, it is possible to obtain information on the depth of the crack, as well as its width.

V. Measurement of Stress

During the last twenty years, several investigators have used acoustic techniques to measure stress. The work is based on the use of the acoustoelastic effect in which the change of acoustic velocity, due to stress of an acoustic wave passing through a solid material, is proportional to the stress. By making longitudinal wave transducers which can be moved over the surface of a material in water, it is possible to map out the principal value of an applied stress field. There are difficulties with this technique due to variations of thickness of the material which can affect the transit times of the acoustic waves, variations in material properties

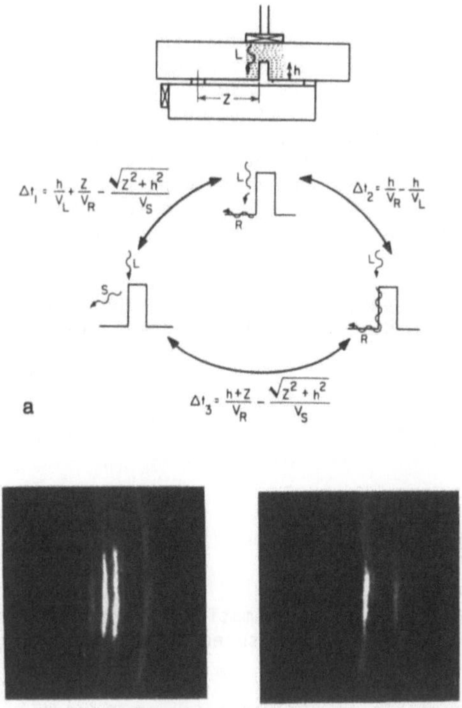

a b c

Fig. 8. Pitch/catch imaging. (a) Schematic of the transmitter/receiver geometry and diagram of the three waves scattered from the crack. (b) Pitch/catch image of a long 7 mm deep slot. (c) Pitch/catch image of a 10 mm long fatigue crack.

which also affect the transit time of the acoustic waves, and difficulties in determining the direction of stress. When there is residual stress present, the problem is even more difficult because the use of a longitudinal wave passing through a slab of material only gives an integrated effect. With residual stress, this integrated effect may well be zero.

Acoustic surface wave techniques have been brought to bear on this problem by HUSSON ET AL [27] and KINO [28], among others. By using surface acoustic waves, it is possible to measure the near-surface stresses on both sides of a slab of metal. Typically, by operating at a frequency of the order of 2 MHz, the penetration depth of the acoustic waves is of the order of 1 mm. Here, the use of surface waves is preferable to the use of x-rays, which only penetrate a few microns and are affected by damage caused by machining effects at the surface.

Several difficulties are present in such measurements. First, the velocity of the acoustic surface wave is affected by stress in the direction of propagation, as well as perpendicular to the direction of propagation along the surface. Husson worked out (KINO [28]) a detailed perturbation theory to predict these effects. HUSSON ET AL [27] used calibration techniques, which we shall describe below to measure the two components of stress. Second, there are problems associated with the contact between a wedge or edge-bonded transducer and the material being measured. Since the change in velocity associated with even very large stresses may well be less than 0.1% , very accurate measurements must be made of acoustic wave velocity or transit time. This is usually carried out by measuring the phase change of the wave from a transmitting to a receiving transducer. If the pair of wedge transducers is moved along the surface of a material, so as to be able to measure stress variations, the contact between the transducers and the sample will introduce phase changes which are quite comparable to those which need to be measured. A different transducer technique must therefore be employed.

HUSSON ET AL [27] and KINO [28] were interested in measuring the residual stress near a weld in a thirteen inch diameter stainless steel pipe, illustrated in Fig. 9.

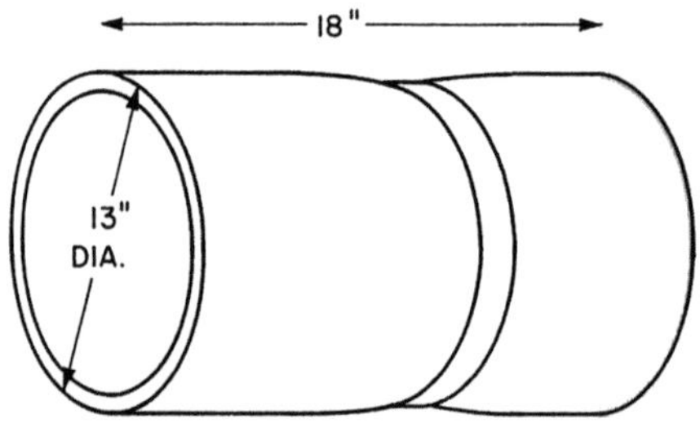

Fig. 9. A schematic of the stainless steel pipe, measured by Husson.

As we have suggested, it was not possible to use wedge or edge-bonded transducers. Instead, an electromagnetic acoustic transducer (EMAT) was used. This device works on the principle that, in the presence of a magnetic field parallel to the surface of a conductor, an rf current flowing near the surface of the conductor will cause the material to move up and down, thus generating a surface acoustic wave. As illustrated in Fig. 10, they made a transmitting transducer consisting of a meander line with several fingers, one-half wavelength apart, to excite a surface acoustic wave as efficiently as possible. In combination with this device, they used two single-strip receiving transducers as receivers. Both sets of transducers had employed permanent magnets to produce the static magnetic field required. The advantage of using these devices is that they do not have to contact the surface and, if the distance between the transducers and the surface varies, the amplitude of the output will change, but the phase will not. The disadvantage of these devices is that they are extremely inefficient, so they need to be excited with as much power as possible. An extremely sensitive receiver must therefore be used. Special care had to be taken to avoid direct electromagnetic pick-up between the receiving and transmitting transducers. To avoid spurious phase changes as much as possible, the phase difference between two receiving transducers with fixed spacing was measured.

The next problem to be tackled was how to calibrate the system. The material properties and texture of commercial materials of this kind were liable to vary from point to point, especially if heat had been applied to the material, and the relationship between velocity and stress would not be accurately known for the particular material being measured. In addition, it was necessary to know the stress coefficients for acoustic waves propagating parallel and perpendicular to the stress. A calibration measurement

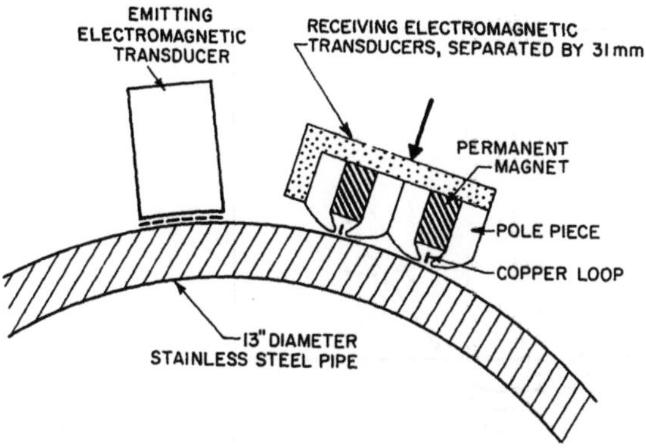

Fig. 10. Emitting and receiving transducers on a curved surface.

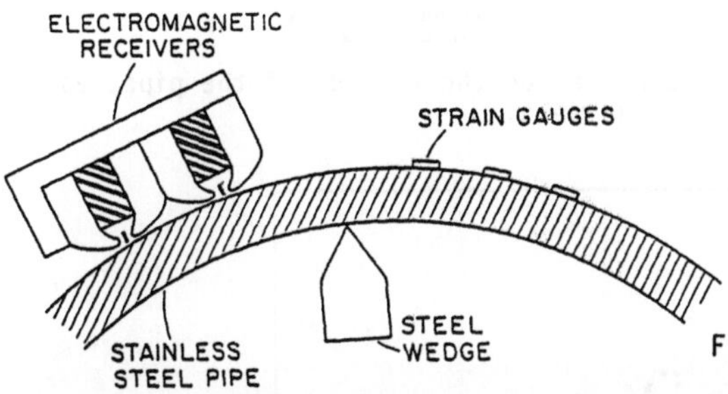

Fig. 11. Calibration measurement.

system using a steel wedge pushed against the inside of the pipe by a car jack was used, as illustrated in Fig. 11. Strain gauges were attached to the surface of the material so that the stresses in each direction applied by the steel wedge could be measured. Then, the surface wave measuring system was used to calibrate the system, both with the wave propagating around the circumference of the metal and in the axial direction.

After calibration, residual stress measurements were made in the axial and circumferential directions at different distances from the weld. As will be seen from Figs. 12 and 13, which were taken on the outside and inside surfaces of the pipe, respectively. There was a great deal of scatter in the measurements, so that some statistical averaging was needed. Nevertheless, it can clearly be seen that, as expected, the stresses tended to reverse in sign on the two sides of the pipe and increase towards the weld, reaching values near the yield stress. An additional, somewhat unexpected, phenomenon was observed; the stresses near the end of the pipe that had been cut increased. This appears to be due to machining damage, and to the fact that the pipe had not been chosen to be long enough. The sample measured was a piece cut from a much longer pipe.

Some confirmatory measurements were made by a hole-drilling trepanning technique, but the calibration methods employed had even more scatter than the surface wave measurements. Suffice it to say that this appears to be an important technique in which the only other far more accurate method involves neutron scattering. Such techniques can only be employed near a nuclear reactor source, so they are very useful for calibration but not for use in the field.

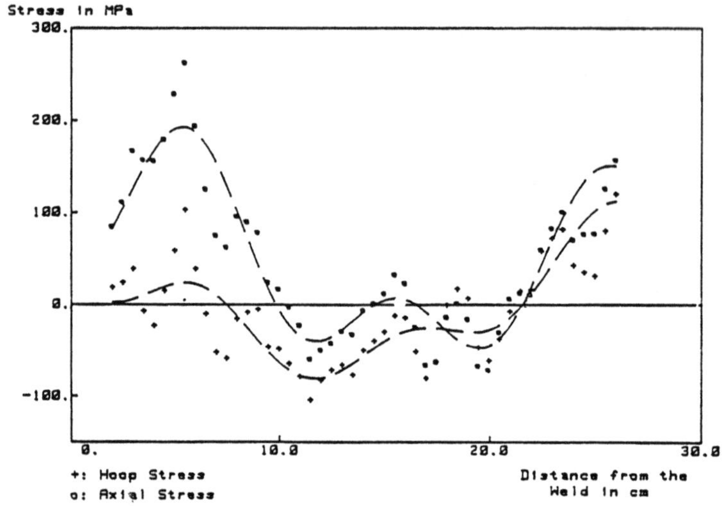

Fig. 12. Measured axial and hoop stresses on the outside of the pipe, as a function of distance from the weld.

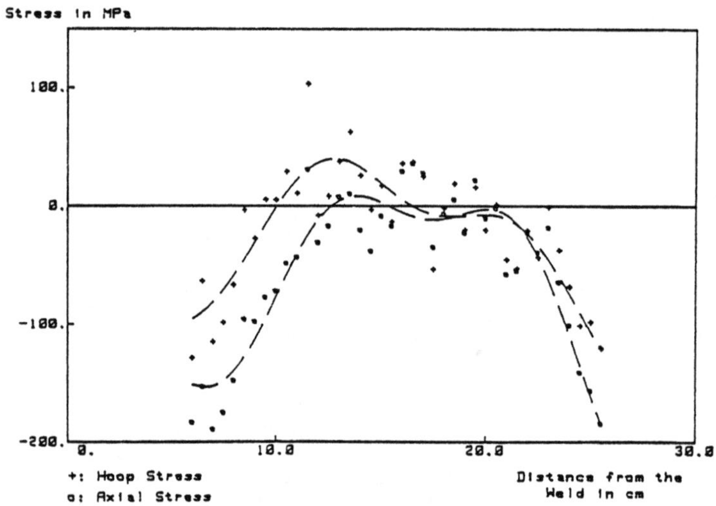

Fig. 13. Measured axial and hoop stresses on the inside of the pipe, as a function of distance from the weld.

VI. Optical Measurements of Surface Acoustic Waves

As we have already mentioned, WICKRAMASINGHE ET AL [7] devised optical probes to measure the velocity of an acoustic surface wave and its scattering from defects. Their basic technique, which was closely related to others used on acoustic surface wave devices, was to allow a focused light source to be incident on a reflecting surface and measure the phase of the reflected light. Since the phase of the reflected light is proportional to the displacement of the surface, this phase measurement makes it possible

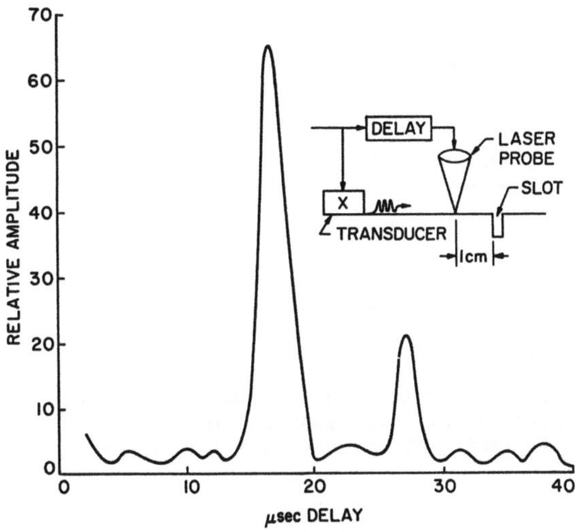

Fig. 14. Time domain measurement of the acoustic reflection coefficient from a slot in an aluminum sample. The time delay between the acoustic tone burst and the laser gating is used to separate the incident and reflected pulses. The reflection coefficent is 0.41 . The acoustic frequency is 5.5 MHz and the slot depth is 0.3 mm . The theoretical reflection coefficent is 0.40 .

to determine both the amplitude and the phase of the acoustic surface displacements on the material. JUNGERMAN ET AL [29] used a fiber-optic probe, which has a two-micrometer light beam definition at the surface of the substrate for this purpose. They used this device to measure the Rayleigh wave reflection coefficient from a crack, as illustrated in Fig. 14. A short tone burst was excited with a wedge transducer at a frequency of 5.5 MHz , and both transmitted and reflected signals were detected separately with a fiber-optic probe. By varying the delay between the gating of the probe and the input signal, it was possible to measure the transmitted and reflected signal separately, and hence the reflection coefficient in the near-field of a slot or crack. As shown in Fig. 14, the measured reflection coefficient was 0.41 , which compared with the theoretical reflection coefficient of 0.4 , calculated by HIRAO ET AL [30]. A similar set of measurements were repeated for a range of slot depths and compared to Hirao's theory. The results are shown in Fig. 15.

For cracks closed at the top by residual stresses, the results would be different. There is a pronounced dip in the frequency characteristic in the region where the crack is of the order of a wavelength deep. Such effects have been calculated theoretically by ACHENBACH ET AL [31], and have been used for direct surface wave measurements with excitation from water by KHURI-YAKUB ET AL [32].

As one more example of the power of this technique, shown in Fig. 16, the fiber-optic probe has been used to probe the near field of a crack, which is excited by a surface wave with a wavelength much longer than the crack depth. In this case, quasi-static theory can be used to predict the profile of the normal surface displacement in the neighborhood of the crack. It is possible to predict from this profile the slot or crack depth (STEELE AND YOO [33]).

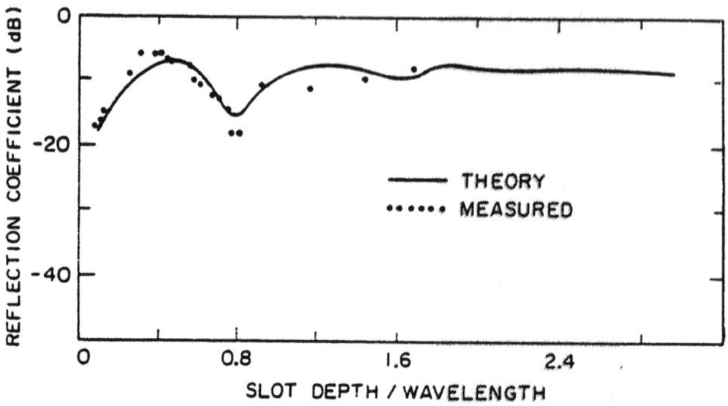

Fig. 15. Rayleigh wave reflection coefficent as a function of the slot depth-acoustic wavelength ratio. Experimental laser probe measurements compared to finite difference theory of Hirao et al [30].

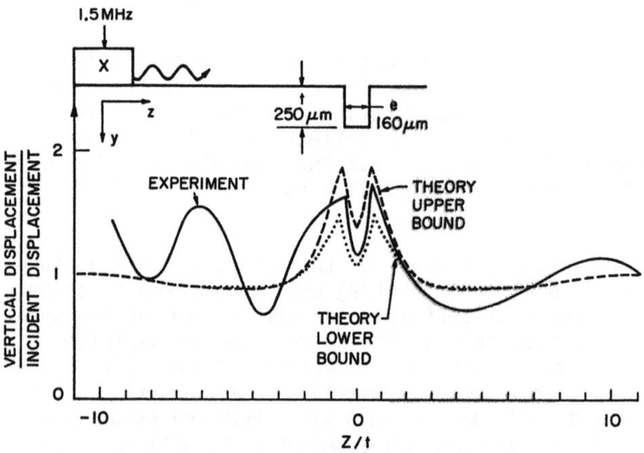

Fig. 16. Vertical displacements in the near field of a 250 μm deep slot with 1.5 MHz Rayleigh wave excitation. The static theory provides an upper bound (dashed) and a lower bound (dots) for theoretical displacements. The half-width d is used to predict a slot depth of 230 μm .

VII. Applications of the Acoustic Microscope

The acoustic microscope has turned out to be a very powerful tool for exciting acoustic surface waves on a solid substrate and imaging effects due to surface acoustic waves propagating along the substrate. The basic acoustic lens configuration of an acoustic microscope is illustrated in Fig. 17 (LIANG ET AL [34]). The focused beam is reflected from a solid surface and received back at the microscope transducer. The signal output is used to vary the intensity of a cathode ray tube, and the lens is scanned mechanically in a raster form, in synchronism with a spot of a cathode ray tube screen, thus forming a picture. When the lens is focused so that its focal point is below the surface of the substrate, as shown in Fig. 17, two signals are received at the lens, one a specularly-reflected wave associated

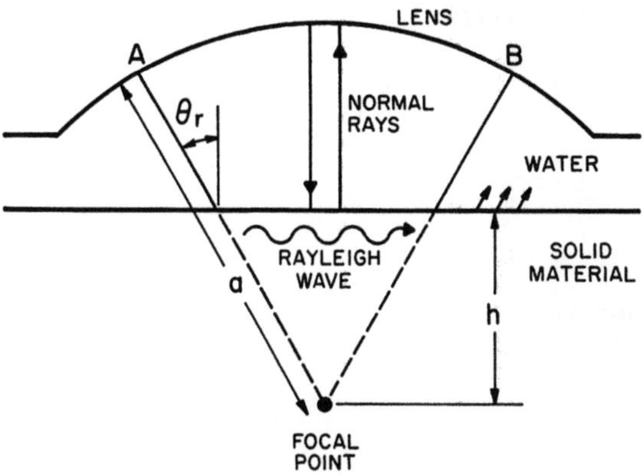

Fig. 17. The acoustic lens configuration of an acoustic microscope. Either quartz or sapphire is employed as the material for the lens.

with the near-axis components of the beam, and the other a wave excited at the Rayleigh angle by the lens which propagates along the surface of the substrate and reradiates at the Rayleigh angle. A signal is picked up on the opposite side of the lens at a point symmetrical to where it is excited.

Therefore, two signal components are picked up by the lens, the specularly-reflected signal and the Rayleigh wave signal. Since the Rayleigh wave signal is delayed as it passes along the substrate, it may be in or out of phase with the specularly-reflected signal, depending on the distance of the focal point below the surface of the substrate. As this distance z is varied, the voltage output from the transducer $V(z)$ varies.
The spacing between the maxima and minima of $V(z)$ is characteristic of the Rayleigh wave velocity and can be used to measure it. More important still, as the lens is scanned over the surface of a substrate, whose material properties are inhomogeneous, bright or dark regions are observed. It is the interference between the two waves that provides the basic contrast mechanism of the acoustic microscope for solid materials.

As an example of the power of this technique, Fig. 18, taken by Quate and his coworkers (BRAY ET AL [35]), shows an image of a metal film on a glass substrate. The optical image shows no changes over the metal-covered region. However, the acoustic images show distinct changes near the corners. These changes correspond to the lack of adhesion of the metal film with the glass, which affects the Rayleigh wave velocity. Depending on the position z of the lens, the contrast mechanism may itself be reversed, but the effects of adhesion show up very clearly in this picture. The $V(z)$ effect itself was discovered by WEGLEIN [36], and explained in detail by ATALAR [37] very clearly.

We will give one more example of the power of the $V(z)$ measurement technique. LIANG ET AL [34,38] used their phase measurement technique to measure both the amplitude and phase of $V(z)$. Extending the work of ATALAR [37], they derived a formula to invert the data by Fourier transform techniques and from it determine the reflection coefficient of a plane wave from a solid as a function of the angle of incidence. The results for aluminum are shown in Figs. 19 and 20. In Fig. 19, the original $V(z)$ amplitude and phase curves are shown. In Fig. 20, the computed results for the reflection coefficient as a function of angle are given, along with

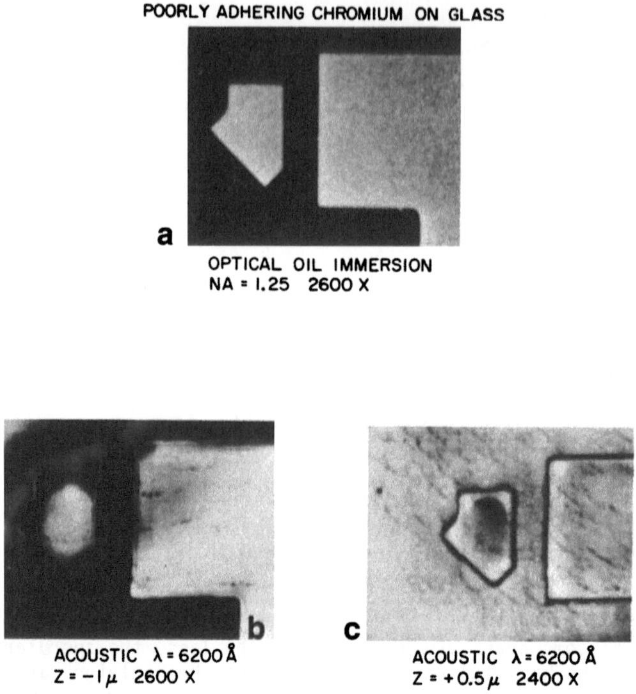

Fig. 18. (a) The optical images of a metal film on a glass substrate. (b) An acoustic image with Z = -1 micron. (c) An acoustic image with Z = +0.5 micron.

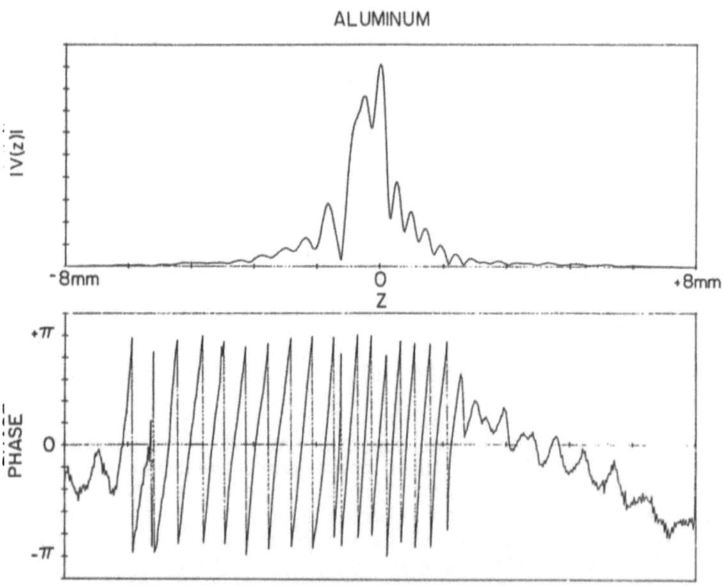

Fig. 19. Experimental V(z) of a water-aluminum interface at 10.17 MHz .

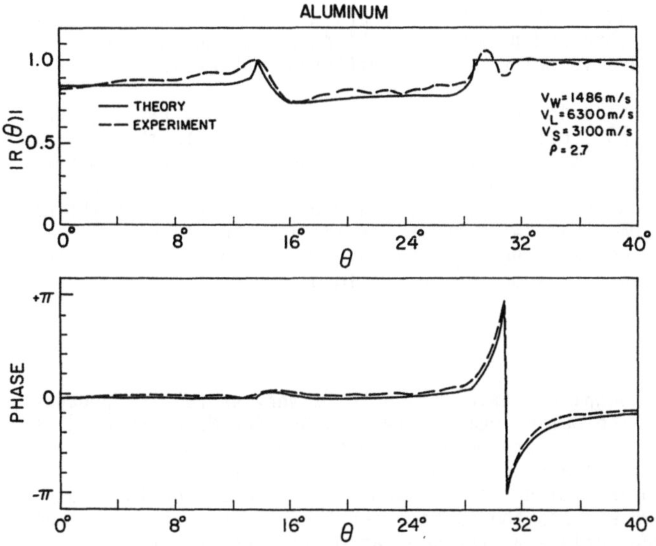

Fig. 20. Inversion of V(z) plot in Fig. 19 to find the reflection coefficient of a plane wave as a function of angle.

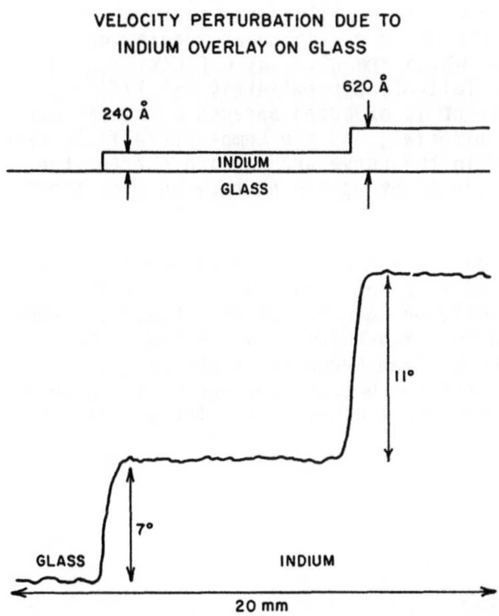

Fig. 21. Measurement of velocity perturbation due to indium film on glass.

theoretical plots for the same material. It will be seen that the Rayleigh wave velocity can be measured very accurately, along with other information on shear and longitudinal wave velocities. By using techniques such as this one, it is therefore possible to obtain a great deal of information on material characteristics, since the surface acoustic wave velocity and attenuation are very dependent on material properties. Very detailed results of this kind have been obtained by KUSHIBIKI ET AL [39], who have measured the surface acoustic wave velocity for many different materials.

An alternative technique is to measure the velocity of the Rayleigh wave directly. LIANG ET AL [34] used a short tone burst to excite the acoustic microscope, thus receiving two tone bursts back, one corresponding to the Rayleigh wave, the other to the specularly-reflected wave. By measuring the phase difference between these waves, they could determine directly the Rayleigh wave velocity. One example of their results is shown in Fig. 21. In this case, using a 60 μm wavelength Rayleigh wave, they were able to measure changes in phase corresponding to a film only 240 Å thick.

VIII. Photoacoustics

Surface acoustic wave techniques have proven to be very useful for photoacoustic applications. The basic idea carried out by STEARNS ET AL [40] is illustrated in Fig. 22. A surface acoustic wave is excited on a substrate by a wedge or edge-bonded transducer and received on a similar transducer. A modulated laser beam is incident on the substrate between the two transducers. The light source heats the surface and there is therefore a temperature modulation of the surface at the modulation frequency of the light. This temperature modulation changes the effective density and elastic constant, and hence the phase of the acoustic wave. The rf modulation of the acoustic wave can be detected as a phase-modulated signal so that very small phase changes can be determined. A simple result taken on a silicon nitride ceramic, shown in Fig. 23, shows the amplitude and phase of the modulated signal output relative to the phase of the input modulation (STEARNS ET AL [40]). These results, which are given as a function of modulation frequency, show an amplitude fall-off approximately as $1/f^{1/2}$, where f is the frequency. This effect is expected because a thermal diffusion wave is being excited in the material, and the temperature falls off as $1/f^{1/2}$. The calculations shown in the curve are carried out for the exact surface wave fields, and show excellent agreement between experiment and theory.

The technique can be used for measuring the profile of an acoustic beam by moving the light beam across the surface, as shown in Fig. 24 (STEARNS ET AL [41]). This technique is a simple one and is not as subject to problems with surface roughness as would be the methods of using the light beam as a detector. Another example given by them shows the scattering around a crack. Here the problem is more complicated because the scattering problem involves both acoustic and thermal wave scattering. The effects that are

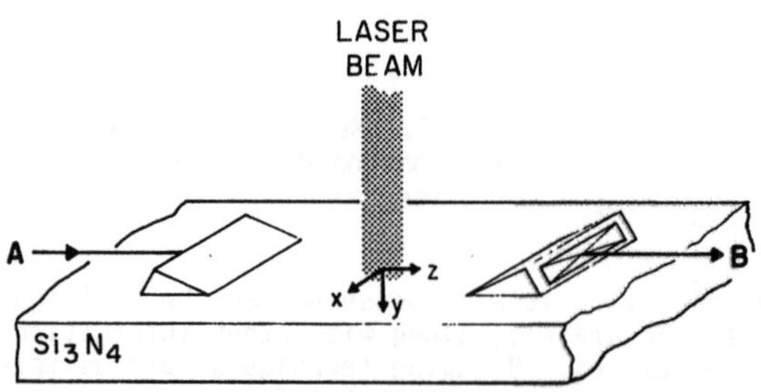

Fig. 22. Experimental set-up for measuring the perturbation of a Rayleigh wave by a laser beam.

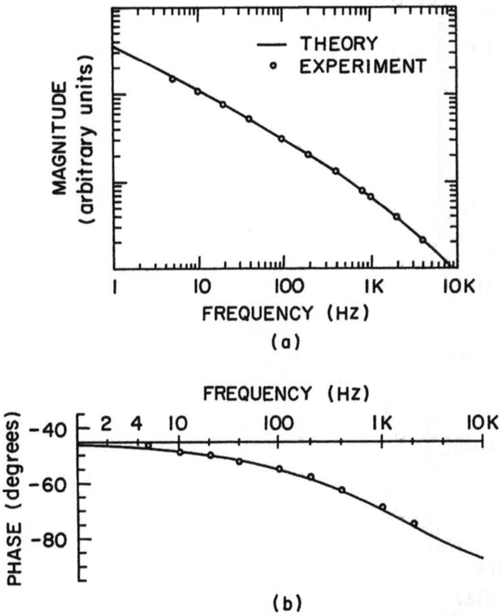

Fig. 23. The theoretical and experimental amplitude and phase of the detected signal with laser heating of Si_3N_4 ceramic.

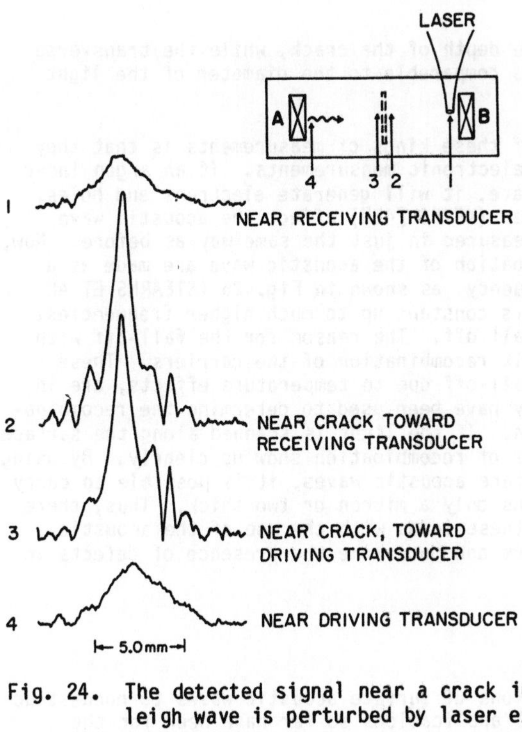

Fig. 24. The detected signal near a crack in a Si_3N_4 ceramic when a Rayleigh wave is perturbed by laser excitation.

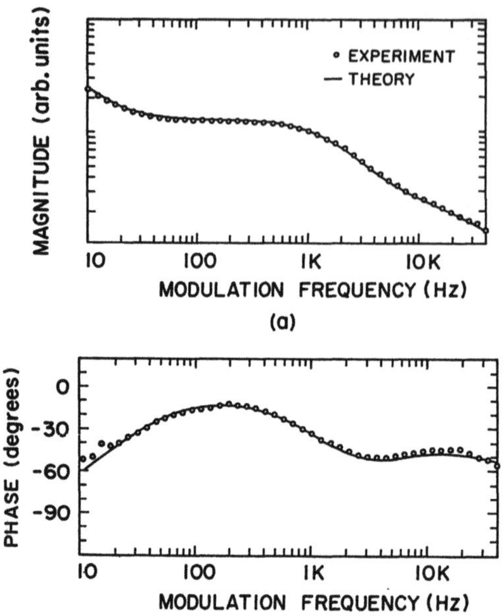

(b)

Fig. 25. The perturbation of a Rayleigh wave by generation of electrons and holes with a 5100 Å argon laser beam.

measured depend radically on the depth of the crack, while the transverse definition of the measurement is comparable to the diameter of the light beam.

A very interesting feature of these kinds of measurements is that they are beginning to be useful for electronic measurements. If an argon laser is incident on a silicon substrate, it will generate electrons and holes. The generated electrons and holes (KEYES [42]) affect the acoustic wave velocity. This effect can be measured in just the same way as before. Now, when measurements of the perturbation of the acoustic wave are made as a function of the modulation frequency, as shown in Fig. 25 (STEARNS ET AL [43]), the phase modulation stays constant up to much higher frequencies, and then eventually begins to fall off. The reason for the fall-off with frequency is associated with bulk recombination of the carriers. These curves, which have an initial fall-off due to temperature effects, are in good agreement with theory; they have been used to determine the recombination rate of carriers in silicon. If the light is scanned along the surface regions, defects and rapid rates of recombination show up clearly. By using sufficiently high-frequency surface acoustic waves, it is possible to carry out these measurements in regions only a micron or two thick. Thus, there are possibilities of combining these ideas with the use of the acoustic microscope to scan semiconductors and determine the presence of defects in silicon.

IX. Conclusion

We have discussed many applications of surface acoustic waves to nondestructive testing. The most powerful applications so far have been for the investigation of cracks. Stress measurements may be a very important use of Rayleigh waves; however, a great deal remains to be done to make such techniques repeatable, practical, and quantitative in the field. Imaging with surface waves is very powerful and will no doubt be pursued further in the future.

Optical and photoacoustic techniques are still in their infancy, and we have only mentioned a few examples of these methods. There are very powerful techniques available for exciting surface acoustic waves with optical beams and receiving them on the same substrate, thus making a noncontacting diagnostic tool (JEN ET AL [44]). Surface acoustic wave techniques can also be used for measuring surface hardness. As these methods are put on a quantitative basis, they should prove very powerful for making quantitative measurements.

The measurement of electronic effects, which we have discussed at the end of this paper, may become very important, for it provides one of the few methods for nondestructive, noncontacting ways of determining surface recombination. If the technique can also provide information on doping, it would be very powerful.

The basic idea of Lord Rayleigh has led to many unexpected results. No doubt there will be many more in the future.

X. References

1. Lord Rayleigh, Proc. London Math. Soc. 17, 4 (1885).
2. Viktorov, I. A., Rayleigh and Lamb Waves (Plenum Press, New York 1967).
3. Pohlman, R., Z. Instrumentenk 71, 322 (1963).
4. R. M. White, Proc. IEEE 58, 1238 (1970).
5. W. R. Smith, H. M. Gerard, J. H. Collins, T. M. Reeder, and H. J. Shaw, IEEE Trans. on Microwave Theory and Techniques MTT-17, 856 (1969).
6. H. K. Wickramasinghe and E. A. Ash, Elect. Lett. 9, 327 (1973).
7. H. K. Wickramasinghe and E. A. Ash, Optical and Acoustical Micro-Electronics, 413, (Ed.: J. D. Fox, Polytechnic Press of the Polytechnic Institute of New York, 1975).
8. H. L. Bertoni and T. Tamir, IEEE Trans. Sonics and Ultrasonics SU-22, 415 (1975).
9. J. Fraser, B. T. Khuri-Yakub, and G. S. Kino, Appl. Phys. Lett. 32 (11), 698 (1 June 1978).
10. F. W. Smith, A. F. Emery, and A. S. Kobayashi, J. Appl. Mech. 34 (4), 953 (December 1967).
11. G. S. Kino, J. Appl. Phys. 49 (6), 3190 (June 1978).
12. B. A. Auld, Wave Motion 1 (1), 3 (January 1979).
13. B. Budiansky and J. R. Rice, J. Appl. Mech. 45 (2), 453 (June 1978).
14. M. T. Resch, B. T. Khuri-Yakub, G. S. Kino, and J. C. Shyne, Appl. Phys. Lett. 34 (3), 182 (Februay 1979).
15. J. J. W. Tien, B. T. Khuri-Yakub, G. S. Kino, A. G. Evans, and D. B. Marshall, J. NDE 2, 219 (1981).
16. B. Budiansky and R. J. O'Connell, Ing. J. Solid Structures 12, 81 (1976).
17. D. B. Marshall and B. R. Lawn, J. Matls. Science 14, 2001 (1979).
18. D. B. Marshall, B. R. Lawn, and P. Chantikul, J. Matls. Science 14, 2225 (1979).
19. D. B. Marshall, A. G. Evans, B. T. Khuri-Yakub, J. W. Tien, and G. S. Kino, Proc. Royal Soc. A385, 461, London, England (1983).
20. B. T. Khuri-Yakub, Proc. of the Fifth International Conference on Nondestructive Testing Methods, Bordeaux, France (May 1983).

21. M. T. Resch, D. V. Nelson, J. C. Shyne, and G. S. Kino, Advances in Crack Growth Measurement, 473 (Ed.: C. J. Beevers, Materials Advisory Services, Ltd., West Midlands, U.K., 1982).
22. M. G. Silk, NDT International 10, 290 (1976).
23. R. J. Hudgell, L. L. Morgan, and R. F. Lumb, Brit. J. NDT 16, 144 (1974).
24. V. Domarkas, B. T. Khuri-Yakub, and G. S. Kino, Appl. Phys. Lett 33 (7), 557 (1 October 1978).
25. S. Ayter and B. A. Auld, Proc. DARPA/AFML Review of Progress in Quantitative NDE, 498 (AFWL-TR-80-4078, Ed.: D. O. Thompson, York, 1979)
26. G. S. Kino, D. K. Peterson, and S. D. Bennett, New Procedures in Nondestructive Testing, 113 (Ed.: P. Holler, Springer-Verlag, New York, 1983).
27. D. Husson, S. D. Bennett, and G. S. Kino, J. Materials Eval. 43 (1), 92 (January 1985).
28. G. S. Kino, Measurement of Stress with Surface Acoustic Waves (Final Report to the Electric Power Research Institute, Palo Alto, CA., NP-3523-LD, Research Project T107-5, 1984).
29. R. L. Jungerman, B. T. Khuri-Yakub, and G. S. Kino, Materials Eval. 42 (4), 444 (April 1984).
30. M. Hirao, H. Fukoaka, and Y. Miura, J. Acoust. Soc. Am. 72, 602 (1982).
31. J. D. Achenbach and R. J. Brind, J. Sound and Vibration 76 (43) (1981).
32. B. T. Khuri-Yakub and L. R. Clarke, Review of Progress in Quantitative Nondestructive Evaluation, 1133 (Eds.: Donald O. Thompson and Dale E. Chimenti, Plenum Press, New York, 1985).
33. C. R. Steele and S. H. Yoo, Appl. Phys. Lett. 44, 857 (1984).
34. K. K. Liang, S. D. Bennett, B. T. Khuri-Yakub, and G. S. Kino, IEEE Trans. on Sonics and Ultrasonics SU-32 (2), 266 (March 1985).
35. R. C. Bray and C. F. Quate, Thin Solid Films 74, 295 (1980).
36. R. D. Weglein, Scanned Image Microscopy, (Ed.: E. A. Ash, Academic Press, London, 1980).
37. A. Atalar, J. Appl. Phys. 49, 5130 (October 1978).
38. K. K. Liang, G. S. Kino, and B. T. Khuri-Yakub, IEEE Trans. on Sonics and Ultrasonics SU-32 (2), 213 (March 1985).
39. J. Kushibiki and N. Chubachi, IEEE Trans. on Sonics and Ultrasonics SU-32 (2), 189 (March 1985).
40. R. G. Stearns, B. T. Khuri-Yakub, and G. S. Kino, Appl. Phys. Lett. 43 (8), 748 (15 October 1983).
41. R. G. Stearns, G. S. Kino, and B. T. Khuri-Yakub, Appl. Phys. Lett. 42 (8), 659 (15 April 1983).
42. R. W. Keyes, Solid State Phys. 20, 37 (1967).
43. R. G. Stearns, B. T. Khuri-Yakub, and G. S. Kino, Proc. IEEE Ultrasonics Symposium, Vol. 1, IEEE Catalog No. 84CH2112-1, 983 (1984).
44. C. K. Jen, P. Cielo, J. Bussiere, F. Nadeau, and G. Farnell, Appl. Phys. Lett. 46 (3), 241 (1 February 1985).

Part VII

Optical Guided Waves and Rayleigh Waves

Optical Waveguides for Application in Acousto-Optic Components

C.W. Pitt

Department of Electronic and Electrical Engineering,
University College London, Torrington Place, London WC1E 7JE, U.K.

1 Introduction

Although optical planar waveguiding layers had been fabricated and tested in the early 1960s [1], the subject entered a period of rapid expansion in the later years of that decade when the researchers [2,3] at Bell Telephone Laboratories coined the phrase 'integrated optics'. This was arguably the most potentially far reaching monolithic system concept to emerge since the Silicon integrated circuit in the late 1950s.

The proposed component functions in the integrated optics of that era often resembled those of the microwave devices developed during the previous two decades; allowing for geometry modification to cope with the dielectric propagation medium, and the much higher frequency of operation, they were indeed similar. The materials systems used in the early waveguiding experiments were usually those that were available as transmission coatings for bulk optical components [4,5,6], as were many of the waveguiding film deposition processes [7,8,9,10]. However the potential of newer processes, such as ion-implantation and epitaxial deposition, were quickly realised [11,12,13] and acted upon. Certainly the need for very low loss films, to carry optical energy over guided path-lengths that were 10^5 times that of the optical transmission path in the antireflection film of a camera lens, was identified in the first few years of the component research programmes.

The original justification for the integrated optical circuit was as a sub-system for optical fibre communications networks [14,15]. The application emphasis began to change in the late 1970s, when it appeared that it would be possible to reduce the optical fibre loss [16] to the point at which signal regeneration circuitry in the communications link could either be eliminated, or economically replaced by optical-to electronic signal convertors and electronic amplifiers. Research shifted towards signal processing concepts, although the components that had been developed for the communications-orientated circuits were often still employed [17]. The new direction accelerated a trend which had emerged soon after the inception of passive waveguide experiments - that of active components. The early active devices frequently used the electro-optic effect in ferroelectric crystalline Lithium Niobate [18] or the weaker effect in Gallium Arsenide [19].

Another form of active device had emerged in the mid-1960s, albeit in bulk form [20,21]: this was the acousto-optic interaction cell, in which laser light could be modulated by microsound waves. The translation of this device into a thin film Bragg cell, using the established surface acoustic wave technology and planar optical waveguides, was achieved in 1970 [22]. Although the device was improved [23], and various other versions (using colinear acousto-optic interaction [24] and optical mode-mode coupling by surface acoustic wave [25]) were proposed and fabricated, these devices were considered to be too slow in operation to be useful as switches or modulators for communications systems. It was not until the emergence of interest in optical signal processing, particularly in large information volume parallel processors, that the waveguiding acousto-optic cell was developed as a high frequency device. The device that stimulated the interest and subsequent development was the integrated acousto-optic spectrum analyser for radar application [26].

This article reviews the limitations of optical waveguide technology in general, and in particular considers the guiding layer requirements of current acousto-optic cells. In the following paper by STEWART & BENNION [27], the potential applications of acousto-optic interaction devices are examined rather more closely.

2 Principles and Limitations of Planar Guides

Figure 1 illustrates in simplified form the principle of operation of a planar optical waveguide [5].

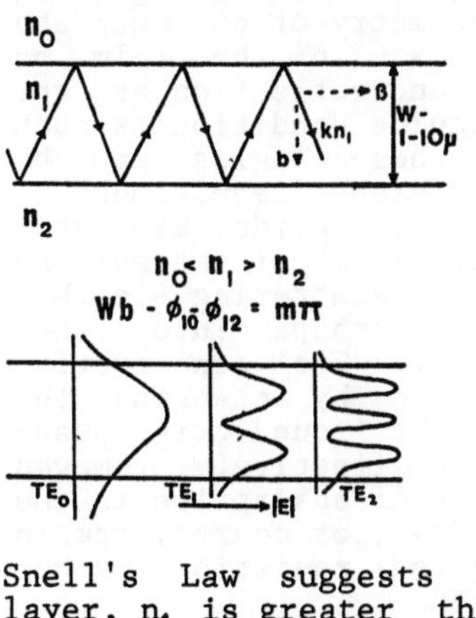

Fig 1. A section through a dielectric layer and surrounding media, showing a light ray reflected at the interfaces: the conditions for guiding the light in a thin layer and the standing electric-field pattern are also indicated.

Snell's Law suggests that if the refractive index of the layer, n_1 is greater than that of the semi-infinite half spaces on either side of the film, then laser light introduced into the film will be reflected within a continuum of angles from both substrate and superstrate interfaces.

However, if the film is made rather thin (thickness, W, of the order of the exciting wavelength or so), then the ray effectively appears as a broad-fronted beam within the film. This leads to interference between the incident and reflected wavefronts at each reflection from the interfaces. The condition for waveguiding then becomes

(a) that the refractive index of the film shall be greater than that of the surrounding media

and

(b) that the interference between the succesive reflections in the film must be constructive.

The latter condition can be expressed in an alternative form - that the round-trip phase change of the plane-normal component, b, of the wave phase-propagation vector, kn_1, must be an integer number of π radians, when the Fresnel reflections coefficients, ϕ_{10} and ϕ_{12}, of the interfaces are accounted for. This condition, known as the waveguide equation, is given in fig. 1; m is then the propagable mode order, and simply gives the number of 'reflection angles' that can be sustained by the guiding film. Under these conditions the optical wave is essentially contained within the guide and propagates at a group velocity of β.

The interferences of the light in the guide sets up standing electric field patterns, as shown for various propagating modes. Note that an evanescent field extends beyond the interfaces - a condition which becomes important when considering defective surfaces in waveguiding components. There is a further effect of the thinness of the film; modes may only propagate if the electric or magnetic field vector lies parallel to the plane of the film. Thus, only transverse-electric (TE) or transverse-magnetic (TM) waves may propagate - there are no hybrid modes in such films.

We now turn to the limitations of such planar structures. Clearly there are several ways in which the waveguiding conditions might be abrogated. If the geometry of an interface is modified, for example by roughness in the film or substrate, then the angles of incidence and reflection at the interfaces will be changed. The waveguide condition is such that only very small variations in these angles can be tolerated, before the constructive interference requirement is impaired and the light is scattered from the guide. Similarly an inhomogeneous region in the refractive index of the guiding film will change the mode angles and cause scattering - either into the superstrate or substrate or perhaps into other propagable modes within the guide. The constructive interference condition may also be broken by changing the thickness of the thin film, altering the 'round-trip' phase change of the wave. Optical energy can be effectively removed from the guide by another mechanism, that of absorption in the thin film material system. The energy does, of course, remain in the structure but usually as infra-red radiation rather than at the desired excitation wavelength.

Many of the early waveguide layers produced for integrated optics experiments suffered severely from these defects [3 - 10] and the devices produced, both passive and active, were often degraded in performance [6,8,24]. Considerable effort has been expended over the past decade-and-a-half in modelling the effect of defects [28, 29], in measuring the effect of them [30, 31, 32], and in attempting to minimise the degredation of the component performance.

Even without defects, the optical energy in a planar guide spreads out in the plane of the film, as the result of diffraction. This can be overcome by machining the planar guide into a stripe on top of the substrate, or by forming the guide as a stripe buried within the substrate. Unfortunately the effect of defects, such as wall roughness is equally as severe as in the case of the planar guide - see Fig 2.

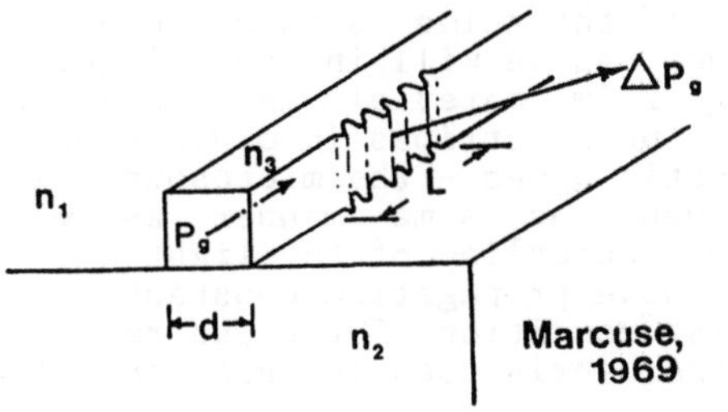

$$\Delta P_g / P_g = \frac{D^2 k^4 L}{2\pi l} (n_3^2 - 1) \int F(n_3, n_1, d, k, l) d\beta$$

l = correlation length of deviation

D = rms deviation of wall

Figure 2. The fraction of the guided optical power that will be scatterd from a stripe guide by wall roughness depends on the wall-perturbation parameters, and also rather strongly on the refractive index of the stripe.

In order to minimise the effect of unavoidable wall roughness, perhaps due to photolithographic and etching processes, the film material should be of relatively low refractive index. Unfortunately, this conflicts with the requirement for a high refractive index to keep light confined as the stripe undergoes a change of direction round a bend - see Fig 3.

The attenuation of a stripe bend has two components [33], as indicated in the figure - the conversion loss as the

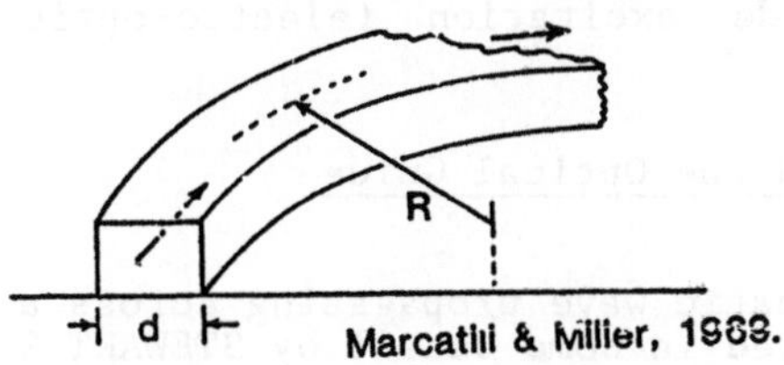

curve radiation loss : $c_1 \exp(-c_2 R)$

mode conversion loss : $(\alpha_g - \alpha_c)(c_3 d^3/R)$

c_1, c_2, c_3, - geometry & index dependent constants

α_g - attenuation of straight guide mode

α_c - attenuation of curved guide mode

Figure 3. The simplified expressions for the curved guide loss indicate that the loss depends on the tightness of the bend and the refractive index of the stripe. There is also some dependence on the scattering losses in the material of the guide.

propagation constant of the straight stripe is gradually converted to that of the normal mode of the guide bend, and the radiation loss as the guided ray interacts with a continually changing guide-superstrate wall angle round the bend. In order to effect tight confinement of the optical energy, the refractive index of the stripe bend must be high - leading to severe loss if wall roughness is present.

This difficulty is exacerbated if the guide is to be used in an acousto-optic device since, as we will indicate later, a high refractive index guiding film material is preferred. There is yet another problem in using stripes in devices which also propagate surface acoustic waves - the microsound wave ripples the guide surface in much the same manner as wall roughness; this leads to some decoupling of the light unless the guided wave and free-space wave propagation constants are carefully set to avoid this condition. For these reasons, stripe waveguides are only infrequently used in acousto-optic interaction components.

The last imperfection to be considered appears in its most severe form in electro-optic film guides. Optical damage [34, 35] is manifest as a change in propagation constant, β, of the guide and as an increase in in-plane scatter in the layer. It occurs under conditions of relatively low power, prolonged period, excitation of the guide or when the guide is carrying a high energy density (usually some $MW.cm^{-2}$) even for a short period (a small fraction of a second). We have also observed that the effect appears to be enhanced by elevated temperature operation. The fundamental causes are ill-understood, and the reduction in optical damage threshold in commercially available materials, compared with stoichiometric materials prepared in the laboratory, is usually attributed to 'photorefractivity'. This generally seems to suggest that the causes are a combination of enhanced optical absorption due to dopant atoms in the material lattice, electronic absorption due to free charges in the material and refractive index perturbation due to free charge relocation under internal electric fields from the waveguide excitation (electro-optic effect).

3 Acousto-optic Interaction and the Optical Guide Requirements

The effect of a surface acoustic wave propagating across a planar optical guide is considered in some detail by STEWART & BENNION [27], but figs 4 and 5 briefly illustrate the main features of the interaction, so that the additional waveguiding requirements for a-o interaction components, etc, may be identified.

The generalised effect of an acoustic wave travelling across the surface of a material, and guided by that free surface, is to create a compression-rarefaction wave and a surface ripple. Both of these effects contribute to a 'grating perturbation' in the effective refractive index of the material as encountered by a guided optical wave confined near the surface

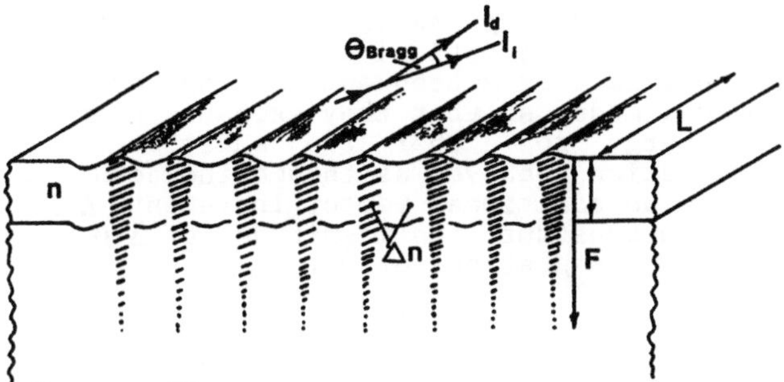

Figure 4. The acousto-optic interaction at the Bragg angle is used to illustrate the effect of a surface acoustic wave on the substrate with optical waveguide overlay.

Bragg condition:-

$$I_d/I_i = \sin^2\left[(\pi/\lambda_0)\cos\Theta_s \cdot L\overline{\Delta n}|F|\right]$$

$$\overline{\Delta n}|F| = |\Delta n_{ao} + \Delta n_{eo} + \Delta n_{sr}|$$

$$\Delta n_{ao} \propto n^3 \qquad \Delta n_{eo} \propto n^3 \qquad \Delta n_{sr} \propto (n^2 - 1)/n$$

of the material. If the material is both electro-optic and piezo-electric, then the acoustic wave will displace lattice ions, creating localised electric fields and initiating a refractive index grating due to the electro-optic effect.These three effects are all strongly dependent on the refractive index of the guiding film - the higher the refractive index, the greater the diffraction efficiency. However, in order to achieve a strong interaction, there must clearly be a good match between the optical and acoustic energy fields. In the diagram, F denotes the overlap integral.

Figure 5 is a reminder that the Bragg interaction condition is not the only form of the a-o scattering mechanism.

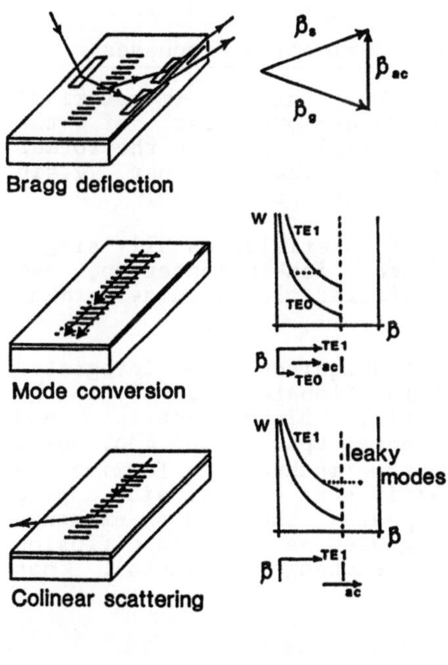

Figure 5. Bragg (single diffraction order), and Raman-Nath (multiple diffraction orders), interaction occurs when the acoustic grating propagation constant (inversely proportional to the grating period, i.e. the acoustic wavelength) couples the incident optical wave to the diffracted wave. Similar coupling can take place colinearly to couple two optical modes together, or couple a guided mode to a superstrate or substrate mode.

If the acoustically formed grating has an appropriate effect on the propagation constant of the optical guided wave, it may cause diffraction to the first, or subsequent grating orders, or it may couple optical modes so that they exchange energy, or it may decouple energy to either the substrate or superstrate. The interaction behaves as though the acoustic surface wave has supplied an additional vector increment, $\Delta\beta$, to the optical propagation constant: the magnitude of $\Delta\beta$ is given by $2\pi/L$, where L is the grating period.

The requirements of the optical guiding film for a-o application can now be defined:

high refractive index - the a-o figure of merit is proportional to the sixth power of the index.

an active layer, i.e. piezo-electric and electro-optic, to enhance efficiency.

good overlap integral - compatibility in thickness between the optical guide and the 1/e depth of the saw penetration.

refractive index uniformity to better than 10^{-6},

thickness uniformity to 10^{-5},

few scattering defects,

low absorption loss, $\ll$1dBcm

- since all of these affect both signal-to-noise ratio and dynamic range in the a-o cell.

4 Optical Waveguide Materials and Fabrication Processes

An examination [36] of the thin film-material/fabrication-process combinations rules out most of the polycrystalline and amorphous materials, because the former suffer from high scatter loss from the columnar grain boundaries, while the amorphous films are not active.

The choice is reduced to ordered materials, usually single crystals, and diffusion, ion-exchange, ion-implantation, and epitaxial layer growth as the preferred planar guide forming processes.

The material that has become almost ubiquitous in integrated optics in the past decade is Lithium Niobate - usually in the form of a poled electro-optic, piezo-electric and pyro-electric crystal of rhombohedral space group R3c, point group 3m. Certainly the most common method of forming an optical waveguide in the material is that of diffusion doping with Titanium [37]. In recent years, the proton-exchange process [38] (more properly designated as a proton diffusion technique, rather than an ion-exchange process), has supplemented Ti:LiNbO(3) as a guiding medium. However there is some evidence that proton doping adversely affects the electro-optic, and possibly the piezo-electric, properties of Lithium Niobate; further discussion of this material system for acousto-optic application will therefore be postponed.

The process for the diffusion-doping of Lithium Niobate with Titanium is similar to that used for doping semiconductor wafers. The crystal wafers are cut from the boule to the desired orientation, lapped, and then diamond-paste or Syton-fluid polished to remove work damage and produce a flat smooth surface. After careful cleaning, a thin layer of Titanium (100 - 1000Å) is electron-beam evaporated or sputtered onto the polished surface. The wafer is then placed in a furnace in an Argon, or sometimes Oxygen, atmosphere for periods ranging from 1 to 30 hours. The resulting guide has an enhanced refractive index at the diffused face, and the index profile extends into the crystal, closely following the dopant profile [39] - see figure 6.

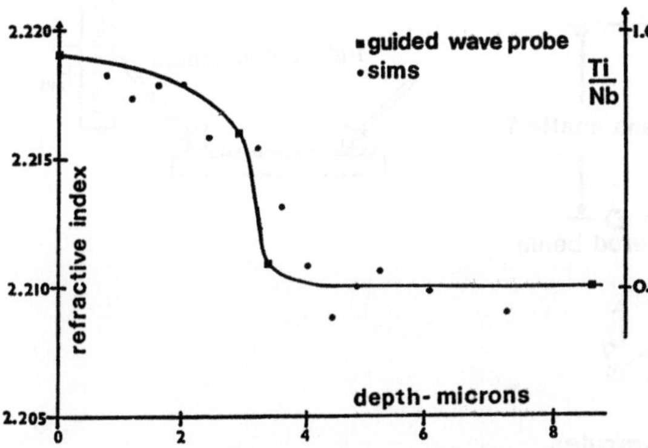

Figure 6. The refractive index profile, produced from the guided mode propagation constant measurements, closely matches that of the Titanium dopant/lattice Niobium ratio measured by Secondary Ion Mass Spectrometry.

The refractive index profile, maximum index value and depth of penetration can all be adjusted to provide a satisfactory waveguide for an acousto-optic interaction cell. The optical loss of the intrinsic material is also low across the visible and near infra-red spectrum - and it does not degrade severely when Titanium doping is added, or even when mixed Lithium Niobate/Tantalate films are fabricated [40] - see figure 7.

The major problem encountered in using Ti:LiNbO(3) waveguides in a-o cells is that of optical damage. There is no satisfactory method of quantifying the reduction in optical damage threshold of Lithium Niobate due to fractional doping with Titanium; however figure 8 demonstrates the effect on a device such as a integrated Bragg-cell spectrum analyser [41]. The experiment is illustrated at the top of the diagram. A He-Ne laser beam was coupled into the waveguide and decoupled

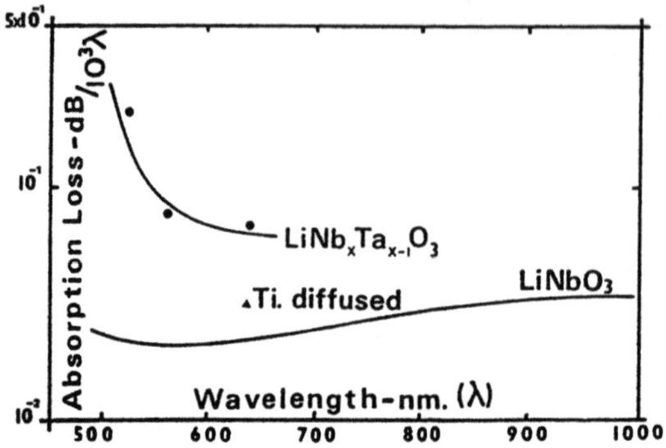

Figure 7. The optical loss in Lithium Niobate; the effect of doping with Titanium and also the increased loss resulting from the formation of a mixed Lithium Niobate/Tantalate guide by melt-phase epitaxy.

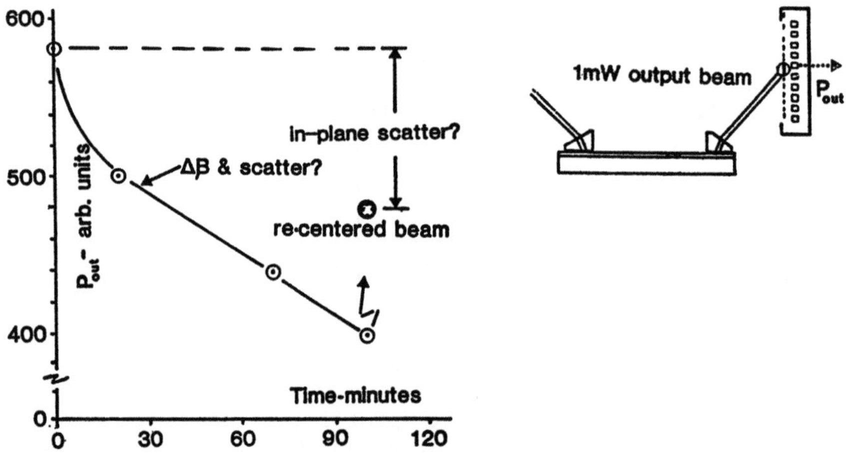

Figure 8. An experiment to illustrate the effect of optical damage in causing the mode output beam to drift, and the guided energy to be scattered in the guide.

to a charge-coupled diode detector array. The main mode spot was centred on one diode in the array, and the output beam power increased to circa 1mW. The detected energy decreased rapidly with time, partially as the result of drift in the guided mode propagation constant and partially, it is thought, as the result of increased in-plane scatter deflecting light from the main mode spot to the scattered light lobes of the output. The latter contention is supported by the reduced main spot power when the spot was re-centered on the detector element at the end of the experiment.

5 Alternative Processes and Materials

The severity of the optical damage in the Ti:LiNbO(3) system is such that considerable effort has been expended to find an alternative material system or process without this limitation - hence the interest in proton exchange [38] and the renewed interest in ion-implantation of Lithium Niobate [42].

There are several other approaches, as yet unproven in the device context, but which may offer some advantages in comparison with existing technology. The refractive index increase resulting from the deliberate reduction in the oxygen stoichiometry of metal-oxide films [43] had been noted in some of the early work on waveguides. A considerable increase in index could be obtained without increasing the optical absorption loss penalty.

A recently developed process [44], which utilises this effect to deposit a waveguide by growing an oxygen deficient homo-epitaxial layer of Lithium Niobate, is illustrated in figure 9. The process follows the principles of molecular beam epitaxy; the layer is deposited on a carefully cleaned substrate (5) in an ultra-high vacuum pumped environment (9,15) by mixing elemental beams of Lithium, Niobium (3) and Oxygen (10) at the substrate. The metal fluxes are supplied by electron-beam guns, while the gas is directed at the substrate by sonic nozzle. The right-hand chamber (14) is used for cleaning and diagnostic evaluation of the substrate prior to growth and in-situ evaluation of the deposited film. The cleaning cycle involves both Ar ion bombardment (11) and thermal annealing on a substrate heater (4); the contamination levels are monitored by Secondary Ion Mass Spectrometry (SIMS) (2) and the crystallinity of the crystal surface is checked by Low Energy Electron Diffraction (LEED) (12,13). The heated

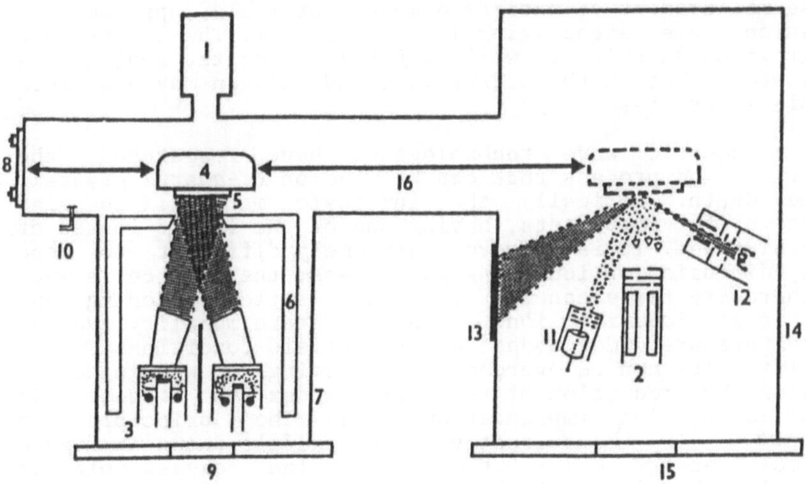

Figure 9. Schematic of the MBE equipment used for the deposition of epitaxial Lithium Niobate.

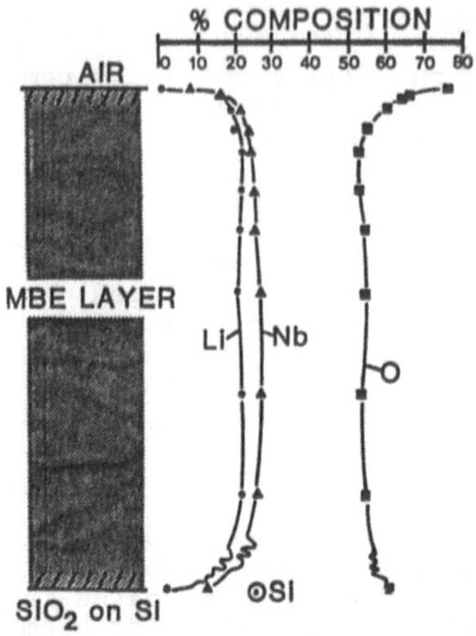

Figure 10. The composition of a layer of LiNbO(3) deposited on oxidised Silicon and measured by Auger Electron Spectroscopy. Note that the Oxygen content is fractionally less than required for full stoichiometry.

substrate is then transferred (16) to the growth chamber (7,8), where preset flux beams, analysed by mass spectrometer (1), are admitted via a cryo-shield (6) onto the growth surface.

The composition of a typical layer - grown on oxidised Silicon for convenience in the Auger Electron Spectroscopy (AES) measurements - is shown in figure 10. The layers have been shown to be single crystal, slightly enhanced in refractive index c.f. stiochiometric bulk-grown crystal, and capable of optical waveguiding; the composition is extremely uniform – producing a well-controlled refractive index layer. Although the current films need to be improved in crystal quality for precise measurement of optical damage threshold, preliminary indications are encouraging in this respect. The layers have been grown on Sapphire as well as Lithium Niobate, and it is anticipated that Lithium Tantalate and Silicon may also form suitable substrates.

Surface acoustic wave technologists have researched the properties and process that can be used on α-quartz crystal, in some depth. Optically the intrinsic material is also attractive for components, having one of the lowest losses of any metal-oxide. It is, however, extremely difficult to dope by any diffusion or ion-exchange process; the lattice is such that there are few vacancies for substitutional doping and interstitial impurity ions tend to retain mobility even at room temperature under modest electric-field conditions.
This difficulty can be overcome by forming a waveguide by utilising the reduction of refractive index when α-quartz is amorphised by ion implantation [45]. The principle is illustrated in the refractive index profile shown in figure 11. A multi-energy (0.7 - 0.8MeV) He^+ implant caused minimal damage in the immediate sub-surface layer (to a depth of approximately 1.5 micron), since the ion energies were so high that only interaction between the lattice electrons and the

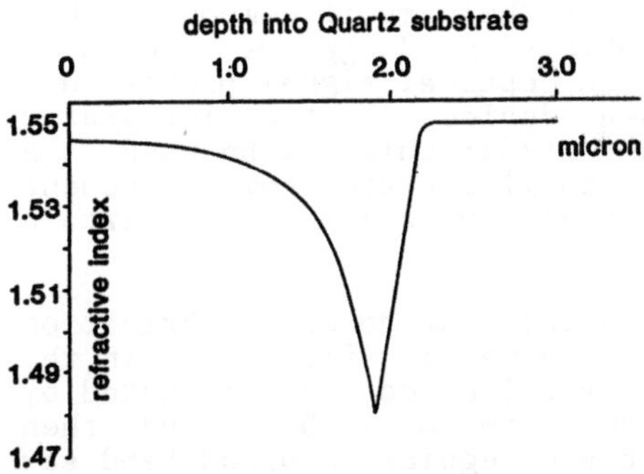

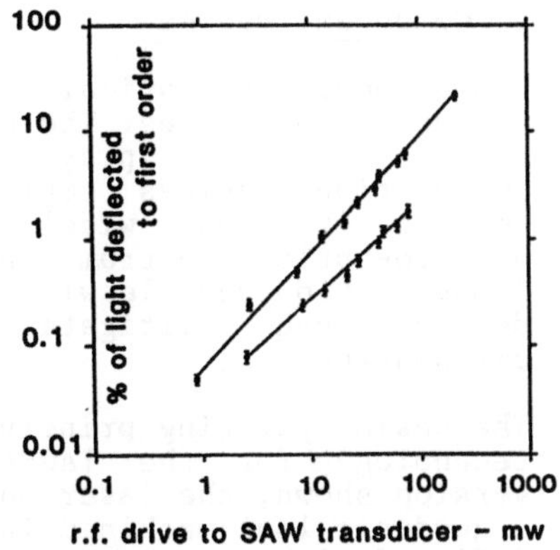

Figure 11. The computed refractive index profile in crystal Quartz resulting from implantation with MeV He^+ ions.

Figure 12. The acousto-optic interaction performance of an α -quartz Bragg cell fabricated by ion-implantation.

incident ions could occur. However, when sufficient energy had been lost by the bombarding ions from this interaction, then the He^+ ions interacted strongly with the lattice ions causing amorphisation to a depth of approximately 2 microns. The reduced index region resulting from the amorphisation, acted as the confining layer for the guided light while the guiding region was the surface region to a depth of 1.5 micron.

An Aluminium film interdigital surface wave transducer (centre frequency 114MHz) was then placed on the modified substrate to form a Bragg cell.The performance is indicated in figure 12.

In this device the substrate was x-cut crystal, the optical propagation was in the z-direction and the saw propagated in the y-direction. The overlap was relatively poor - an acoustic wavelength of 26 micron and an optical guide depth of approximately 1.5 micron, but the interaction length was made long (100 acoustic wavelengths) because the optical layer was very uniform. We achieved 22% optical power deflected to the first order at 240 mW r.f. power input to the transducer, giving an efficiency of 0.25% per mW. The dynamic range was 24dB. Both efficiency and dynamic range were achieved without use of signal recovery techniques.

The a-o figure of merit is rather similar in Quartz and in Lithium Niobate,if the elecro-optic contribution from LiNbO(3) is neglected (7.5 x 10^{-18} , and 6.8 x 10^{-16} respectively) - but the e-o effect does make a very large improvement in practical devices. Although similar efficiency devices can be constructed in both materials systems, the electro-mechanical coupling coefficient of Quartz is 25 times smaller than that of Lithium Niobate, so that Quartz devices are acoustically narrow-band in comparison.

6 Component Design

Improvement in optical materials is not the only route to enhanced a-o devices. The design of the circuit components has a major role to play in optimising Bragg or colinear interaction devices employed in optical signal processing. Perhaps the most widely known device is the integrated acousto-optic spectrum analyser [26]; this may be used as a demonstration vehicle with which to illustrate how component design can mitigate material problems in certain circumstances.

The basic operating principle and the most common fabrication technology for the IAOSA are shown in figure 13. In the version shown, the laser output is collected and collimated by a geodesic lens machined into the Lithium Niobate and then diffused with Titanium to form the waveguide. A broad-band saw trnsducer converts the r.f. frequency signal into an analogous travelling acoustic surface wave grating. The interaction of the broad guided laser beam and the sound wave takes place in the Bragg cell; the diffracted light is collected by the Fourier transform lens and converted into spots on the detector diode array.The frequency components of the original signal, and their magnitude, can then be derived from the position of the light spots on the detector array and the optical power in each spot.

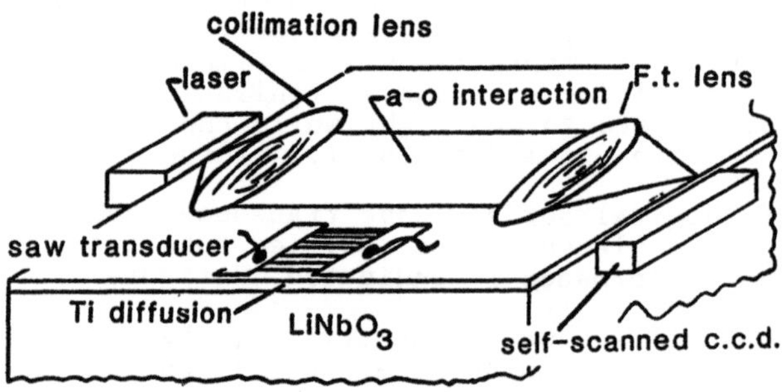

Figure 13. An illustration of an integrated Lithium Niobate acousto-optic spectrum-analyser for use at rf frequencies.

It is extremely difficult to make the geodesic lenses aberration-free, and optical energy displaced both by the aberrations, and by scattering centres in the planar guide and lens regions, is directly collected by the Fourier transform lens. This light, together with the unwanted scattering in the Bragg interaction cell, forms the noise signal in the focal line of the output lens, i.e. at the detector plane. The problems of signal-to-noise ratio and dynamic range are increased by the long path lengths, and proportionally high scatter-noise, in the device; the lens f numbers are such that the total length of the device may be over 10 cms.

An alternative version [46] has been proposed using only thin film components [47]. The top view schematic of the device is

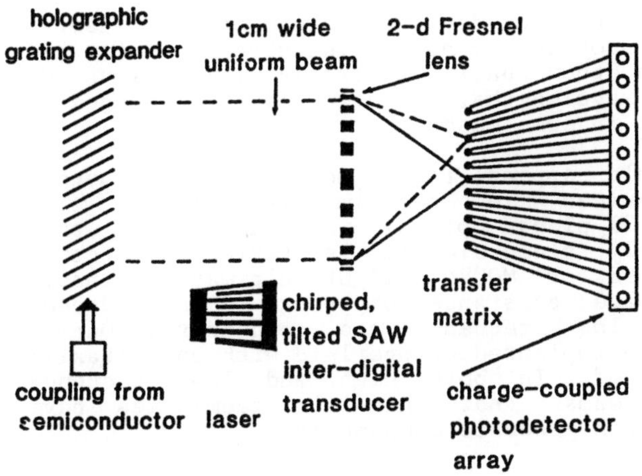

Figure 14. A thin film component version of the IAOSA.

shown in figure 14. The collimation lens has been replaced by a holographic grating etched into the surface of the waveguide and which collimates the laser output by reflection [48,49]. Any scattering in the guide, and scattering levels are low since the grating is narrow (less than 1 mm) and perturbs the planar guide very little, is orthogonal to the interaction region and Fourier lens. The Bragg cell region is reduced in size to accomodate only the saw transducer. The Fourier transform lens has been replaced by a two-dimensional Fresnel lens (30 micron wide)[50]of high efficiency and short focal length, again reducing scattering. The matrix of waveguide stripes at the output enables the focal spot geometry and spacing to be conveniently converted to those of a commercial charge coupled diode array, but it has another advantage if carefully designed. Both version of the device are subject to optical damage at the focal spots if operated in the visible spectrum.

However, by suitable shaping of the stripes in the transfer matrix much of the light scattered by optical damage, caused by prolonged excitation at a single acoustic frequency, can be collected as signal rather than noise. There is further benefit to be gained by using thin film components - they can be defined by high definition lithography techniques, thus reducing non-material scattering dramatically.

7 Concluding Comments

Careful component design may enable many of the limitations of Ti:LiNbO(3) to be overcome in most signal processing applications of a-o cells; the outstanding exception is optical damage, which can be mitigated in the visible spectrum, but has not been completely overcome to date. Fortunately, the optical damage threshold is much less degraded by Titanium doping if the excitation is near-ir rather than visible spectrum. Since many of the applications will require small lasers, probably semiconductor devices instead of gas lasers, the optical damage problem may not prove intractable for a-o circuits. However the other basic materials limitations do still occur at ir wavelengths, although slightly reduced by the longer wavelength in some cases.

In summary, it would appear that most high-frequency, broad-band a-o devices in the immediate future will incorporate Ti:LiNbO(3) planar optical waveguides, but that development of highly piezo-electric epitaxial dielectric layers, either by MBE or MOCVD, might ultimately enable improved optical damage resistance components to be produced. For narrow-band or low frequency a-o components, Quartz substrate, either ion implanted or overlaid with an epitaxial layer, could be employed. In both high and low frequency applications, it seems that thin-film geometries have advantages compared with geodesic components.

Acknowledgements

The author would like to thank all the colleagues who have worked with him on the projects outlined and referenced in this paper.

References

1. H. Osterberg and L.W. Smith: "Transmission of Optical Energy along Surfaces: Pt I & II", J. Opt. Soc. Am., 54, pp1073-1084, 1964.
2. S.E. Miller: "Integrated Optics: An Introduction", BSTJ, 48, pp2133-2160, 1969.
3. J.E. Goell and R.D. Standley: "Integrated Optical Circuits", Proc. IEEE , 58, pp1504-1512, 1970.
4. P.K. Tien, R. Ulrich and R.J. Martin: "Modes of Propagating Light Waves in Thin Deposited Semiconductor Films", APL, 14, pp291-294, 1969.
5. P.K. Tien: "Light Waves in Thin Films & Integrated Optics", App. Opt., 10, pp2395-2413, 1971.
6. R. Ulrich and R.J. Martin: "Geometrical Optics in Thin Film Light Guides", App. Opt., 10, pp2077-2085, 1971.
7. J.E. Goell and R.D. Standley: "Sputtered Glass Waveguide for Integrated Optical Circuits", BSTJ, 48, pp3445-3448, 1969.
8. R. Schubert and J.H. Harris: "Optical Guided-wave Focussing & Diffraction", J. Opt. Soc. Am., 61, pp154-161, 1971.
9. R. Ulrich, H.P. Weber, E.A. Chandross, W.J. Tomlinson and E.A. Franke: "Embossed Optical Waveguides", APL, 20, pp213-215, 1972.
10. C.W. Pitt: "Sputtered Glass Optical Waveguides", El. Letts, 9, 17, pp401-403, 1973.
11. R.L. Hines and R. Arndt: "Radiation Effects of Quartz & Vitreous Silica by 7.5 keV to 59 keV Positive Ions", Phys. Rev., 119, 2, pp623-633, 1960.
12. E.R. Schineller, R.P. Flam and D.W. Wilmot: "Optical Waveguides Formed by Proton Irradiation of Fused Silica", J. Opt. Soc. Am., 58, 9, pp1171-1176, 1968.

13. P.K. Tien, S. Riva-Sanseverino, R.J. Martin, A.A. Ballman and H. Brown: "Optical Waveguide Modes in Single-crystalline $LiNbO_3$-$LiTaO_3$ Solid-solution Films, APL, 24, 10, pp503-506, 1974.
14. E.A. Ash, C.W. Pitt and M.G.F. Wilson: "Integrated Optics for Fibre Communications Systems", Proc. 1st European Conference on Optical Fibre Communications (IEE publication #132), Institute of Elec. Engineers, London, pp 131-134, 1975.
15. F.A. Blum and J.C. Campbell: "Monolithic GaAs Circuit Elements for Integrated Optics", Proceedings of the Topical Meeting on Integrated Optics, Opt. Soc. America, paper MA4, 1976.
16. N. Shibata, M. Kawachi, S. Sudo, T. Edahiro: "Low-loss High-Numerical-Aperture Optical Fibre Fabricated by V.A.D. Method", El. Letts, 15, 21, pp680-681, 1979.
17. T.G. Giallorenzi: "An Overview of Micro-Optic Signal Processing Research", Proceedings of the Topical Meeting on Integrated and Guided Wave Optics, Opt. Soc. America, paper MA1, 1977.
18. J.M. Hammer: "Digital Electro-Optic Grating Deflector and Modulator", APL, 18, pp147-149, 1971.
19. D. Hall, A. Yariv and E. Garmire: "Observation of Propagation Cut-off and its Control in Thin Optical Waveguides", APL, 17, pp127-129, 1970.
20. C.F. Quate, C.D.W. Wilkinson and D.K. Winslow: "Interaction of Light and Microwave Sound", Proc. IEEE, 53, pp1604-1623, 1965.
21. R. Adler: "Interaction between Light and Sound", IEEE Spectrum, pp42-54, 1967.
22. L. Kuhn, M.L. Dakss, P.F. Heidrich and B.A. Scott: "Deflection of an Optical Guided Wave by a Surface Acoustic Wave", APL, 17, pp265-267, 1970.
23. J.M. White, P.F. Heidrich, E.G. Lean: "Thin Film Acousto-optic Interaction in $LiNbO_3$", El. Letts, 10, pp510-511, 1975.
24. F.R. Gfeller and C.W. Pitt: "Co-linear Acousto-optic Deflection in Thin Films", El. Letts., 8, 22, pp549-551, 1972.
25. L. Kuhn, P.F. Heidrich, and E.G. Lean: "Optical-Guided Wave Mode Conversion by an Acoustic Surface Wave", APL, 19, pp428-430, 1971.
26. D.B. Anderson, J.T. Boyd, M.C. Hamilton and R.R. August: "An Integrated Optical Approach to the Fourier Transform", IEEE JQE, QE-13, 4, pp268, 1977.
27. W.J. Stewart and I. Bennion: "Acoustic-Optical Interactions and Wave/Wave Scatter Processing", Rayleigh Wave Centenary, Series on Wave Phenomena, (Springer, Berlin, Heidelberg, 1985).
28. D. Marcuse: "Mode Conversion Caused by Surface Imperfections of a Dielectric Slab Waveguide", BSTJ, 48, pp3187-3215, 1969.
29. D. Marcuse: "Radiation Losses of Dielectric Waveguides in Terms of the Power Spectrum of the Wall Distortion Function", BSTJ, 48, pp3233-3242, 1969.
30. J.H. Harris, D.P. GiaRusso, R. Schubert: "Optical Waveguide Scattering and Griffith Microcracks", Proc. IEEE (Lett.), 59, pp1123-1124, 1971.
31. C.W. Pitt, F.R. Gfeller and R.J. Stevens: "R.f. Sputtered Thin Films for Integrated Optical Components", Thin Solid Films, 26, pp25-51, 1975.

32. D.J. Walter and J. Houghton: "Attenuation in Thin Film Optical Waveguides due to Roughness-Induced Mode Coupling", Thin Solid Films, 52, pp461-476, 1978.
33. E.A.J. Marcatili and S.E. Miller: "Improved Relations Describing Directional Control in Electromagnetic Wave Guidance", BSTJ, 48, pp2161-2188, 1969.
34. G.E. Peterson, A.M. Glass and T.J. Negran: "Control of the Susceptibility of Lithium Niobate to Laser Induced Refractive Index Changes", APL, 19, pp130-132, 1971.
35. A.M. Glass, I.P. Kaminow, A.A. Ballman and D.H. Olson: "Absorption Loss & Photorefractive Index Changes in Ti:$LiNbO_3$ Crystals and Waveguides", Appl. Optics, 19, 2, pp276-281, 1980.
36. C.W. Pitt: "Thin Film Optical Waveguides Fabricated by Ion-Assisted Techniques, Thin Solid Films, 86, pp137-146, 1981.
37. R.V. Schmidt and I.P. Kaminow: "Metal-diffused Optical Waveguides in $LiNbO_3$, APL, 25, pp458-460, 1974.
38. J.L. Jackel, C.E. Rice and J.J. Veselka: "Proton Exchange for High Index Waveguides in $LiNbO_3$", APL, 41, pp607-608, 1982.
39. R.A. Betts, C.W. Pitt, K.R. Riddle and L.M. Walpita: "A Comparative Study of the Dopant Profiles in Diffused Planar Optical Waveguides by SIMS and Guided Wave Probe", Appl. Phys., A31, pp29-35, 1983.
40. A.A. Ballman, H. Brown, P.K. Tien and S. Riva-Sanseverino: "The Growth of Solid Solution $LiNbO_3$-$LiTaO_3$ Thin Films for Optical Waveguides", J. Cryst. Growth, 30, 1, pp37-41, 1975.
41. A.P. Overbury: Private Communication, June 1985.
42. G.L. Destefanis, J.P. Galliard, E.L. Ligeon, S. Valette, B.W. Farmery, P.D. Townsend and A. Perez: "The Formation of Waveguides and Modulators in $LiNbO_3$ by ion Implantation", J.A.P., 50, 12, pp7898-7905, 1979.
43. C.W. Pitt: "Sputtered Glass Optical Waveguides", El. Letts., 9, 17, pp401-403, 1973.
44. R.A. Betts and C.W. Pitt: "The Growth of Thin Film Lithium Niobate by Molecular Beam Epitaxy" - to be published.
45. C.W. Pitt, J.D. Skinner and P.D. Townsend: "Acousto-optic Diffraction of Optical Guided Waves in Crystal Quartz", El. Letts., 20, 1, pp4-5, 1984.
46. V. Neuman, C.W. Pitt, and L.M. Walpita: "An Integrated Optic Spectrum Analyser Using Grating Components", Proc. First European Conference on Integrated Optics, IEE Publication #201, pp89-92, 1981.
47. C.W. Pitt: "Use of Thin Films in Optical Waveguiding Devices - A Case Study", Vacuum, 34, 3-4, pp399-403, 1984.
48. L.M. Walpita and C.W. Pitt: "Beam Expansion in Linear Constant Period Gratings", App. Optics, 23, 19, pp3434-3438, 1984.
49. L.M. Walpita and C.W. Pitt: "Fabrication and Evaluation of Linear Constant Period Beam Expander Gratings", Optics Communications, 52, 4, pp241-246, 1984.
50. C.W. Pitt, G. Trotter and J.D. Skinner: "Computer Simulation of Guided Wave Fresnel Lenses", Optics Communications, 53, 2, pp87-90, 1985.

Acoustic-Optical Interactions and Wave/Wave Scatter Processing

W.J. Stewart and I. Bennion

Plessey Research (Caswell) Ltd, Allen Clark Research Centre, Caswell, Towcester, Northants, NN12 8EQ, England

1. Introduction

The interaction of Rayleigh waves with guided light waves has been most intensively studied for the case of the acousto-optic interaction between a propagating surface acoustic wave and a light wave confined in a planar waveguide. This interaction has formed the basis for a number of devices including modulators, deflectors, scanners, filters, switches and spectrum analysers. This paper offers primarily, therefore, a selective review of these device concepts and the principles underlying their operation. However, acoustic waves are not the only surface waves with which guided optical waves interact and we present, briefly, some extension of the principles to include other wave types.

2. Fundamentals of Acousto-optic Interaction

The basic structure in which guided-wave acousto-optic interactions may be studied is illustrated in Fig. 1. Surface acoustic waves (SAW) are generated by a transducer which is typically an interdigital electrode array. In the case illustrated in Fig. 1, the waveguide is formed in a piezoelectric material - eg $LiNbO_3$ - and the transducer is fabricated directly on its surface: otherwise, a piezoelectric film must be deposited below or above the transducer. The SAW produces a propagating periodic disturbance from which the guided light wave can scatter, and this disturbance has contributions from a number of physical mechanisms. Firstly, the acoustic strain wave modulates the refractive index of the waveguide through the photoelastic effect by an amount Δn_a which may be evaluated, generally, by a relatively lengthy tensor analysis[1]. Secondly, the surface deformation created by the

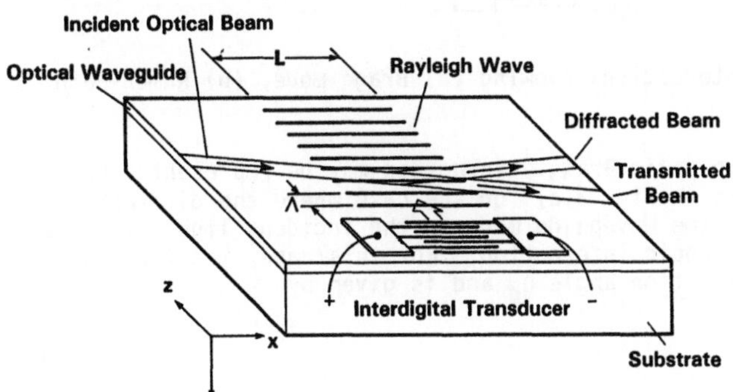

Fig. 1 Basic experimental structure for guided-wave optical interactions with Rayleigh waves

SAW produces a periodic variation in the thickness of the waveguide which, in turn, modifies the effective refractive index by an amount Δn_d. Generally, $\Delta n_a \gg \Delta n_d$, but there are instances where this does not hold true and the Δn_d contribution has to be considered[2]. A third contribution to the overall Δn arises when the waveguide is piezoelectric in which case the SAW is accompanied by periodic electric fields which induce a further refractive index change Δn_e through the electro-optic effect: in strongly piezoelectric materials such as $LiNbO_3$, the contribution from Δn_e can be substantial or even dominant[3].

In the structure of Fig 1, then, the guided optical wave encounters a propagating periodic refractive index given by

$$\Delta n(x,z) = \Delta n_0 \; a(x) \; \exp 2\pi j(f_0 t - \frac{z}{\Lambda}) \qquad (1)$$

where $\Delta n = \Delta n_a + \Delta n_d + \Delta n_e$, Δn_0 is the amplitude of Δn, $a(x)$ the transverse variation of Δn, f_0 and Λ the frequency and wavelength of the SAW, respectively. The optical wave, thus, encounters a moving phase grating from which it can diffract. We distinguish between two modes of diffraction as illustrated by Fig. 2. The Bragg mode, Fig 2(a), is the basis of most acousto-optic beam deflection devices and is characterised by diffraction of the incident optical wave into almost entirely one order. This occurs when the interaction length L is sufficiently long that the grating may be considered "thick", or when the acousto-optic parameter $Q \equiv 2\pi\lambda_g L/\Lambda^2$ is greater than about 4π, with λ_g the guide optical wavelength. When on the other hand $Q < 0.3$, Raman-Nath diffraction occurs as shown in Fig. 2(b): in this condition, the acoustic grating appears to the optical wave as a thin phase grating and a multiplicity of diffracted orders results. In this paper, we are almost solely concerned with the Bragg interaction.

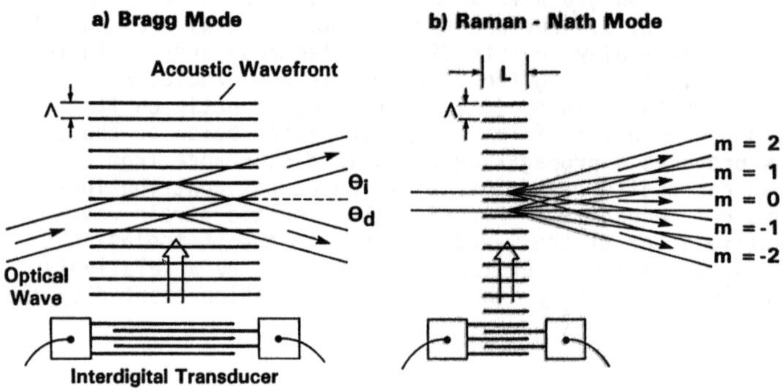

Fig. 2 Acousto-optic interactions showing (a) Bragg mode, (b) Raman-Nath mode

The phase-matched acousto-optic Bragg interaction may be represented by the wave vector diagram shown in Fig. 3(a) for the case where the diffracted light propagates in the same waveguide mode as the incident light: in this case we refer to the isotropic interaction. For this case, the incident angle θ_i equals the diffraction angle θ_d and is given by

$$\theta_d = \sin^{-1} \left\{ \frac{\lambda_g f_0}{2V} \right\} \qquad (2)$$

where $\lambda_g = \lambda_0/N$, N is the effective index of the waveguide mode, and V is the acoustic wave velocity. For acoustic frequencies around f_0, the Bragg condition is not exactly satisfied and the diffracted beam is scanned in angle with reduced intensity following the familiar bell-shaped response curve from which it can be shown that the number of resolvable spots for an acoustic bandwidth Δf corresponding to half-power diffracted intensity is [4,5]

$$NRS = \frac{2\Lambda D}{\lambda_g L} = \Delta f.\tau \qquad (3)$$

where D is the width of the optical beam, and τ = D/V the transit time of the acoustic wave across the optical beam. A complete expression for the diffraction efficiency is a complicated function of the optical and acoustical field distributions and is given by TSAI[6].

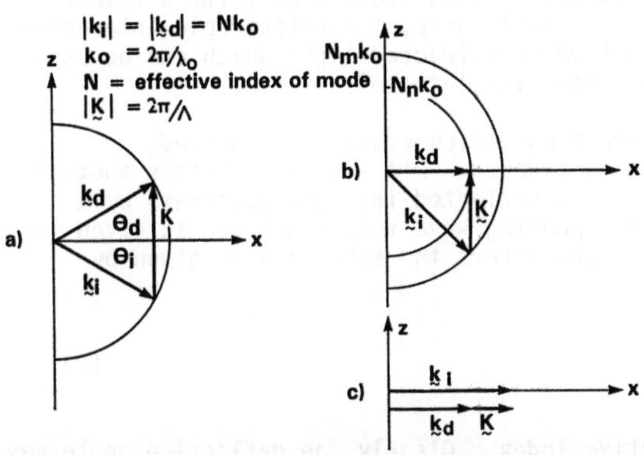

Fig. 3 Wave vector diagrams for acousto-optic interactions: (a) incident and diffracted waves in the same mode; (b) incident and diffracted waves in different modes; (c) collinear interaction

In the first demonstration of guided-wave Bragg diffraction by KUHN et al in 1970[7], the SAW was generated on an α-quartz substrate on which a glass optical waveguide was sputter-deposited. Since then the majority of devices studies have used waveguides in $LiNbO_3$, in particular waveguides formed by titanium diffusion[8] which offers the most attractive combination of optical, acoustic, acousto-optic and piezoelectric properties. Acousto-optic Bragg diffraction in a Ti:$LiNbO_3$ waveguide was first demonstrated by SCHMIDT and KAMINOW in 1975[9] and has since been refined to form the basis for the first true application of integrated optics technology, the RF spectrum analyser[10].

In the guided-wave acousto-optic interaction, it is also possible for the incident and diffracted waves to propagate in different optical waveguide modes, eg TE_m and TE_n, as depicted in Fig 3(b). This interaction may also take place collinearly as shown in Fig. 3(c) and has been considered as the basis for a tunable acousto-optic filter[11,12]. It is further possible for the incident and diffracted waves to be in orthogonally polarised optical waveguide modes, eg TE_m and TM_n. In this case, referred to as the anisotropic interaction, the wave vector diagram is also described by Fig. 3(b). This interaction is significant since it offers high efficiency diffraction over a wider acoustic bandwidth than is achievable in the normal isotropic process and, hence, a larger number of resolvable spots. In the optimised anisotropic interaction, the wave vector of the diffracted wave is orthogonal - or very nearly so - to that of the acoustic wave which has centre frequency chosen to satisfy the condition[5]

$$f_o = \frac{V}{\lambda_o} (N_m{}^2 - N_n{}^2) \tag{4}$$

and angle of incidence

$$\theta_i = \sin^{-1} \left\{ \frac{N_m{}^2 - N_n{}^2}{N_m} \right\} \tag{5}$$

TSAI et al[13] have demonstrated seven-fold increase in bandwidth in Y-cut Ti-diffused $LiNbO_3$ using the anisotropic interaction, albeit for a five-fold increase in acoustic drive power. Significantly, the anisotropic interaction provides reduced background noise from undiffracted light which can be used to advantage in devices requiring high signal-to-noise ratio.

Up to this point, only those interactions which result in a guided, diffracted optical wave have been considered. Fig 4(a) illustrates another case in which a guided optical wave is deflected into the substrate by a counter-propagating SAW. The corresponding wave vector diagram is given in Fig 4(b). The angle θ_d at which light enters the substrate is given by

$$\theta_d = \cos^{-1} \left\{ \frac{N_i - \lambda_o . f_o / V}{n_s} \right\} \tag{6}$$

where n_s is the substrate refractive index. Clearly the deflection angle may be changed by adjusting the SAW frequency. This configuration has received some experimental study[14,15] but tends to suffer from inherently low efficiency[16].

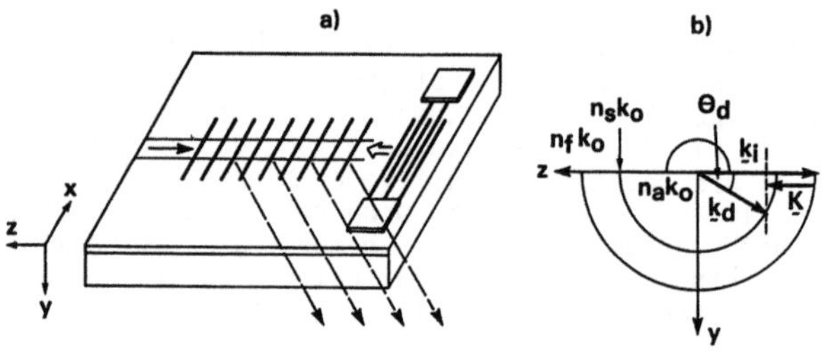

Fig. 4 (a) Waveguide acousto-optic interaction between a guided mode and a radiation mode; (b) wave vector diagram

It is also to be noted that the acousto-optic interaction leads to a shift in frequency of the incident light wave. In the coplanar case illustrated in Fig. 1, the optical frequency is upshifted by an amount equal to the SAW frequency: if the SAW direction were reversed, a corresponding downshift would occur. Some efforts to exploit this effect to produce frequency modulators/SSB modulators is described later.

3. Efficiency of the Phased-matched Bragg Interaction

The major applications of guided-wave acousto-optic devices utilise the Bragg interaction to achieve modulation, deflection, and related signal processing functions. The key interaction parameters have been discussed by several authors[5,6,17] and we summarise here selected results.

For the phase-matched Bragg interaction, the diffracted light intensity, I_d, relates to the incident intensity, I_i, through

$$I_d/I_i = \eta = \sin^2(\kappa L) \tag{7}$$

where κ is a coupling coefficient, the derivation of which is discussed in [5] and [17]. Following [17], we may express eqn (7) as

$$\eta = \sin^2 \left\{ \frac{\pi}{\lambda_0 \cos\theta_i} \cdot I \cdot \left(P_a \cdot \frac{M_2}{2} \cdot \frac{L}{H} \right)^{\frac{1}{2}} \right\} \tag{8}$$

where P_a is the acoustic power, L, H, the acoustic beam width and depth, respectively, I is an overlap integral describing the overlap of the optical and acoustic fields and has value between 0 and 1, M_2 is a material figure of merit given by

$$M_2 = \frac{n^6 p^2}{\rho V^3} \tag{9}$$

with n the refractive index, p the effective photoelastic constant, ρ the density, and V the acoustic velocity. Ideally, then, an efficient interaction occurs for low input acoustic power when L/H is large, M_2 is large, and I should be as near to unity as possible. Clearly, if the

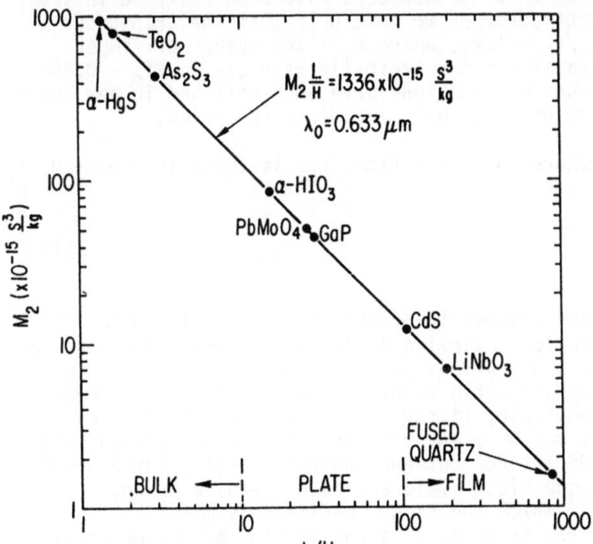

Fig. 5 Acousto-optic figure of merit M_2 plotted against the aspect ratio L/H requiired to achieve 100% diffraction with 150 mW input acoustic power. After E G LEAN et al [17]

incident and diffracted waves are in the same optical waveguide mode, the value of I can be close to 1. On the other hand, interactions between different modes of the same polarisation will have low values of I due to the orthogonality of waveguide modes. Interactions involving orthogonal polarisations can, however, still have values of I close to unity. Fig.5 shows the required relationship between L/H and M_2 to achieve 100% diffraction efficiency for 150 mW of acoustic power with I = 1[17]. Superimposed are the values of M_2 for selected acousto-optic materials. It may be seen that for L/H > 100, readily achievable in optical waveguide geometries, materials of relatively modest M_2 may be used in contrast with the case of bulk acousto-optic devices in which L/H must be <10 necessitating the use of materials with the highest M_2 values. When a material of high M_2 can also form an optical waveguide of good optical quality, then the required P_a can, of course, be correspondingly reduced.

4. Bandwidth and Speed of Operation

The acoustic bandwidth is given by (eqn (3))

$$\Delta f = \frac{2\Lambda V}{\lambda_g L} \tag{10}$$

and is a bandwidth imposed by the interaction length L which is essentially equal to the width of the acoustic beam. L must be minimised to maximise Δf which is accompanied by corresponding reduction in diffraction efficiency through eqn (7). For a given amount of acoustic power, P_a, there is inevitably a trade-off between efficiency and bandwidth.

There is, however, a second contributing factor to the overall device bandwidth that comes from the bandwidth of the interdigital transducer. The frequency response and efficiency of transducers have been analysed in some detail[18] and for our present purposes we note only that the transducer fractional bandwidth $\Delta f_{trans}/f_o \sim 1/N_e$, where N_e = the number of finger electrode pairs. The SAW transducer may typically have $\Delta f_{trans}/f_o \sim 0.35$ although various multiple transducer designs have been utilised in devices to achieve greater overall bandwidth and these are discussed below.

For modulator applications where switching time, T_s, is the prime concern, T_s can be expressed as

$$T_s \simeq \frac{1}{\Delta f_{trans}} + \tau \tag{11}$$

It is clear that the efficiency-bandwidth product of an acousto-optic isotropic Bragg-deflection device employing a simple interdigital transducer is somewhat restricted. For applications requiring maximisation of this product, a number of more sophisticated transducer arrangements have been employed. Parallel-driven, multiple, tilted arrays of simple transducers of staggered centre frequencies may be used, with the tilt angle between each adjacent pair equal to the difference between the Bragg angles corresponding to the respective centre frequencies. This approach yields a broad, composite frequency band over which the Bragg condition is met. Alternatively, the phased-array transducer, shown in Fig. 6, may be used. Here a number of identical transducers are connected together in a stepped pattern such that a phase shift is produced between adjacent sections as the SAW frequency is changed. This results in steering of the acoustic beam,

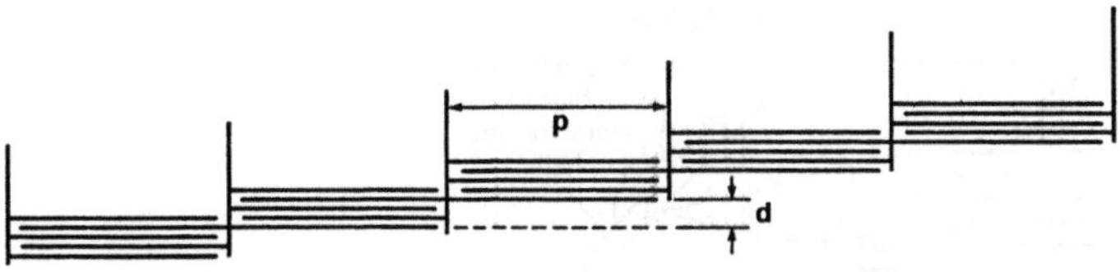

Fig. 6 Phased array beam-steering SAW transducer for wideband acousto-optic interactions

which may track the Bragg condition and render it possible to obtain high diffraction efficiency over a relatively wide frequency band. The phased-array can achieve higher diffraction efficiency than the multiple tilted array since the resulting wide acoustic beam permits a large interaction length to be maintained. On the other hand, the overall bandwidth is restricted in the phased-array device by that of the transducer. An obvious extension of these concepts uses multiple tilted phased arrays[6].

It is noteworthy, finally, that highest centre frequencies require transducer interdigital finger widths that are incompatible with standard photolithographic processing. State-of-art devices utilise electron-beam written transducer patterns. A Bragg cell on Y-cut $LiNbO_3$ employing a phased-array transducer with 0.5 μm finger widths has been operated at a centre frequency of 1.74 GHz with 530 MHz bandwidth[19].

5. Acousto-optic Signal Processing

Although a major application of guided-wave acousto-optics is in switching or modulation of light waves, there are many wider potential applications in signal processing techniques. Therefore, at this point it is appropriate to consider the general requirements for optical signal processing.

The key factor in most such applications is the nature of the optical detector. Almost without exception optical detectors measure power rather than amplitude. This results in a nonlinear (quadratic) relationship between optical amplitude and detected signal. Although this may appear curious since no physical nonlinearity is involved it is vital to obtaining correlation, for example. So in a correlator two diffracted waves add linearly but the detection process leads to a correlation signal. Equivalently, double-diffraction processes give nonlinear power throughput. This kind of effect also produces some strange results in the basic diffraction process, so that spreading the acoustic power over a longer wavefront increases diffracted optical power. This shows in the L/H dependence in the equations above.

Guided-wave acousto-optic correlation may be performed in two modes. The first of these, space-integrating correlation, is depicted in Fig. 7. The received signal $S_1(t)$ is fed to the first transducer of the device which then produces a spatial representation of the temporal variation of the signal. Reference signal $S_2(-t)$ is applied to the second transducer. Collimated, guided light is sequentially diffracted by the two acoustic waves so that the spatial diffraction pattern is the product of those due to S_1 and S_2. The resulting light is spatially integrated on to a photodetector by a focussing lens. This correlator is capable of large range but only limited time-bandwidth product. The second, more widely studied correlator is the

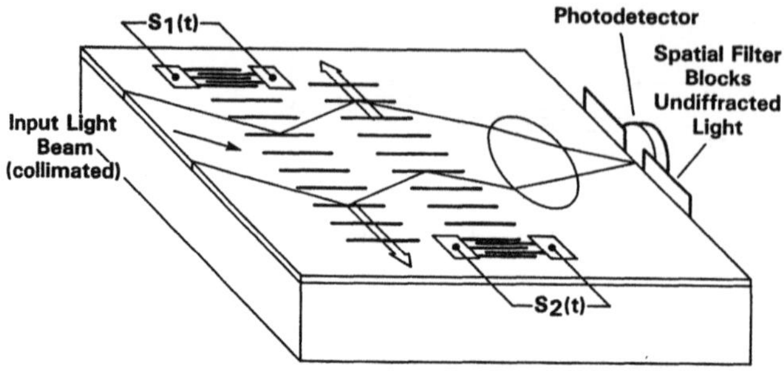

Fig. 7 Guided-wave acousto-optic space-integrating correlator

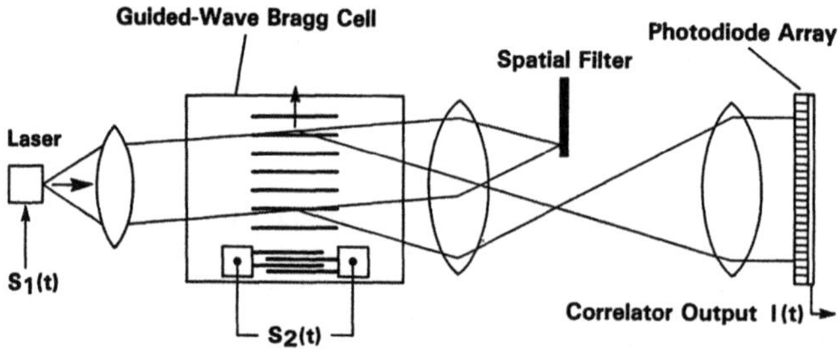

Fig. 8 Acousto-optic time-integrating correlator using guided-wave Bragg cell

time-integrating correlator illustrated in Fig. 8. In this case the incoming signal $S_1(t)$ is used to modulate the laser source which is then collimated and incident on the SAW generated by the transducer driven by the reference signal $S_2(t)$. The diffracted light is focussed, spatially filtered, and relayed to a photodetector array where it forms the correlation signal. This appears spatially but is read out in time using a CCD array. Very large time-bandwidth products may be obtained. Both correlators are guided-wave equivalents of more familiar bulk acousto-optic devices[20], and their full potential has yet to be demonstrated.

The most widely studied signal processing device utilising guided-wave acousto-optics is the RF sectrum analyser[10,21]. The hybrid integrated optical device is shown in Fig. 9. The laser is end-fire launched into a Ti:$LiNbO_3$ waveguide where it is collimated, diffracted, and Fourier transformed before emerging to be detected at a photodiode array. The device illustrated uses geodesic lenses precision machined into the substrate. The integrated optical spectrum analyser has struggled to attain its targetted dynamic range, principally due to difficulties in lens fabrication, despite truly impressive waveguide scatter performance (in-plane scatter -55 dB at 1^0 separation from main peak). However, a linear dynamic range of 46 dB has been measured for a 2 μs integration time for a device with 400 MHz bandwidth centred on 600 MHz[22].

Recently, the acousto-optic interaction has been demonstrated in a Ti:$LiNbO_3$ waveguide formed on a spherical substrate that serves to guide, collimate and focus the light[23]. Spectrum analysis was demonstrated in addition to correlation and convolution of RF signals.

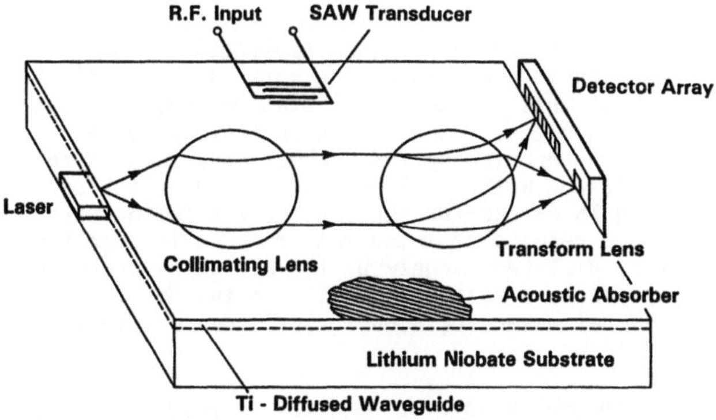

Fig. 9 Integrated optical Bragg cell spectrum analyser

Finally, we note that the potentially attractive possibility of ultimate integration of active and passive components offered by the III-V semiconductor system may even include acousto-optic devices, and wideband diffraction has been demonstrated in a GaAs-GaAlAs waveguide surmounted by ZnO[24].

6. Acousto-Optic Interactions in Channel Waveguides

Up to this point we have considered optical waveguides wherein light is confined in one dimension only. The greatest potential of integrated optics lies with devices in which light is guided by strip or channel waveguides, with confinement in both dimensions perpendicular to the direction of propagation. In this format, a wide range of modulation and switching functions have been demonstrated, primarily using the electro-optic effect, with Ti:$LiNbO_3$ the predominant waveguide system. Acousto-optic interactions in channel waveguides have been little studied by comparison, despite some potentially useful capabilities. We consider these interactions here.

Fig. 10 shows a channel waveguide crossover device studied by TSAI et al[25]. Two identical, multimode Ti-diffused channel waveguides in $LiNbO_3$ cross at an angle equal to twice the Bragg angle for diffraction by the SAW generated by the transducer shown. The double-diffused region where the channel waveguides cross helps to reduce crosstalk between the guides. TSAI et al

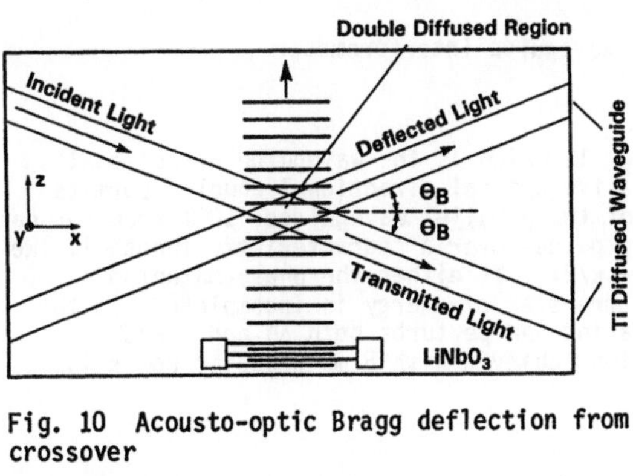

Fig. 10 Acousto-optic Bragg deflection from a SAW in a Ti:$LiNbO_3$ waveguide crossover

performed a number of experiments on this device and demonstrated 97% deflection of light from one input waveguide with the application of 9V. The SAW frequency was 634 MHz corresponding to a waveguide intersection angle of 3^0. Being multimode, operation of this device is somewhat dependent on the relative excitation of the various modes. Experiments on a single-mode version of the same structure were reported by VON HELMOLT et al[26] wherein 6% deflection efficiency was obtained with a 629 MHz SAW and intersection angle 3.5^0. SSB modulation with 16 dB carrier suppression was demonstrated. The low efficiency of this device results from the rather short interaction length, 120 μm yielding Q = 7.6, dictated largely by the waveguide crossover geometry and, indeed, this represents a major difficulty in the design and implementation of practical channel waveguide acousto-optic devices in this form.

Further studies of single-mode channel waveguide devices are reported in [27] using waveguide structures familiar in integrated optics through their applications in electro-optic switching and modulation[28]. Fig 11 shows a waveguide Mach-Zehnder interferometer driven by a SAW. In the interferometer, a single-mode input waveguide divides into two nominally identical, parallel, arms which are recombined to form an output guide. Waves from the two arms combine in the output guide with the field distribution of the guided mode only if they arrive together exactly in phase; otherwise, the output field contains a proportion which is in the next higher-order mode of the guide which is cut-off, and this light is lost by radiation into the substrate. The Δn induced by the SAW is here used to induce phase shifts between waves in each of the two arms of the interferometer which results in intensity modulation. Detailed experimental results on this structure are given in [27].

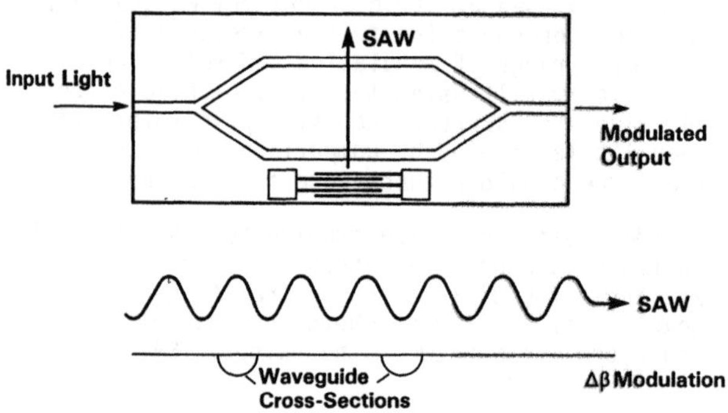

Fig. 11 SAW-driven Mach-Zehnder waveguide interferometer

Another structure is shown in Fig. 12 which is the waveguide directional coupler driven by a SAW. The passive optical directional coupler permits transfer of energy between each of two parallel waveguides: 100% transfer can be achieved between phase-matched guides over a characteristic length l. We define a coupling coefficient $\kappa = \pi/2l$. If either the phase-matching condition or κ is perturbed, the transfer of energy is incomplete. In the structure of Fig. 12, the propagating SAW perturbs both $\Delta\beta$ and κ and modulation occurs. 100% modulation achieved with 8 mW acoustic power is reported in [27].

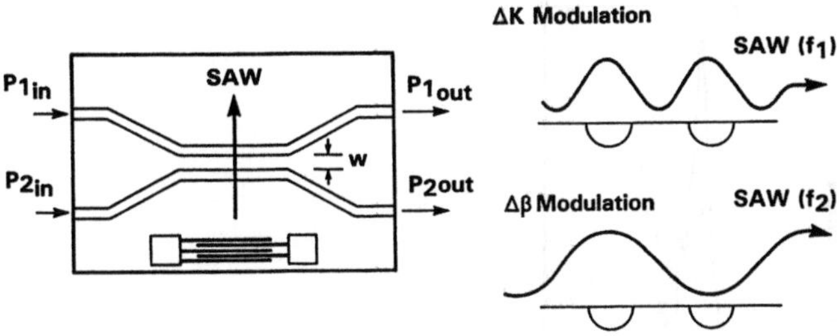

Fig. 12 SAW-driven waveguide directional coupler

7. Thermal Wave Diffraction

All of the above acousto-optical diffraction systems could be described as Brillouin scattering phenomena, but this term is usually reserved for scattering of light waves from thermally induced sound waves. This phenomenon has been observed for surface acoustic waves [29] in the configuration shown in Fig. 13.

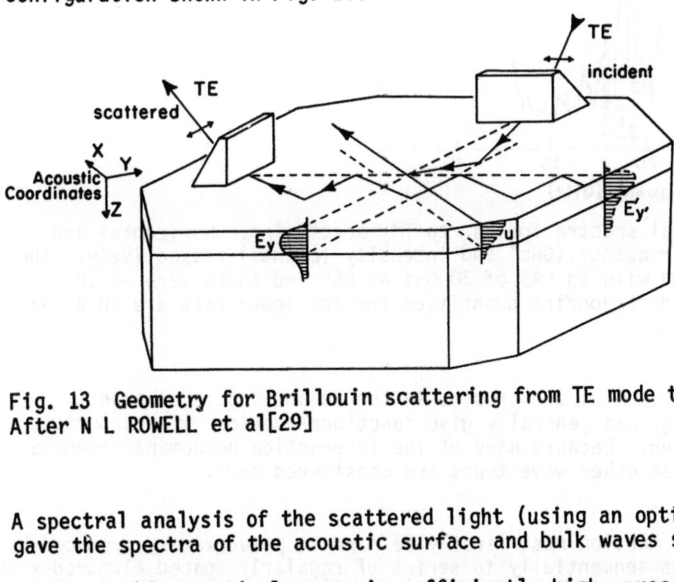

Fig. 13 Geometry for Brillouin scattering from TE mode to TE mode at 90°. After N L ROWELL et al[29]

A spectral analysis of the scattered light (using an optical Fabry-Perot) gave the spectra of the acoustic surface and bulk waves shown in Fig. 14.

If the incident optical power is sufficiently high, processes of this kind can lead to stimulated Brillouin scattering in which power is fed from the optical wave into the acoustic wave. Such a process clearly has potential for nonlinear signal processing with acoustic readout.

8. Surface Wave Interactions

Acoustic surface waves are by no means the only surface waves that can interact with optical guided waves. Indeed, as must be clear from the above discussions, an interaction can generally be obtained subject to a phase matching condition (possible for most systems giving reasonable wavelength matching) and some mechanism which will perturb the optical wave velocity.

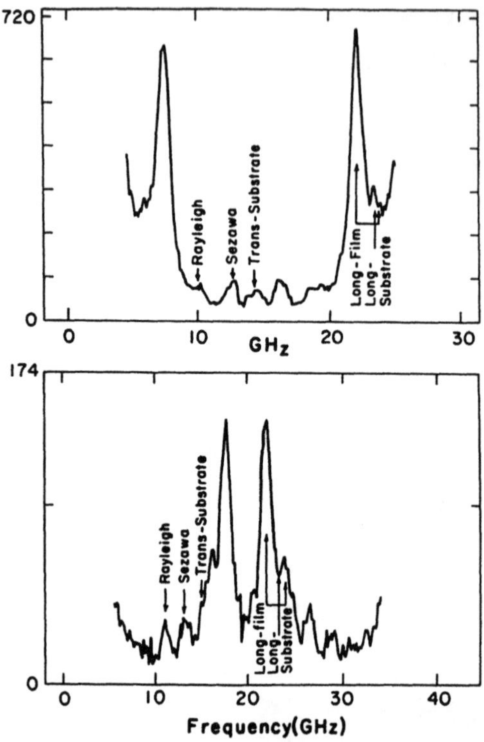

Fig. 14 Experimental spectra for TM_0 to TM_0 scattering. Horizontal and vertical axes are frequency (GHz) and intensity (counts) respectively. The upper data was taken with an FRS of 30 GHz at 86^0 and there were $\sim 5\times 10^7$ counts at 0 GHz. Corresponding quantities for the lower data are 40.8 GHz, 89^0 and 10^7 counts.

Mixed interactions, such as are present in the acoustic case, though partially cancelling, can generally give functional devices if appropriate conditions are chosen. Because many of the interaction phenomena resemble acousto-optics, these other wave types are considered here.

Pseudowaves: Brief mention must firstly be made to pseudowaves, produced by applying AC voltages sequentially to series of regularly spaced electrodes on electro-optic waveguides, which have been used in integrated optical signal processing. The similarity of some structures, for example, the spectrum analyser[30], to their acousto-optic counterparts is pronounced and, indeed, some structures may be used in conjunction with acoustics. A significant advantage which they possess is relative ease of control of the wave 'velocity'.

Magnetostatic Wave Interactions: Magnetostatic waves (Fig. 15), being magnetic spin waves, are very different in characteristics to acoustic waves. A major part of their attraction for potential signal-processing applications lies with the tunability of their velocity with applied field.

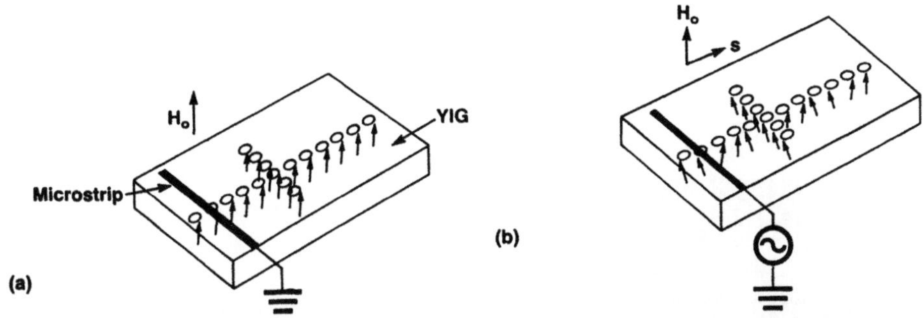

Fig. 15 Magnetostatic waves: (a) with uniform applied field, the magnetic moments in the thin ferrimagnetic film precess at the same rate with the same phase; (b) an applied RF field generates a perturbation which propagates through coupling of adjacent magnetic moments.

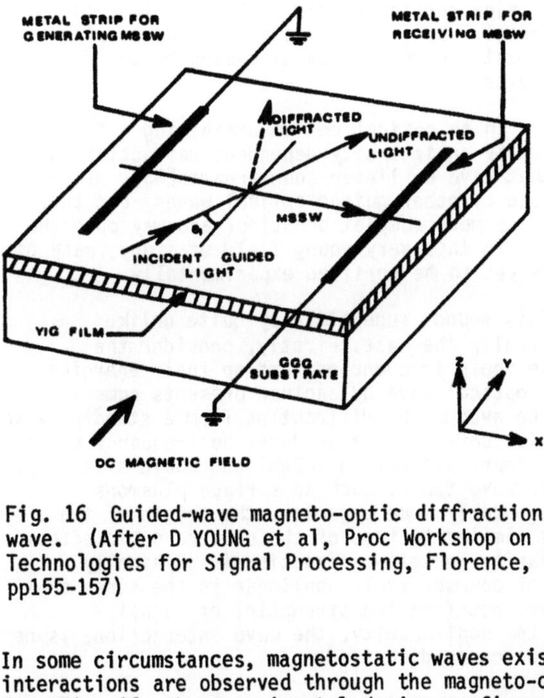

Fig. 16 Guided-wave magneto-optic diffraction from a magnetostatic surface wave. (After D YOUNG et al, Proc Workshop on Integrated Optical and Related Technologies for Signal Processing, Florence, Italy, Sept 10-11, 1984, pp155-157)

In some circumstances, magnetostatic waves exist as surface waves and optical interactions are observed through the magneto-optic effect. As may be seen from Fig. 16, the experimental device configuration closely resembles the acousto-optic case and TM-TE mode conversion has been induced by both codirectional and contradirectional interactions with efficiencies up to 4%[31]. The potential exists for practical devices operating in the 1-20 GHz frequency range.

Surface Plasmons: Surface plasmons closely resemble guided optical waves and exist at the boundary between a metal and a dielectric[32]. The field distribution of the plasmon decays exponentially away from the boundary into both media as shown in Fig. 17, with a wide range of characteristics ranging from waves propagating with the major portion of the field within the

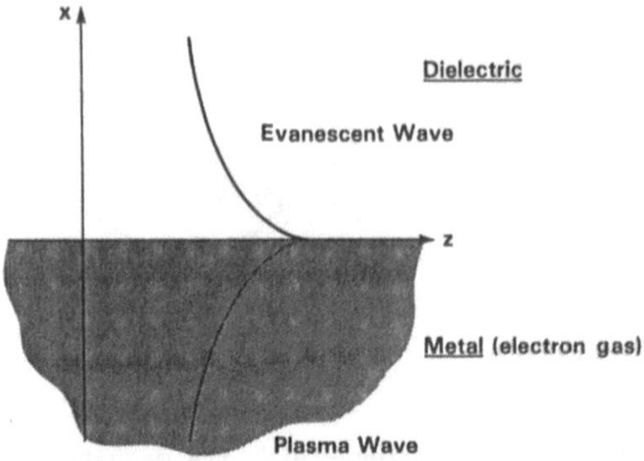

Fig. 17 Surface plasmon fields at a metal-dielectric interface

dielectric to those wherein the major portion of the field comprises the plasma oscillations within the metal. Surface plasmons could be considered for interactions with surface acoustic waves or, as discussed below, for interaction with other optical waves.

Nonlinear Optical Surface Waves: In an optical medium exhibiting a third order nonlinearity, characterised by an intensity-dependent refractive index, some surface wave types exist that have no linear counterparts[33]. They may find applications similar to those of other guided optical waves, but their intensity-dependent nature leads to more complex behaviour and may open the way to new processing functions. In this very young field of study, many of the theoretical predictions have yet to be verified experimentally.

Optical-optical Diffraction: This sounds superficially quite unlike acousto-optics but this is not really the case. Firstly, consider the function of diffraction: because their frequencies, and so their energies, are similar, diffraction of one optical wave by another presents some problems. These can, however, be avoided by diffracting from a standing wave produced by two other waves: this configuration produces no frequency shift. It is widely used in degenerate four-wave mixing (DFWM) and could similarly, in principle, be used with other wave types, such as surface plasmons. Secondly, in certain materials, notably some organics, the origin of the optical nonlinearity is closely similar to that of the electro-optic effect and, thus, the model of nonlinearity as self-diffraction is physically reasonable. Such systems are, of course, truly nonlinear in the sense that they do not permit superposition, but from the standpoint of signal processing the crucial part of the nonlinearity, the wave interaction, is not unlike the other wave diffraction cases discussed.

References

1. D F Nelson and M Lax: "Theory of the photoelastic interaction", Phys Rev B 3 1971 pp2778-2794
2. R Normandin, V C Y So, N Rowell and G I Stegeman: "Scattering of guided optical beams by surface acoustic waves in thin films", J Opt Soc Amer, 69 1979 pp1153-1165
3. J M White, P F Heidrich and E G Lean: "Thin-film acousto-optic interaction in $LiNbO_3$", Electron Lett 10 1975 pp510-511
4. E G Lean: "Acousto-optical interactions" in Introduction to Integrated Optics, ed M K Barnoski, Plenum, New York 1974

5. R V Schmidt: "Acoustooptic interactions between guided optical waves and acoustic surface waves", IEEE Trans Sonics and Ultrasonics SU-23 1976 pp22-33.
6. C S Tsai: "Guided-wave acoustooptic Bragg modulators for wide-band integrated optic communications and signal processing", IEEE Trans Circuits and Systems, CAS-26 1979 pp1072-1098.
7. L Kuhn, M L Dakss, P F Heidrich and B A Scott: "Deflection of an optical guided wave by a surface acoustic wave", Appl Phys Lett 17 1970 pp265-267
8. R V Schmidt and I P Kaminow: "Metal-diffused optical waveguides in $LiNbO_3$", Appl Phys Lett 25 1974 pp458-460
9. R V Schmidt and I P Kaminow: "Acoustooptic Bragg deflection in $LiNbO_3$ Ti diffused waveguides", IEEE J Quantum Electron QE-11 1975 pp57-58
10. D Mergerian, E C Malarkey, R P Pautienus, J C Bradley, G E Marx, L D Hutcheson and A L Kellner: "An integrated optical radio frequency (rf) spectrum analyser", Proc SPIE 239 Guided-Wave Optical and Surface Acoustic Wave Devices, Systems and Applications, 1980 pp121-127
11. L Kuhn, P F Heidrich and E G Lean: "Optical guided wave mode conversion by an acoustic surface wave", Appl Phys Lett, 19 1971 pp428-430
12. M S Chang:" "Tolerance to the phase matching condition for the integrated acousto-optic filter" Appl Opt 13 1974 pp1867-1871
13. C S Tsai, I W Yao, B Kim and L T Nguyen: "Wideband guided-wave anisotropic acoustooptic Bragg diffraction in $LiNbO_3$ waveguides", Proc Int Conf on Integrated Optics and Optical Fibre Communication, Tokyo Japan July 18-20 1977 pp57-60
14. F R Gfeller and C W Pitt: "Collinear acoustooptic deflection in thin films", Electron Lett 8 1972 pp549-551
15. I P Kaminow: "Optical waveguide modulators", IEEE Trans Microwave Theory Techn MTT-23 1975 pp57-70
16. D Marcuse: "Light scattering from periodic refractive-index fluctuations in asymmetric slab waveguides", IEEE J Quantum Electron QE-11 1975 pp162-168
17. E G Lean, J M White and C D W Wilkinson: "Thin-film acousto-optic devices", Proc IEEE, 64 1976 pp779-788
18. R W Smith, H M Gerard, J H Collins, T M Reeder and H J Shaw: "Design of surface wave delay lines with interdigital transducers", IEEE Trans Microwave Theory Techn MTT-17 1969 pp865-873
19. C Stewart, W J Stewart and G Scrivener: "500 MHz bandwidth guided wave L-band Bragg cell", Electron Lett 17 1981 pp971-973
20. N J Berg and J N Lee, ed Acousto-Optic Signal Processing: Theory and Implementation, Marcel Dekker, New York 1983
21. A Dawar, R M De La Rue, G F Doughty, N Finlayson, S M Al-Shukri and J Singh: "Acousto-optic techniques in integrated optics", Proc SPIE 517 Integrated Optical Circuit Engineering 1984 pp64-73
22. D Mergerian, E C Malarkey and R P Pautienus: "High dynamic range integrated optical RF spectrum analyser", Proc 4th International Conf Integrated Optics and Optical Fibre Communication, Tokyo, Japan, June 1983 pp260-261.
23. Q Li, C S Tsai, S Sottini and C C Lee: "Light propagation and acousto-optic interaction in a $LiNbO_3$ spherical waveguide", Appl Phys Lett 46 1985 pp707-709
24. C J Lii, C C Lee, O Yamazaki, L S Yap, K Wasa, J Merz and C S Tsai: "Efficient wideband acoustooptic Bragg diffraction in GaAs-GaAlAs waveguide structure", Proc 4th International Conf Integrated Optics and Optical Fibre Communication, Tokyo, Japan, June 1983 pp252-253
25. C Tsai, C L Chang, C C Lee and K Y Liao: "Acousto-optic Bragg deflection in channel optical waveguides", Tech Digest Topical Meeting Integrated and Guided-Wave Optics, Incline Village, Nevada, Jan 1980 paper PD-7

26. C H Von Helmolt, R T Kersten, W Auch and W Steudle: "Bragg switch/SSB-modulator with integrated optic single-mode waveguides", Second Europ Conf Integrated Optics, Florence, Italy, October 1983 pp132-135.
27. C H Von Helmolt: "Integrated optic strip waveguides driven by surface acoustic waves", J Opt Commun 2 1981 pp142-154
28. R C Alferness: "Guided-wave devices for optical communication", IEEE J Quantum Electron QE-17 1981 pp946-959
29. N L Rowell, V C-Y So, R Normandin and G I Stegeman: "Brillouin scattering in thin film optical waveguides.III:Experiments", Can J Phys 60 1982 pp1821-1837
30. L Thylen and L Stensland: "Electro-optic approach to an integrated optics spectrum analyser", Appl Opt 20 1981 pp1825-1832
31. A D Fisher, J N Lee, E S Gaynor and A B Tveten: "Optical guided-wave interactions with magnetostatic waves at microwave frequencies", Appl Phys Lett 41 1982 pp779-781
32. H Raether: "Surface plasma oscillations and their applications", in Physics of Thin Films Vol.9 ed G Hass, M H Francombe and R W Hoffman, Academic Press, 1977
33. C T Seaton, G I Stegeman, W M Hetherington and H G Winful: "Nonlinear third order integrated optics", Proc 3rd Europ Conf on Integrated Optics, Berlin, Germany, May 1985 pp178-195.

Index of Contributors

Subject Index

MIX
Papier aus verantwortungsvollen Quellen
Paper from responsible sources
FSC® C105338

If you have any concerns about our products,
you can contact us on
ProductSafety@springernature.com

In case Publisher is established outside the EU,
the EU authorized representative is:
Springer Nature Customer Service Center GmbH
Europaplatz 3, 69115 Heidelberg, Germany

Printed by Libri Plureos GmbH
in Hamburg, Germany